Fabrication and Application of Nanomaterials

Fabrication and Application of Nanomaterials

Sulalit Bandyopadhyay, Ph.D.

McGraw Hill

New York Chicago San Francisco
Athens London Madrid
Mexico City Milan New Delhi
Singapore Sydney Toronto

Library of Congress Control Number: 2019937651

Fabrication and Application of Nanomaterials

1 2 3 4 5 6 QVS 23 22 21 20 19

ISBN 978-1-260-13223-6
MHID 1-260-13223-4

This book is printed on acid-free paper.

Sponsoring Editor
Summers Scholl

Editorial Supervisor
Stephen M. Smith

Production Supervisor
Lynn M. Messina

Acquisitions Coordinator
Elizabeth M. Houde

Project Manager
Touseen Qadri, MPS Limited

Copy Editor
Mohammad Taiyab Khan, MPS Limited

Proofreader
Heather Mann, MPS Limited

Indexer
Edwin Durbin

Art Director, Cover
Jeff Weeks

Composition
MPS Limited

About the Author

Sulalit Bandyopadhyay earned a Ph.D. and M.Sc. in chemical engineering from the Norwegian University of Science and Technology (NTNU). He is a post-doctoral researcher in the Department of Chemical Engineering at NTNU and a visiting researcher at TU Delft and IHE Delft Institute for Water Education. He has been actively teaching chemical engineering courses at the graduate and post-graduate levels at NTNU since 2013. Dr. Bandyopadhyay's research interests include synthesis, characterization, and functionalization of nanoparticles; development of nanoparticle-based hydrological tracers; drug delivery; modeling of nanosystems; and recycling of Li-ion batteries.

Contents

Preface

Nanotechnology is an advancing field generating new innovations every day. Synthesizing nanomaterials using solution-based approaches offers a plethora of advantages for tuning their functional properties to make them apt for a multitude of applications. However, there are very few documented works that cover the full span from synthesis, functionalization, and characterization to specific applications, and more so at the level of a textbook that can intrigue readers to venture further into this exciting world. This formed the basis for taking up this ambitious project!

The passion to write a book on this topic generated from teaching a course (Fabrication and Application of Nanomaterials) in the Department of Chemical Engineering, Norwegian University of Science and Technology (NTNU), Trondheim, Norway. I have taught this course since 2013 and the contents of the course have developed with the emergence of new concepts. What must not be forgotten while using this book is that the theoretical developments and generalized concepts stem from fundamental sciences, but the topics have been covered from a chemical engineering background!

The book provides in-depth knowledge on synthesis, characterization, and functionalization of nanomaterials using solution-based bottom-up approaches developed from fundamental principles of nucleation and growth. It is written as a textbook for graduate students pursuing chemical engineering, nanotechnology, materials science, biotechnology, chemistry, nanomedicine, biophysics, and related disciplines. The reader will master the techniques for synthesis and application of fabricated nanomaterials through concise theory, numerical problems, and recent examples from industry and academia.

The book relies on developing reading material for graduate students interested in fabrication of nanomaterials, with the overall objective of applying chemical engineering principles to the field of nanoscience. Thus, the reader will be exposed to limitless possibilities of designing smart nanomaterials and subsequently utilizing them in various engineering fields.

The author is a chemical engineer with specialization in the field of polymers, colloids, and nanomaterials and with 10 years of academic experience. This book is meant to be used as a primary textbook or supplementary reading material for subjects such as fabrication and application of nanomaterials, nanomedicine, nanochemistry, chemical methods for synthesis and characterization of nanomaterials, functional materials, advanced characterization techniques, and the like.

The book is an excellent tool for Ph.D. students, scientists, and researchers pursuing nanotechnology-based fundamental research as well as applications of nanomaterials. It provides not only a generic overview of the world of nanoparticles but also a comprehensive summary starting from the fundamentals and ending at the potential industrial applications of specific nanoparticles. Professors and lecturers at various institutes can use this book as reading material for innovative research applying state-of-the-art industrial techniques, in addition to utilizing the book content for the development of new specialization courses and modules related to the subject matter.

Classical concepts of nucleation and growth are used to explain the synthesis of inorganic and polymeric nanomaterials. These are then functionalized toward specific applications including targeted drug delivery, hydrological tracing, water purification, and catalysis. Characterization of the physicochemical properties is illustrated using state-of-the-art techniques, both *in situ* and *post situ* modifications. The first eight chapters prepare the reader with hands-on experience to select from a wide array of nanomaterials, the ideal candidates suited for one of the specific applications highlighted in the next four chapters. Although only four applications have been discussed in detail in this book, the reader will get a holistic overview of available alternatives to synthesize nanoparticles in solution, tailor them toward a specific application, characterize them, and design smart nanomaterials for tailored applications. The chapters have been supplemented with concept-based questions, most of which are built on direct knowledge interpretation from the respective chapter, and research-based questions that are a result of experiments performed in laboratories and therefore require application of the knowledge gained from the respective chapter. With a mix of theoretical and numerical problems, the exercises are expected to add dimension to overall understanding of the chapters.

This book wouldn't have seen the light of the day without the support from all the chapter contributors, colleagues, Ph.D. and masters students in nanoresearch, friends, and family! Kudos to all of you.

Dr. Sulalit Bandyopadhyay
Post-Doctoral Researcher, Department of Chemical Engineering
Norwegian University of Science and Technology (NTNU)
Trondheim, Norway
Visiting Researcher, Department of Water Management,
TU Delft, The Netherlands

Fabrication and Application of Nanomaterials

CHAPTER 1
Introduction

Nanotechnology is at the forefront of science in today's world. The credit for coining the word *nanotechnology* is given to a renowned American theoretical physicist, Richard P. Feynman. The term *nano* refers to a billionth of a meter. For visualization, it can be imagined that if one nanoparticle (NP) were the size of a football, then a doughnut would be as big as New Zealand! During the annual meeting of the American Physical Society in December 1959 at Caltech, Feynman delivered a famous lecture, "There's plenty of room at the bottom." In his lecture, he talked about rearrangement of atoms to synthesize materials at the atomic level.[1]

> But I am not afraid to consider the final question as to whether, ultimately—in the great future—we can arrange the atoms the way we want; the very atoms, all the way down! What would happen if we could arrange the atoms one by one the way we want them (within reason, of course; you can't put them so that they are chemically unstable, for example).
>
> —Richard P. Feynman

At that time, these words were considered a mere science fiction. However, we see the insight of Feynman in the new millennia, when research in the field of nanoscience has upsurged exponentially.

1.1 Importance of Nanoscale

Bulk materials are macroscopic and possess continuous physical properties. However, as we go lower in dimensional scales, the principles of classical physics are rendered insufficient in describing the behavior of particles, such as movement, energy, and thermal and electrical conductivity. The principles of quantum mechanics come into play at such atomic scales. One of the most studied elements is gold, which shows completely different optical, mechanical, and electrical properties at nanoscale as compared to bulk materials. The color of a gold NP varies with its size, ranging from intense red for sizes less than 100 nm to dirty yellowish for larger particles. The colors of Au NPs are remarkably different from the color of bulk gold material.

The following sections of the chapter will give you an overview of nanoscience and NPs, starting with the different types of NPs, their functionalization, and their characterization, followed by their specific applications. The chapter is designed in a way to acquaint you with the structure of the rest of the textbook, which will equip you to synthesize nanomaterials, and ascribe functional groups in order to tailor them toward specific applications.

1.2 Inorganic Nanoparticles

Inorganic materials such as gold (Au), silver (Ag), iron (Fe), silicon (Si), platinum (Pt), and nickel (Ni) predominantly comprise inorganic NPs. Depending on composition, these can be categorized as metallic, semi-conductor, and metal oxide NPs. These materials exhibit exciting properties as their size approaches nanoscale. Recent research aims at exploiting the high surface to volume ratios of these NPs for advanced and diverse applications ranging from biosensing to magnetic recording to energy harvesting and water management. The two categories of inorganic NPs are discussed in brief here but will be detailed throughout the book.

1.2.1 Plasmonic Nanoparticles

One of the most widely studied plasmonic materials in NP research are the gold NPs. Also known as *colloidal gold*, Au NPs find their existence since the Roman times when they were used for staining of glass for decorative purposes. However, the present scientific assessment of Au NPs began in the 1850s when Michael Faraday noticed that their colloidal properties varied from the bulk. Au NPs can be synthesized in different shapes and sizes, each category possessing distinguished optical properties. These interesting optical characteristics of colloidal gold stem from a unique phenomenon known as *localized surface plasmon resonance* (LSPR).

The LSPR results from the interaction of light with the electrons of the conduction band.[2] Coherent localized plasmon oscillations with a resonant frequency result from this interaction and is dependent on composition, size, geometry, dielectric properties of particles, and solvent and particle–particle separation of the NPs. Metals such as Au and Ag display LSPR in the visible range of electromagnetic spectrum due to the energy levels of d-d transitions. Of all the metals, Ag exhibits the sharpest and the strongest bands. However, Au is preferred over Ag for biomedical applications owing to biocompatibility, inert nature, and strong thiol affinity for immobilization of biomolecules.

As already mentioned, the properties of NPs vary from those of the bulk material. Thus, the minute dimensions of these particles place them in quantum size regime. At these magnitudes, continuous electron bands turn discrete and properties including plasmon resonance are enhanced with the size of NPs. The oscillating electrical field causes coherent oscillation of conduction electrons upon plane-wave excitation of small particles for which $R/\lambda < 0.1$ (R is the radius of the NP and λ is the wavelength of the incident light). This effect results in the accumulation of polarization charges on the NP surface.[3,4] The electron cloud is displaced relative to the nuclei owing to the oscillations of electrons in the conduction band, giving rise to an opposing Coulombic attractive force, as shown in Fig. 1.1. This force by the nuclei controls the oscillation of the electron cloud at a specific frequency. If the frequency of the oscillating electrons matches that of the incident electromagnetic wave (light), a resonance condition occurs. The resonantly enhanced field inside the NP generates a dipolar field at its surface that is responsible for the intensification of scattering cross section and absorption along with strengthened electromagnetic field in the immediate vicinity of the NP surface. Typically, Au NPs exhibit LSPR at around 520 nm, though the absorption is highly determined by the shape and size of electron distribution, density of electrons, and effective electron mass.[5,6]

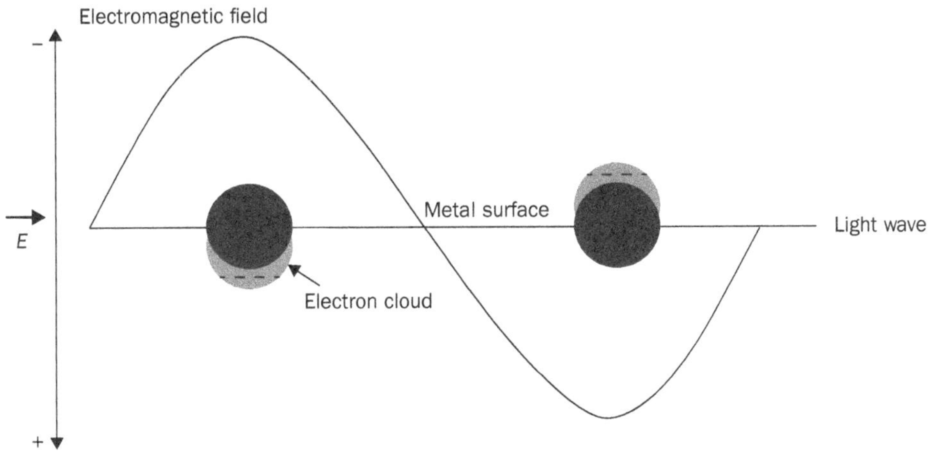

Figure 1.1 Schematic showing localized surface plasmon resonance (LSPR) in a metallic nanoparticle.[7]

1.2.2 Magnetic Nanoparticles

Surface and finite-size effects govern the magnetic behavior of the individual NPs at nanoscales. These NPs possess extraordinary magnetic properties such as extremely low coercivities, high magnetic saturations, and high magnetic susceptibilities among others. The motion of particles having mass and carrying an electric charge generates magnetic effects. The spinning of an electrically charged particle creates a magnetic dipole known as *magneton*. Such magnetons are associated in groups for a ferromagnetic material. A Weiss domain constitutes all the magnetons that are aligned in a particular direction by the exchange forces. The structure of this domain defines the size dependence of its magnetic behavior. As we go down in dimensions, the material is reduced to a single domain below a critical value of <15 nm (for typical materials).

The magnetic domain theory states that factors such as magnetic saturation, strength of crystal anisotropy and exchange forces, surface or domain-wall energy, and particle size influence the critical size of the single domain. The interaction of magnetic materials with the applied magnetic field is described by hysteresis, characterized by remanence and coercivity. The latter is the more intriguing property when dealing with small particles. Coercivity increases up to a maximum upon reduction in size and reduces sharply to zero as the size approaches that of a single domain.

Particles are deemed superparamagnetic at zero coercivity, which is mostly a result of thermal effects. These particles magnetize in the presence of an applied magnetic field, but demagnetize with zero hysteresis upon removal of the external field. The thermal variations are sufficient to spontaneously demagnetize already saturated assembly. Thus, the "active" behavior of the particles is avoided. Some of the different kinds of magnetic materials based on the relative magnitude and direction of constituting dipoles are shown in Fig. 1.2.

Such magnetic NPs find several applications as trigger substances in biological systems due to their magnetization and rapid demagnetization in and without the presence of an external magnetic field, respectively. Crystalline materials such as Fe, Co,

FIGURE 1.2 Schematic depicting different kinds of magnetic materials based on magnitude and direction of magnetic dipoles represented by arrows. (Adapted from Krahne et al.[8])

and Ni exhibit ferromagnetism, but magnetite (Fe_3O_4) finds major use as it is most magnetic among all the naturally occurring minerals.[9,10]

1.3 Organic Nanoparticles

Besides inorganic NPs, organic NPs have gained considerable importance, especially in the field of medicine due to their biocompatibility. One of the classes of organic NPs is liposomes. These are spherical shaped concentric vesicles comprising at least one lipid bilayer enclosing an aqueous volume. They can be synthesized from natural and synthetic phospholipids. Biodegradable polymeric NPs, such as poly lactic-co-glycolic acid (PLGA), poly lactic acid (PLA), and polycaprolactone (PCL), find applications in tissue engineering and targeted drug delivery as well as in the development of vaccine strategies.[11] To avoid toxicity-related effects, natural polymers, namely chitosan, and gelatin among others have gained importance. These NPs offer improved and effective administration of drugs over the conventional oral and intravenous methods.

As an alternative carrier system to liposomes and polymeric NPs, solid lipid NPs (SLNs) have emerged as potential colloidal delivery systems. SLNs are submicron-sized lipid emulsions in which the liquid lipid (oil) is substituted by a solid lipid. Another category of organic NPs widely studied is dendrimers. They consist of a highly branched polymeric network of macromolecules formed by either convergent or divergent synthesis routes. NPs offer an extremely symmetric, well-defined structure with monodisperse size distribution and high surface functionality.

The uniqueness of inorganic and polymeric NPs can be appreciated owing to the fact that they not only exhibit different characteristics at smaller sizes but also possess high surface-to-volume ratios. This enables them to be functionalized (or coated) with various chemical moieties that enhance their physico-chemical properties. The surface layer can be organic in nature, such as polymeric coatings, or inorganic in nature, such as core-shell NPs like Fe core and Au shell NPs. Such functionalized NPs can be utilized in building novel hierarchical structures by exploiting fundamental complexities in bonding, interfacial interactions, and structure between particles; undoubtedly NPs represent the future of nanoscience research as they offer multiple functionalities.

1.4 Synthesis

Depending on the end-product and desired functionality, there are numerous ways to synthesize NPs. These fabrication techniques aim at constructing nanostructures having dimensions lower than a few hundred nanometers. These methods can be broadly classified under top-down and bottom-up approaches. These have assisted in integrating engineered nanomaterials into multifunctional devices opening up for a wide array of applications.

In the top-down approach, one starts with a larger dimension and breaks down the structure into smaller particles until the desired dimension and shape is achieved. On the contrary, in case of bottom-up approach, nanomaterials are fabricated starting from the atomic or molecular scale upwards into a higher complex nanoscale assembly. The advantage of the latter technique is that the optimized process can lead to no wastage of precursor materials. This book primarily illustrates the bottom-up synthesis approach of nanomaterial fabrication. However, a few top-down approaches will also be discussed in this chapter, along with the bottom-up techniques, as other possible synthesis routes.

1.4.1 Top-Down Synthesis

Milling is one of the oldest top-down techniques used for synthesis of nanomaterials. It is a high-energy synthesis process that breaks down larger particles into smaller ones or causes reactions in mixtures due to enormous mechanical forces. It is widely used to fabricate amorphous, crystalline, and composite NPs from elemental powders and compounds. NPs of Fe, Co, FeCo, $SmCo_5$, and NdFeB have been successfully synthesized using ball milling, producing particles of sizes less than 30 nm with a narrow size distribution.[12]

Various lithography techniques have been used for synthesizing NPs with determined final feature resolutions. A mask covers the required material while the unwanted material is etched using chemical, UV light, x-ray, or electron beam methods. Introduced in the 1960s, optical photolithography has since then been frequently used in the microelectronics industry. Optical lithography techniques make use of a collimated light passing through a quartz plate and a photoresist-coated semiconductor wafer to generate patterns on a surface. The light-sensitive photoresist can be either positive or negative. Using a positive photoresist causes the areas exposed to light to etch away, hence leaving the pattern as that of the mask. However, in the case of negative resist, the areas unexposed to light are etched away leaving a pattern complementary to the mask pattern. Resolutions of 1 μm can be achieved by conventional photolithography techniques.[13]

Extreme UV lithography uses higher wavelength x rays, thereby creating nanoscale patterns by utilizing reflected light marks and Bragg scattering.[14] Resolution augmentation through photo-induced deactivation (RAPID) is a new technique in which two simultaneous pulsed lasers of 800 nm cause cross-linking in the polymeric photoresist material due to multi-photon absorption and deactivation of photopolymerization due to single-photon absorption. The phase shaping of the deactivation beam enables one to synthesize features with resolutions as low as 40 nm.[15,16] A more superior, mask-less lithography approach is the electron beam (e-beam) lithography that exposes the top surface of the resist-coated material to the beam. It involves two-dimensional scanning of the surface of the material. The electron projection lithography permits a greater penetration depth than normal e-beam lithography and has been used successfully for fabrication of nanoscale devices.[17]

An alternative approach to photolithography is soft lithography that is inexpensive and allows transfer of thin, molecular patterns (even wet) onto a surface. This method utilizes a microscale replica of a polymer such as poly(dimethylsiloxane) (PDMS) and may be classified into microstamping, stencil, and microfluidic patterning. Stamping enables biomolecules such as peptides, proteins, and polysaccharides to be transferred onto a surface using a stamp. It is a single-step process allowing fabrication of large-scale patterns. The patterning uses PDMS moulds to create micro-sized channels on the substrate that are later filled by the fluid material. Controlled deposition of cells has been carried out using this technique with applications in tissue engineering.[18] Other lithography techniques include scanning probe lithography allowing to pattern nanometer scale features. Several approaches fall under this category including scanning tunneling microscopy (STM), atomic force microscopy (AFM), near-field scanning optical microscopy (NSOM), and so on.[13]

1.4.2 Bottom-Up Synthesis

Self-assembly of atoms and molecules, resulting in multifunctional nanomaterials, is the ultimate objective of this approach without the need to waste or eliminate parts from the final product as in the case of top-down approach. Even though bottom-up method results in smaller sized NPs in addition to a better shape and size control, there remains some challenges. These include, for instance, control of impurities, lattice site uniformity, quality of reactants, surface preparation, and conditioning for controlled deposition of atoms. Synthesis of nanomaterials using bottom-up approaches will be the primary focus of this book.

Nucleation is the first step in the formation of NPs in solution. The driving force for nucleation is governed by supersaturation. Once the stable nuclei are formed, growth of the nuclei sets in and this determines the final shapes of the particles. Concepts of nucleation and growth will be discussed in detail in Chaps. 2 and 3, respectively.

An ideal NP synthesized using solution-based processes consists of a core and passivating/stabilizing ligand, as shown in Fig. 1.3. Both the core and the ligand layer taken together contribute to the material properties and hence the final applications of NPs. For a brief introduction to bottom-up synthesis procedures, some of the most

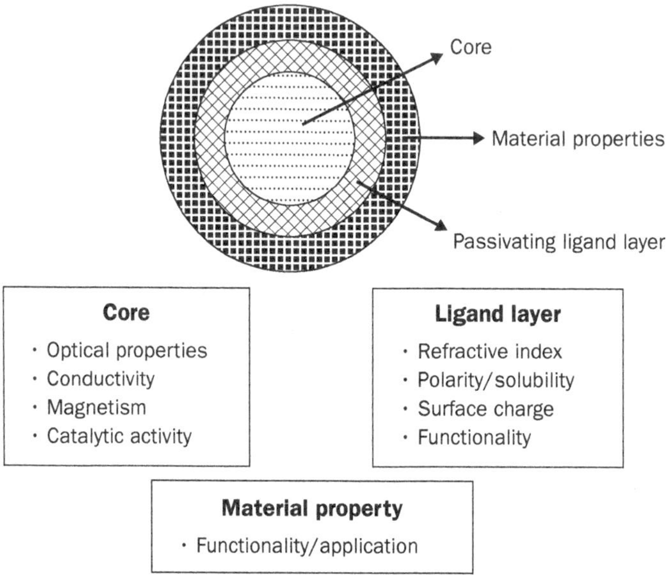

Figure 1.3 Schematic of nanoparticles synthesized using bottom-up approach showing various properties from the core and ligand layers.

commonly used techniques will be highlighted here and will be further detailed in Chap. 4.

Chemical reduction technique typically involves reducing metallic precursor salts by a reducing agent followed by stabilization of the particles with a passivating ligand. Many organic as well as inorganic reagents are used as reducing agents; sodium citrate, ascorbate, sodium borohydride, Tollens' reagent, *N,N*-dimethylformamide (DMF), and block polymers of poly-(ethylene glycol) are the common reducing agents for Au and Ag NP synthesis. As a rule of thumb, the generation of metallic element from the reduced salts further results in agglomeration into oligomeric clusters. These clusters are stabilized by ligands that prevent aggregation of formed NPs.

Sol-gel synthesis process is undoubtedly the most widely used process for fabricating metal oxides and alloys due to its cost-effective nature. Metal precursors, from a solution, are deposited on appropriate substrates and subsequently heat-treated to cause oxidation or sintering of the products. However, major focus has been put on hydrolysis and condensation of molecular precursors for the growth of various inorganic materials. Superhydrophobic surfaces and films utilizing silica have been synthesized using sol-gel techniques through dip coating, spin coating, spraying, or layer-by-layer deposition.

One of the recent methods for synthesizing and surface engineering complex nanomaterials is atomic layer epitaxy (ALE) or atomic layer deposition (ALD). It is based on the principle of sequential self-terminating surface reactions of gaseous precursors. ALD provides excellent control over the deposited layer thickness at atomic levels resulting in extremely thin, high-quality conformal films.[19] Core-shell NPs as well as particles with hollow shell morphologies can be fabricated by this technique in addition to obtaining crystalline surfaces thereby enhancing their physico-chemical properties sought after in applications such as chemical nanoreactors, enhanced emissions, targeted drug delivery, and imaging.[19,20]

Another synthesis method that does not rely on the external interactions is molecular self-assembly. Self-assembly occurs quite spontaneously in nature, for instance, the formation of lipid bilayer in cell membranes. Attempts have been made to mimic the forces of nature to control the intermolecular forces and their intensities between molecular systems. This could result in novel synthesis pathways to create unique nanomaterials. However, the processes of nature are relatively complex, thus making molecular self-assembly an empirical approach whose mechanism is not yet completely understood. In fabricating self-assembled monolayers, the layers are chemically synthesized and are bound together by weak van der Waals forces. Thiols and silanes are the most frequently used molecules for monolayer self-assembly. The presence of weak van der Waals forces allows reversible solution processing of self-assembled monolayers and results in economic fabrication of various nanostructures at an industrial scale. Mimicking of biological materials has enabled the use of this technique to generate nanomaterials for a large number of biological applications.

1.5 Functionalization

After the synthesis of NPs and nanostructures, they need to be appropriately functionalized in order to make them suitable candidates for various applications. There are two ways to achieve this: (1) *in situ* functionalization and (2) *post situ* functionalization. In *in situ* functionalization, NPs are functionalized with appropriate ligands during the synthesis process, while in *post situ*, functionalization is carried out after the synthesis of the NPs.

The functionalization of metallic NPs will be discussed in detail in Chap. 5. The adsorption of polymeric ligands on the metallic surface is often explained using Langmuir adsorption isotherm, a topic that would be dealt with in Chap. 5. Langmuir adsorption is one of the most used approaches to study macromolecule binding to metal surfaces and NPs, although the underlying assumptions are often not rigidly followed. For the discussion of assumptions and deviations refer to Chap. 5.

The binding ligands can also be stimuli-sensitive, which means they can respond to external stimuli such as temperature, pH, light, chemical environment, and so on. Instead of adsorbing such polymeric molecules on the surfaces of NPs, they can be grown atop the NPs. Synthesis and physico-chemical properties of such polymers or polymeric NPs will be discussed in Chaps. 7 and 8.

1.6 Characterization

Validation of the synthesized hybrid NPs is equally crucial before going to their applications to visually or analytically observe the consistency of the final products. Let us take the example of shopping for a dining table. Before deciding on the right table to buy, you will look at the size, color, surface texture, and practicality of fitting the table in the available space of the dining room, among other features. Same approach is applicable for the NPs, which makes characterization an essential component from synthesis to application.

There are numerous ways to characterize NPs. The *in situ* or solution-based methods include dynamic light scattering (DLS), UV-visible spectrometry (UV-Vis), differential scanning calorimetry (DSC), spectrofluorometry, and electrophoretic mobility measurements, to name a few. A few of the dry techniques used for NP characterization are transmission electron microscopy (TEM), x-ray photoelectron spectroscopy (XPS), Fourier transform infrared spectroscopy (FTIR), x-ray diffraction (XRD), energy-dispersive x-ray spectroscopy (EDX), and others. Atomic force microscopy (AFM) is a technique that can be used in dry as well as solution modes. Some of these techniques will be the subject matter of Chap. 8.

1.7 Applications

NPs have characteristics such as biocompatibility, environmentally friendly, ease of synthesis, and high surface functionality. It is because of these characteristics nanomaterials have found applications in various fields, as shown in Fig. 1.4. One of the fastest growing use of nanomaterials is in the biomedical industry. As already stated, NPs have shown their importance as potential targeted drug delivery systems in association with various drugs. Their small size along with their ease of administration has elevated their position as the carrier devices. Magneto-plasmonic NPs possess theranostic advantages as they can be used for imaging, diagnostics, and therapeutics simultaneously, especially in tackling the issue with cancer cells.

The high surface-to-volume ratios of these small NPs provide an extremely large number of reaction sites to undergo physical and chemical reactions, thereby opening doors into the field of catalysis. Nanoporous materials have been used as supports for catalytic applications as well as membranes for separation processes. These catalysts

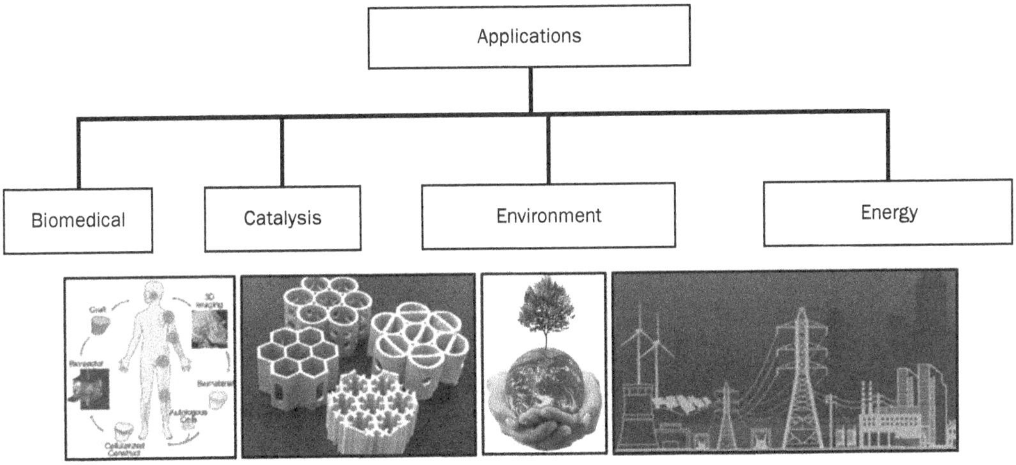

FIGURE 1.4 Different fields of applications of nanomaterials that will be covered in the book.[21]

are used not only in chemical processing but also in environment conservation, such as in cleaning up of oil spills and removing pollutants from air and groundwater. NPs are used for fabricating novel materials. Exceptional light absorption properties of plasmonic NPs facilitate their use in solar cell fabrication.

Innovative approaches have led to the use of NPs as energy harvesting devices in electrostatics, piezoelectrics, and electromagnets. Thus, they can be used as sensing devices, as devices for environmental monitoring, as personal electronic devices, and as defense equipment. It might be possible to efficiently harvest energy from winds and oceans in the future. Thus, there are innumerable application potentials offered by the field of nanoscience and nanotechnology. Some important applications of nanomaterials are in catalysis, drug delivery, water management, and energy harvesting, which will be discussed in Chaps. 9, 10, 11, and 12, respectively.

Countless NPs exist, which may offer limitless application possibilities. By exploring novel synthesis routes and techniques, it is possible to create NP devices based on the desired target end use. Hybrid NPs open a new regime of smart materials, which will take nanoscience to new heights in innovation and technology at an unprecedented pace. Although there is still a long way to go in understanding the underlying mechanisms of NP synthesis and physico-chemical properties, you shall unlock these unanswered perceptions as the characterization techniques improve in the future allowing us to better understand the world of nanomaterials.

1.8 Study Questions

1.1 Compare and contrast the bottom-up and top-down approaches for nanomaterial fabrication.

1.2 What is meant by functionalization of NPs? What are the different roles functionalization can serve?

1.3 List two examples each of top-down and bottom-up approaches that may be used for synthesizing metallic NPs.

1.4 Mention some characteristic features of NPs that make them important candidates in the biomedical industry.

1.5 Mention some important characterization techniques that can be used to determine the physico-chemical properties of NPs in solution.

1.9 References

1. Feynman, R. P., There's plenty of room at the bottom. *Engineering and Science.* **1960,** *23*(5), 22–36.
2. Willets, K. A.; Van Duyne, R. P., Localized surface plasmon resonance spectroscopy and sensing. *Annual Review of Physical Chemistry.* **2007,** *58*, 267–297.
3. Anker, J. N.; Hall, W. P.; Lyandres, O.; Shah, N. C.; Zhao, J.; Van Duyne, R. P., Biosensing with plasmonic nanosensors. *Nature Materials.* **2008,** *7*(6), 442–453.
4. Haes, A. J.; Van Duyne, R. P., A unified view of propagating and localized surface plasmon resonance biosensors. *Analytical and Bioanalytical Chemistry.* **2004,** *379*(7–8), 920–930.
5. Maier, S. A.; Atwater, H. A., Plasmonics: localization and guiding of electromagnetic energy in metal/dielectric structures. *Journal of Applied Physics.* **2005,** *98*(1), 10.
6. Kelly, K. L.; Coronado, E.; Zhao, L. L.; Schatz, G. C., The optical properties of metal nanoparticles: the influence of size, shape, and dielectric environment. *Journal of Physical Chemistry B.* **2003,** *107*(3), 668–677.
7. Zhang, J. Z.; Noguez, C., Plasmonic optical properties and applications of metal nanostructures. *Plasmonics.* **2008,** *3*(4), 127–150.
8. Krahne, R.; Morello, G.; Figuerola, A.; George, C.; Deka, S.; Manna, L., Physical properties of elongated inorganic nanoparticles. *Physics Reports.* **2011,** *501*(3–5), 75–221.
9. Rockenberger, Joerg; Scher, E. C.; Alivisatos, A. P., A new nonhydrolytic single-precursor approach to surfactant-capped nanocrystals of transition metal oxides. *Journal of the American Chemical Society.* **1999,** *121*(49), 11595–11596.
10. Dabbousi, B. O.; Rodriguez-Viejo, J.; Mikulec, F. V.; Heine, J. R.; Mattoussi, H.; Ober, R.; Jensen, K. F.; Bawendi, M. G., (CdSe) ZnS core–shell quantum dots: synthesis and characterization of a size series of highly luminescent nanocrystallites. *Journal of Physical Chemistry B.* **1997,** *101*(46), 9463–9475.
11. Gutjahr, A.; Phelip, C.; Coolen, A.-L.; Monge, C.; Boisgard, A.-S.; Paul, S.; Verrier, B., Biodegradable polymeric nanoparticles-based vaccine adjuvants for lymph nodes targeting. *Vaccines.* **2016,** *4*(4), 34.
12. Chakka, V.; Altuncevahir, B.; Jin, Z.; Li, Y.; Liu, J., Magnetic nanoparticles produced by surfactant-assisted ball milling. *Journal of Applied Physics.* **2006,** *99*(8), 08E912.
13. Gates, B. D.; Xu, Q.; Stewart, M.; Ryan, D.; Willson, C. G.; Whitesides, G. M., New approaches to nanofabrication: molding, printing, and other techniques. *Chemical Reviews.* **2005,** *105*(4), 1171–1196.
14. Wu, B.; Kumar, A., *Extreme Ultraviolet Lithography.* New York: McGraw-Hill, 2009.
15. Li, L.; Gattass, R. R.; Gershgoren, E.; Hwang, H.; Fourkas, J. T., Achieving $\lambda/20$ resolution by one-color initiation and deactivation of polymerization. *Science.* **2009,** *324*(5929), 910–913.
16. Schmid, G. M.; Miller, M.; Brooks, C.; Khusnatdinov, N.; LaBrake, D.; Resnick, D. J.; Sreenivasan, S.; Gauzner, G.; Lee, K.; Kuo, D., Step and flash imprint lithography for manufacturing patterned media. *Journal of Vacuum Science & Technology B:*

Microelectronics and Nanometer Structures Processing, Measurement, and Phenomena. **2009,** *27*(2), 573–580.
17. Berger, S.; Gibson, J.; Camarda, R.; Farrow, R.; Huggins, H.; Kraus, J.; Liddle, J., Projection electron-beam lithography: a new approach. *Journal of Vacuum Science & Technology B: Microelectronics and Nanometer Structures Processing, Measurement, and Phenomena.* **1991,** *9*(6), 2996–2999.
18. Betancourt, T.; Brannon-Peppas, L., Micro-and nanofabrication methods in nanotechnological medical and pharmaceutical devices. *International Journal of Nanomedicine.* **2006,** *1*(4), 483.
19. Kim, H.; Maeng, W.-J., Applications of atomic layer deposition to nanofabrication and emerging nanodevices. *Thin Solid Films.* **2009,** *517*(8), 2563–2580.
20. King, D. M.; Johnson, S. I.; Li, J.; Du, X.; Liang, X.; Weimer, A. W., Atomic layer deposition of quantum-confined ZnO nanostructures. *Nanotechnology.* **2009,** *20*(19), 195401.
21. http://homepage.ntu.edu.tw/~pgchao/ortho.htm
http://jxhxhg.com/view24-25.html
http://itemmom.com/index.php/making-a-difference-one-plant-at-a-time/world-environment-day-image/
http://energywindowmedia.com/en/electricity-had-greater-share-of-investment-two-straight-years-iea/

CHAPTER 2
Nucleation

Crystallization is a commonly used bottom-up approach in the synthesis of inorganic NPs. It is a first-order phase transition process where matter is transformed from a disordered state with high free energy to a crystal phase associated with a low free energy. For crystallization to occur from vapor, melt, or solution, the solute-containing media have to be supersaturated, i.e., in a non-equilibrium state, which creates the excess free energy, and thus the driving force for crystallization. Synthesis of crystals from vapor or melt media is seldom carried out, whereas solution crystallization is used for a broad range of industrial and laboratory processes. Therefore, the discussion in this and next chapters will mainly refer to solution crystallization.

Nucleation is the first stage of crystallization that represents the birth of the new phase within the bulk. Since it is important to control nucleation for many crystallization applications, it is of primary importance to acquire a fundamental understanding of nucleation pathways. The classical nucleation theory (CNT), which originated in the 19th century, has provided the basic theoretical framework for understanding nucleation on a thermodynamic basis. It describes the nuclei formation via the dynamic and stochastic association of monomeric units that overcome a free energy barrier at a critical size and result in a phase separation in the system.[1] According to the CNT, the so-called monomeric units can be ions, atoms, or molecules depending on the nature of the forming material, which come together and part again until reaching a critical size that forms a stable cluster in solution, the nucleus.

The CNT has proven to be useful in analyzing experimental data related to nucleation kinetics and parameters related to it; however, the error margin for kinetic data can have high orders of magnitude. In addition, it does not yet provide information about the structural characteristics of clusters or comprehensive insight into the pathways leading from the solution to the precipitated solid.[2,3] Rather, the CNT withholds certain assumptions regarding the clusters that simplify the thermodynamic framework it is built on, such as neglecting interactions between clusters and size-dependency of surface energy.[4] The CNT has been criticized on these bases and questions raised on some experimental observations related to complex cluster structures, which have led alternative hypotheses for nucleation to emerge. However, it still stands as the most commonly used theoretical framework on nucleation and with further research, its apparent limitations can be overcome via more extensive and elaborate interpretations. This chapter covers the fundamental background of the CNT and its implications in detail, and also mentions the alternative hypotheses in order to better comprehend the strengths and weaknesses of this theory.

2.1 Classical Nucleation Theory

According to the CNT, the thermodynamic background for nucleation lies on the reduction of free energy of the system by the formation of solid phase, and supersaturation is a prerequisite for spontaneous crystallization reactions. It is often easy to generate supersaturation in a solution by applying any kind of process that brings the solution concentration above the solubility limit of the material of interest. For solids that dissolve endothermically, cooling is a common method used to generate supersaturation. Alternatively, in a solution, concentration of the solute can be increased in a controlled manner via *in situ* chemical reactions; for example, reduction from their salt is a commonly employed method to prepare supersaturated solutions for metallic solids. Once the solubility limit is crossed, solution is said to be supersaturated and the system enters the metastable zone where crystallization can occur. Most industrial and laboratory crystallization processes are carried out at the metastable zone since further increase in supersaturation can lead to abrupt precipitation without the ability to exert any control on it at the labile zone. Although the solubility diagram in Fig. 2.1 gives a clear idea of concentration-based solubility and supersaturation concept, it has very limited applicability and huge deviations from reality in terms of the thermodynamic quantification the CNT is based on. The CNT employs *activity-based supersaturation* as the fundamental driving force for the spontaneous nucleation to occur, which deduces from the chemical potential difference between the metastable and stable states of a system.

2.1.1 Activity-Based Supersaturation

In solution crystallization, the fundamental driving force is the difference between the chemical potential of a given substance in the solution (state 1) and in the solid phase (state 2):[5]

$$\Delta\mu = \mu_1 - \mu_2 \tag{2.1}$$

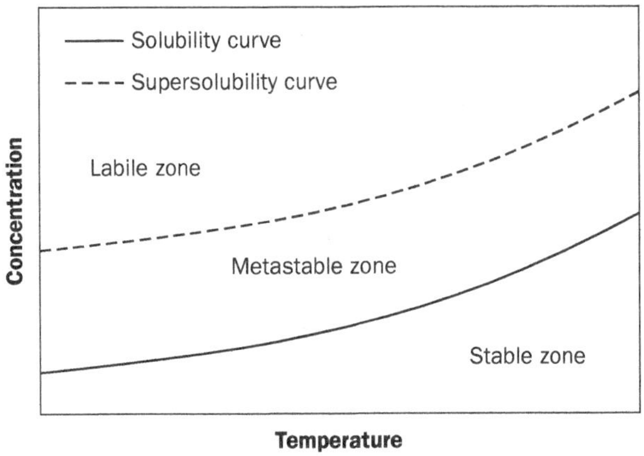

Figure 2.1 Concentration versus temperature graph for a solid that dissolves endothermically showing the stable, metastable, and labile zones separated by solubility curves.

The chemical potential, μ, can be expressed in terms of the standard potential, μ_0, and the activity, a, as:

$$\mu = \mu_0 + RT\ln a \tag{2.2}$$

where, T is the temperature and R is the universal gas constant.

The activity of a solute is a measure of its *effective concentration* in solution and it deviates from concentration (C) by a factor of its activity coefficient, γ, in a non-ideal solution [Eq. (2.3)]. Activity coefficient depends on the ionic strength of the solution, charge, and size of each component due to the interactions between chemical species.

$$a = C\gamma \tag{2.3}$$

By combining Eqs. (2.1) and (2.2), the driving force for the crystallization can be expressed as:

$$\Delta\mu = RT\ln\left(\frac{a}{a^*}\right) \tag{2.4}$$

where a is the activity in an arbitrary state and a^* is the activity of the same solute at equilibrium. The activity ratio given in Eq. (2.4) is defined as supersaturation, S, and it is a measure of the tendency of a system to precipitate. The difference in the chemical potential must be larger than zero for crystallization to occur spontaneously, which provides $S > 1$.

$$S = \left(\frac{a}{a^*}\right) \tag{2.5}$$

For electrolytes that dissociate upon dissolution, the mean ionic activity, $a_\pm$, should be used to calculate supersaturation, which is the multiplication of the activity of all anions and cations in solution that forms the solid. In this case the numerator in Eq. (2.5) is defined as the ionic activity product (IAP), and the denominator is named as the solubility product, K_{sp}, which gives the ionic activity product of the cations and anions at equilibrium state for the corresponding solid. Then, supersaturation can be defined in the same manner as given in Eq. (2.5):

$$S = \left(\frac{a_\pm^\nu}{a_\pm^{*\nu}}\right)^{\frac{1}{\nu}} = \left(\frac{IAP}{K_{sp}}\right)^{\frac{1}{\nu}} \tag{2.6}$$

where v is the number of moles of ions in 1 mole of electrolyte.

The definition of supersaturation can vary between different sources, although the point of origin for any defined concept is the chemical potential difference between two phases, or in other words the driving force for crystallization. While it is possible to represent the driving force for crystallization and supersaturation in multiple ways, great caution should be taken to clearly present the employed terminology.

Once the supersaturation is established in a solution, the metastable system allows for nucleation, which represents the appearance of the new phase within the bulk solution. Nucleation events can be classified as primary and secondary nucleation, where primary nucleation processes represent all cases of nucleation in systems that do not contain the forming crystalline matter and secondary nucleation occurs when parent crystals catalyze the formation of a new crystalline phase of the same substance. Secondary nucleation mechanisms are evaluated under two categories in general as originating from the parent crystal or from the concentrated liquid layer adjacent to the crystal surface.[6] The first mechanism suggests breeding of new nuclei via breaking of fast-growing parent crystals, whereas the second mechanism proposes enhanced possibility of nucleation within the concentrated adhesion layer upon its removal into the solution.[7] However, both of these mechanisms call for shear forces being active on the particles, which require a micron size range. At smaller size, particles do not experience the shear forces and consequently, the secondary nucleation mechanisms become irrelevant for the formation of NPs. Thus, this chapter will cover only the primary nucleation mechanisms.

2.1.2 Primary Nucleation

Primary nucleation can further be divided into homogenous and heterogeneous nucleation. Homogenous nucleation occurs spontaneously in the system, whereas heterogeneous nucleation is induced by the presence of a solid surface that can be the surface wall of the vessel, a foreign particle, or an impurity in the medium.

2.1.2.1 *Primary Homogeneous Nucleation*

In the case of primary homogeneous nucleation, where nuclei emerge from solution on its own accord, the total Gibbs free energy change, ΔG, can be expressed as the sum of the free energy change for the phase transformation, ΔG_v, and the free energy change for the formation of a surface ΔG_s [Eq. (2.7)].[5]

$$\Delta G = \Delta G_v + \Delta G_s \tag{2.7}$$

The first term on the right hand side of Eq. (2.7) expresses the excess free energy of the metastable solution over the solid deposition. This term is negative and proportional to the volume of the crystallite. The second term expresses the energy cost of creating a solid–liquid interface, thus it is positive and increases proportional to the surface area of the crystallite. Thus, ΔG can be formulated to:

$$\Delta G = V\Delta G_v + A\gamma \tag{2.8}$$

where V is the volume and A is the surface area of the nucleus, ΔG_v is the free energy change of transformation per unit volume, v_m, and γ is the surface free energy of the solid–liquid interface. Then, the overall free energy change due to nucleation can be expressed for a solid body of any geometric shape with a known dimension. For the ease of notation here, ΔG is given for a spherical nucleus with the radius r:

$$\Delta G = \frac{4}{3}\pi r^3 \Delta G_v + 4\pi r^2 \gamma \tag{2.9}$$

where ΔG_v represents the chemical potential difference that induces the phase transformation and scales with supersaturation as follows. k_B is the Boltzmann constant.

$$\Delta G_v = -\frac{\nu k_B T \ln S}{v_m} \tag{2.10}$$

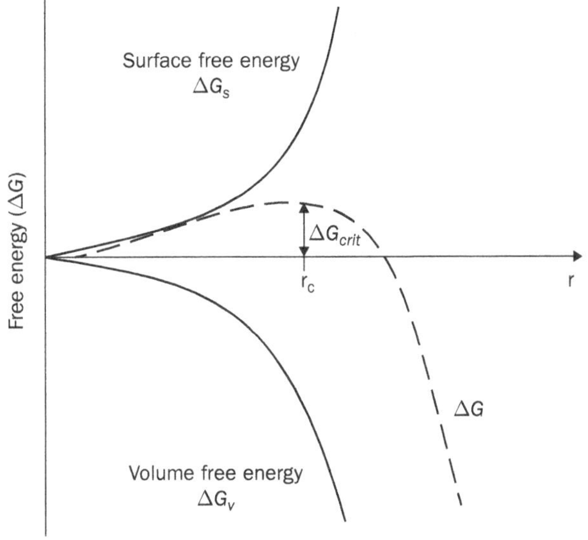

FIGURE 2.2 The change of free energy as a function of radius for a spherical nucleus.[8]

According to the CNT, a stable nucleus can form at a critical cluster size when its excess free energy balances the energetic costs of creating a surface. Figure 2.2 shows the dependence of the total Gibbs free energy change in a nucleating system to the cluster size with the contributions of volume and surface free energies. At small sizes, the area-dependent surface energy term surpasses the volume-dependent term and dominates the total free energy change. Thus, stabilization of these clusters becomes energetically unfavorable in the system due to the overall increase in the free energy. A stable nucleus can form only when a critical size is reached so that the slope of the ΔG curve attains a negative value.

The point where the first derivative of Eq. (2.9) with respect to r equals zero represents the critical radius size, r^*, and denotes the energy barrier, ΔG^*, which must be overcome to form a spherical nucleus from solute molecules:

$$r^* = -\frac{2\gamma}{\Delta G_v} = \frac{2\gamma v_m}{\nu k_B T \ln S} \tag{2.11}$$

$$\Delta G^* = \frac{16\pi\gamma^3}{3\Delta G_v^2} = \frac{16\pi\gamma^3 v_m^2}{3k_B^3 T^3 (\nu \ln S)^2} \tag{2.12}$$

Equations (2.11) and (2.12) clearly illustrate the effective parameters on the critical size and energy barrier for primary homogeneous nucleation as temperature, interfacial energy, and supersaturation.

In practice, calculation of the volume or surface area of solids are often based on methods that measure one dimension of the particles such as equivalent diameter.

Then, the same deductions can be made for any geometry to express ΔG^* by using the volume and surface shape factors, k_v and k_a, that relate the volume and surface areas of a particle to its measured dimension, L:

$$V = k_v L^3 \tag{2.13}$$

$$A = k_a L^2 \tag{2.14}$$

The general expression for ΔG can be given as below and ΔG^* is deduced from the first derivative of ΔG with respect to L:

$$\Delta G = k_v L^3 \Delta G_v + k_a L^2 \gamma \tag{2.15}$$

$$\Delta G^* = \frac{4k_a^3 \gamma^3}{27k_v^2 \Delta G_v^2} \tag{2.16}$$

2.1.2.2 Primary Heterogeneous Nucleation

In the case of heterogeneous nucleation, Eqs. (2.11) and (2.12) remain intact to define the critical size and the energy barrier for nucleation. However, when nucleation occurs on a foreign surface, the interfacial free energy term needs to be modified to include contributions of both the interfacial energies between the crystal and solution and the other between the crystal and the substrate. Since the interfacial energy between a crystal nucleus and a solid substrate is typically lower than that of the crystal in contact with the solution, the presence of foreign bodies results in reducing the effective interfacial free energy, and thus the energy barrier for nucleation.[9] The extent of this reduction depends on the affinity between the substrate surface and the forming nucleus, i.e., wettability.

In the presence of a substrate surface, the interfacial energies between the solution, substrate, and nucleus are balanced at a contact angle, θ, between the nucleating phase and the substrate surface as expressed by Young's equation, as given in Eq. (2.17) (Fig. 2.3):

$$\gamma_{ns} + \gamma_{nl} \cos\theta = \gamma_{sl} \tag{2.17}$$

The contact angle is a measure of the affinity between the two surfaces, and the stronger bonding between the nucleus and the substrate leads to lower values of θ, which denotes a lower interfacial energy. The overall energy barrier for heterogeneous nucleation can be expressed by taking into account this reduction in surface energy by a factor, ϕ, which depends on the contact angle:

$$\Delta G_{het}^* = \phi \Delta G_{\text{hom}}^* = \left(\frac{(2+\cos\theta)(1-\cos\theta)^2}{4} \right) \Delta G_{\text{hom}}^* \tag{2.18}$$

Equation (2.18) shows that when θ is less than 180°, the energy barrier for nucleation is lower than that of the homogeneous nucleation. Since ϕ is a factor of θ and independent of the nucleus radius, r, the critical radius does not change for heterogeneous nucleation (Fig. 2.3D).

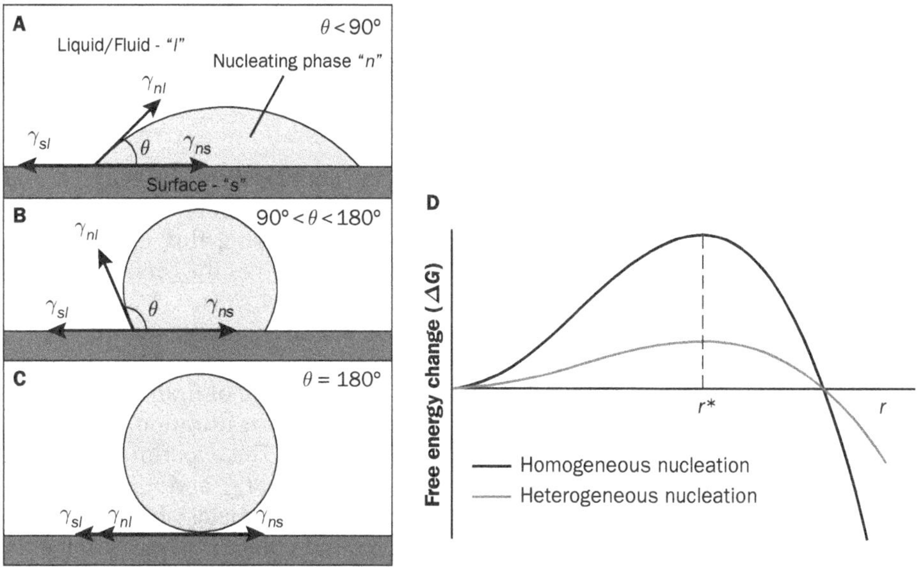

FIGURE 2.3 Illustration of the equilibrium described by Young's equation. The interfacial tensions γ_{sl}, γ_{ns}, and γ_{nl} between substrate surface (s), nucleus (*n*), and solution (*l*) are balanced at a contact angle θ between the nucleating phase and the surface. (A) Favorable nucleus–surface interactions result in $\theta<90°$. (B) Unfavorable nucleus–surface interactions (poor wetting) result in $90°<\theta<180°$. (C) A complete absence of wetting where $\Delta G^*_{het} = \Delta G^*_{hom}$.[10] (D) The different energy barriers for homogeneous and heterogeneous nucleations.

The energy barrier for homogeneous nucleation is usually too high to enable precipitation and it is difficult to eliminate the presence of foreign particles in a precipitating system. Thus, most of the primary nucleation events happen via heterogeneous pathways in practice.[8] In this case, the resulting number of nuclei that forms via heterogeneous nucleation equals the expected number of foreign particles in regular laboratory/industry operations, which is estimated at 10^6–10^8 particles per cubic centimeter. Homogeneous nucleation can be achieved only under very high supersaturation, which in turn results in the formation of much higher numbers of particles.

2.1.3 Nucleation Rate

Once the energy barrier for nucleation is defined, it can be used to express the nucleation rate in the form of the Arrhenius reaction rate equation. The nucleation rate, *J*, given in Eq. (2.19) expresses the number of stable nuclei formed per unit time per unit volume.

$$J = A\exp\left[-\frac{\Delta G^*}{k_B T}\right] \tag{2.19}$$

The magnitude of the free energy barrier to nucleation with respect to the thermal energy, $k_B T$, is a critical factor determining the number of particles produced; thus, the rate of nucleation is efficiently influenced by the thermodynamic parameters defining ΔG^*. The pre-exponential factor, A, deals with the kinetic factors of nucleation and is given as:

$$A = zf^*C_0 \tag{2.20}$$

The Zeldovich factor, z, corrects for the fraction of critical size nuclei that dissolve rather than grow; f^* represents the attachment frequency of building units to nucleus and can be represented as the multiplication of incoming flux of monomers to the nucleus surface and the nucleus surface area; and C_0 gives the concentration of nucleation sites.[11,12]

Direct measurement of nucleation rates is not an easy task and in practice nucleation rates are mostly determined in relation to induction time. Between the establishment of supersaturation in a system and the appearance of first stable nuclei, a time period usually elapses depending on the level of supersaturation, state of agitation, viscosity of the solution, presence of impurities, etc. The experimentally observable value of this time period is referred as induction time, t_{ind}, and it is a measure of the ability of a supersaturated solution to remain in this metastable state. The rate of nucleation, J, in a system can be related to the induction time measurements by the following expression:

$$J \approx \frac{N}{t_{ind}} \tag{2.21}$$

where N is the number of stable nuclei formed. When the appearance of the stable nuclei is responsible for bringing the solution out of the metastable equilibrium, and the experimental technique allows for counting N effectively, induction time can be given as:[2]

$$t_{ind} = \frac{N}{JV} \tag{2.22}$$

where V is the solution volume.

In most applied cases, however, the induction time measurements depend on monitoring of certain variables in the system such as pH, turbidity, conductivity, etc., until a detectable change occurs.[13] Although induction time cannot be regarded as a fundamental property of a system, it has been shown to provide useful information about the nucleation rate and mechanism. Combining Eqs. (2.19) and (2.21) establishes the relationship between J and t_{ind} [Eq. (2.23)], which allows for deduction of information on nucleation mechanisms and interfacial energy (γ), as shown in Fig. 2.4.

$$\ln t_{ind} \sim \frac{\gamma^3}{T^3(\ln S)^2} \tag{2.23}$$

In practice, the breakdown of the solution metastability is mostly due to progressive appearance of many nuclei and their overgrowth to macroscopically large sizes. Therefore, experimentally determined induction times can be considered as the sum of the time lapse until formation of the stable nuclei and the time lag until the onset of a

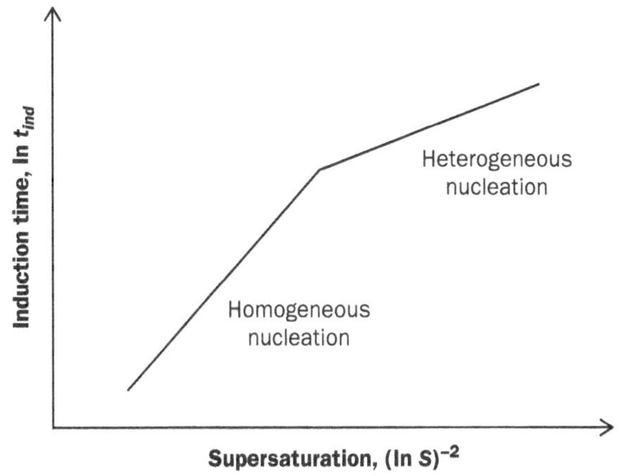

FIGURE 2.4 Logarithmic plot of induction time as a function of initial supersaturation. Change of slope indicates a transition between homogeneous and heterogeneous nucleation mechanisms. Note that slope of the lines relates to interfacial energy.

significant detectable change in the system by crystal growth.[14] Then, the induction time can be given as:

$$t_{ind} = \sqrt[4]{\frac{3\alpha_v}{\pi R^3 J}} \tag{2.24}$$

when nuclei grow as spheres with a time-independent growth rate, R, and α_v is the detectable volume (or mass) fraction of the new phase.[15] Hence, the use of Eq. (2.24) is valid when the experimental method used detects the induction time based on the appearance of a certain volume (or mass) of many nucleating crystallites.

2.1.4 Phase Stability and Phase Transitions

For a defined set of experimental conditions (e.g., temperature, pressure, and composition), the solid phase resulting in the minimum free energy in the system is thermodynamically the most stable one and all other phases are metastable with respect to this phase under the same conditions. The classical theory of crystallization assumes formation of the thermodynamically most stable phase out of a supersaturated solution. However, it does not dictate that the supersaturated system transforms directly into the most stable phase. Instead, although metastable, intermediate phases can still appear and exist in the system for relatively long periods of time.

When a solution is supersaturated with respect to multiple phases, each phase can nucleate at a rate given by Eq. (2.19). Although the value of supersaturation will be the largest for the most stable phase with the lowest solubility, the interfacial energy of metastable phases are typically lower. As a result, metastable phases can have lower energy barriers of formation and initially appear in a supersaturated system as a coaction of thermodynamic and kinetic factors.[16] The interfacial energy contributions can further become comparable to the phase stability differences and gain more importance on determining the precipitation pathways with small particle size in nanometer

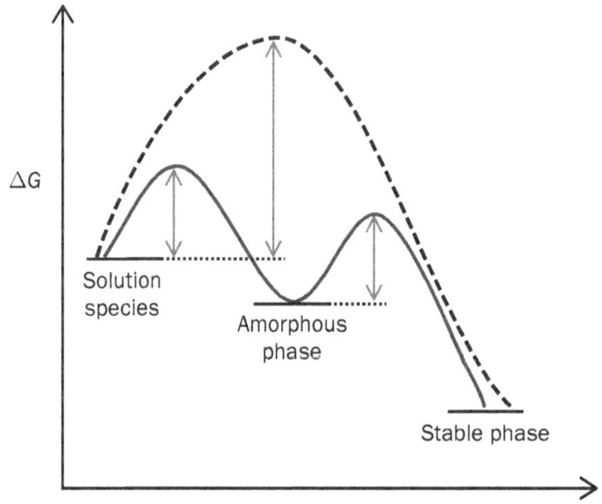

FIGURE 2.5 Reaction coordinate diagram demonstrates alternative pathways to lower Gibbs free energy: direct transformation to stable phase (dashed line) or transformation via metastable intermediates (solid line).

range.[17,18] Given the sufficient time and the free energy of activation, they eventually transform into the more stable phase by dissolution and re-precipitation reactions. This stepwise formation behavior, as shown in Fig. 2.5, is denoted as Ostwald's rule of stages and particularly favorable in systems where the final stable phase has a high activation energy barrier for formation. Observation of amorphous phases prior to crystallization in many systems stands as an example of this rule where crystalline phases have high interfacial energies and large unit cell volumes.[19,20]

Metastable phases are characterized by local minima in the energy landscape and are potentially capable of transforming to the stable phase. The driving force for this transformation process is minimization of the total free energy of the system that originates from the solubility difference between the metastable and stable phases.[21] In a solution in equilibrium with the metastable solid and supersaturated with respect to stable phase, the solvent-mediated transformation initiates by the appearance of the nuclei of the stable phase. As these nuclei grow, solution becomes undersaturated with respect to the metastable phase. Thus, the metastable solid dissolves and produces supersaturation for the continuing growth of the stable phase. This dissolution-growth process continues until metastable phase disappears and the transformation is complete.[22]

2.2 Alternative Nucleation Hypotheses

The classical nucleation theory presents a robust theoretical background and useful quantitative expressions, yet there are some questions unanswered concerning the nucleation phenomenon as mentioned in the beginning of this chapter. Despite vigorous research in this field, in-depth information has stayed unknown about nucleation mainly due to the limitations of analytic techniques to monitor such small-scale processes in real time and *in situ*. The attempts devoted to overcome this challenge and advancement of both experimental [wide-angle x-ray scattering (WAXS), small-angle x-ray scattering (SAXS), cryo-transmission electron microscopy (cryo-TEM)]

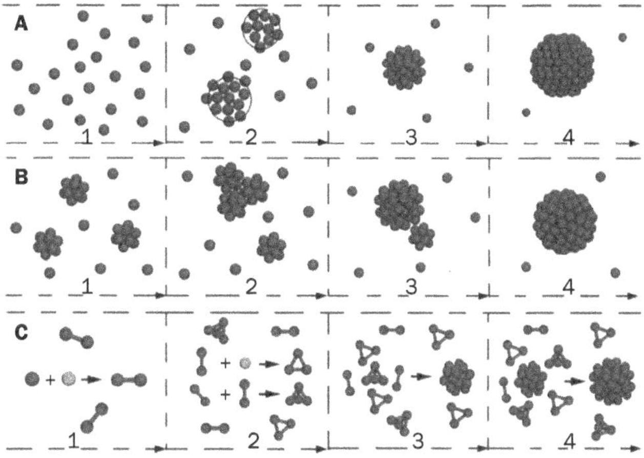

Figure 2.6 Schematic illustrations of nucleation routes for metallic nanocrystals: (A) classical nucleation pathway via monomer addition, (B) cluster-aggregation-mediated nucleation, and (C) nucleation through aggregation of complexes.[23]

and theoretical (molecular dynamic simulations, density functional theory calculations) techniques have delivered new observations and complementary alternative hypotheses on the nucleation pathways.[23] In contrast to nuclei formation by monomer attachment model proposed in the CNT, particle-based pathways for nucleation via aggregation and reorganization of solution species have been suggested in an increasing number of studies during the last decade. Generally referred to as pre-nucleation clusters (PNCs), the aggregates are described to comprise multi-ion complexes or clusters with a disordered nature. According to these alternative hypotheses, clusters that form prior to nuclei do not lead to immediate growth, instead, through further aggregation and internal reorganization, they aid in nucleation. For example, in the case of metallic NPs, recent research has intensively proposed aggregation of clusters or ionic complexes to mediate the nucleation rather than nuclei formation by monomer attachment model (Fig. 2.6A).[23] The cluster aggregation pathway suggests formation of small size clusters initially, which later interact with each other to form the stable nuclei via aggregation (Fig. 2.6B).[24] According to the complex-mediated nucleation, the metallic ions that are not fully reduced first form complexes via ionic interactions, and then, they further aggregate and reduce to form nuclei (Fig. 2.6C). Both of these pathways suggest prior steps to nucleation through transient species with dynamic structures.

Experimental observations of intermediate steps in various crystallization systems and their endorsement via dynamic simulations have further encouraged the suggestion of the multistep mechanism via particle aggregation and reorganization as a universal mechanism underlying most crystallization processes from solution.[25] The driving force behind the formation of intermediates has been assigned as resulting in a local or global minimum in free energy, which gives the PNCs either a metastable or stable nature, respectively, with respect to the monomers in solution.[9] It is important to note here, the very nature of these clusters initially defines whether the proposed hypothesis clearly contradicts to the CNT or can be assessed as a supplementing idea to the classical concept by taking free energy landscapes and kinetic contributions into account. Since the free energy landscape of a system can possess multiple local minima associated with different structural states and varying energy barriers, nucleation can follow a complex pathway including the evolution of the separating phase.[26,27] Several studies have shown such an approach to evaluate the presence of metastable pre-nucleation species within the classical concept via introducing small extensions to the CNT to account for the manifold of nucleation and pre-nucleation phenomena known today.[28-31] The notion of stable PNCs is clearly contradictive to the CNT as their formation is proposed to be entropically driven without any energy barrier to overcome and they lower the free energy of the system.[32] However, strong criticism has been raised against the stable PNCs as nucleation promoters. Molecular simulations have suggested that the energetically favorable nature assigned to them gives rise to a higher energy barrier for the transition of stable PNCs to crystalline nuclei than the energy barrier for nucleation to take place within the bulk solution.[30] Other studies have shown that stable PNCs decrease the effective supersaturation, thus the probability of nucleation in a system.[28,29] These contradicting results display the lack of quantitative models for particle-based nucleation pathways, which arise another discord between classical and non-classical theories while defining the effective variables on nucleation kinetics. The CNT shows successful implementations in defining the effective parameters on nucleation kinetics as interfacial free energy, temperature, and supersaturation, where the latter is the most dominant variable. Although there are no unified nucleation rate equations proposed for aggregation mechanisms, reorganization of clusters is a suggested rate determining step for particle-based nucleation. However, the nucleation rate then becomes independent of solution supersaturation, which shows clear contradiction to the CNT.

2.3 Concluding Remarks

Nucleation is the first stage of a crystallization process; therefore, the ability to exert any control over nucleation is an important step toward fabrication of particles with desired properties. The classical nucleation theory offers a fundamental framework to nucleation and identifies the effective parameters acting on nucleation kinetics. Once recognized, these parameters can be used to manipulate important properties of the final particles such as crystal phase, particle size, and size distribution.

Nucleation stage is highly deterministic on the final size and size distribution of particles since crystal size is determined by the competing mechanisms of nucleation and growth over the consumption of supersaturation. High nucleation rates result in formation of many particles with small size, which can be achieved by tuning the effective parameters of temperature, interfacial free energy, and most dominantly, supersaturation. In quantitative terms, the nucleation rate can be regulated most effectively as a function of supersaturation. However, while doing so, special concern should be given to stepwise precipitation pathways and phase transitions that can be induced under changing driving force due to both thermodynamic and kinetic effects. Using additives is another method to influence nucleation rate via effects of additives either on interfacial free energy or on kinetics of nucleation. Promoted nucleation rates can be observed if the additives decrease the energy barrier for nucleation by offering heterogeneous nucleation sites and reducing the interfacial energy parameter. However, the presence of an additive might interfere with the kinetic factors of nucleation that are expressed in the pre-exponential factor, A, in the nucleation rate equation [Eq. (2.19)].

In the cases where additives cause repulsion forces, interfere with the incoming flux of monomers to the nuclei surface, or block the nucleation sites, their presence can lead to decreased nucleation rates.

Along with size, uniform size distribution can also be an important property in many applications of NPs. A commonly used strategy to ensure formation of small particles with narrow size distribution is burst nucleation, which refers to an experimental scheme that allows for intense nucleation during a short period of time without significant change in supersaturation, subsequently followed by growth. When nuclei are formed under constant conditions that provide a constant critical size and are subjected to similar growth conditions in terms of supersaturation and time of growth, a uniform size distribution is expected.

As presented in this chapter, the CNT offers qualitative and quantitative insight into the complex behavior of nucleating systems, and a means to control particle properties. The ongoing progress in nucleation studies, however, demonstrates the need for the CNT to be expanded to cover the variety of nucleation pathways, which require considering both thermodynamic and kinetic factors.[9]

2.4 Study Questions

2.4.1 Concept-Based Questions

2.1 What is the generic condition necessary for spontaneous crystallization to occur in a solution?

2.2 What is the fundamental driving force for nucleation?

2.3 Explain the relationship between the activity and concentration of a solute in solution.

2.4 Define the two opposite energy contributions to the total Gibbs free energy change for nucleation.

2.5 What does the "critical size" refer to in classical nucleation theory?

2.6 Define homogeneous and heterogeneous nucleations and compare them in terms of their energy barriers and critical sizes for nucleation.

2.7 Name three dominant variables that determine the nucleation rate (J) in a solution.

2.8 Explain the relationship between the nucleation rate (J) and induction time.

2.9 What does the Ostwald's rule of stages refer to?

2.10 Derive an expression for the homogeneous nucleation rate. Based on that expression, explain how the interfacial tension of a precipitating solid can be estimated and which experimental data you would need.

2.11 The nucleation rate is difficult to measure experimentally. An indirect way is to analyze measurements of the induction time:

$$t_{ind} = \sqrt[4]{\frac{3\alpha_v}{\pi R^3 J}}$$

The induction time was measured as 10 seconds for $CaCO_3$ precipitated from solution at 25°C and a supersaturation of $S = 5$. The overall rate of crystal growth is calculated as $1.6 \cdot 10^{-10}$ m/s.

a. What is the nucleation rate if $\alpha_v = 1.10^{-8}$?

b. How many particles will be produced if all the crystals are nucleated within the induction time without any reduction in supersaturation?

c. What is the most likely nucleation mechanism in this case?

d. Suggest a way to reduce the particle size at these conditions.

2.4.2 Research-Based Questions

2.12 In their study, Lu et al. synthesized nickel (Ni) particles and they tailored the particle size by forming palladium (Pd) particles in the reaction solution prior to Ni precipitation via reduction from its salt.[33] There was a dramatic decrease in Ni particle size accompanied with higher numbers of particles with increasing amounts of Pd, as shown in the figure below. Suggest a possible explanation on how the concentration of Pd effects the final size and number of Ni particles.

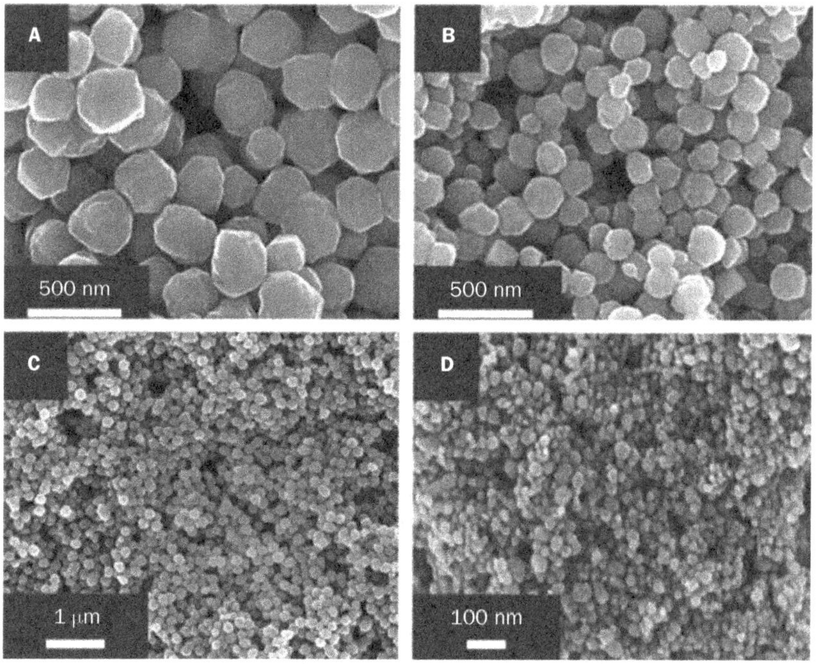

Dispersed Ni particles of different diameter precipitated in boiling propylene glycol process in the presence of different amounts of Pd (added as $[PdCl_4]^{2-}$ ions): (A) 0 ppm Pd, (B) 100 ppm Pd, (C) 1000 ppm Pd, and (D) 2500 ppm Pd.

2.13 Wang et al. synthesized gold NPs by reduction of tetrachloro auric acid (TCAA) with ascorbic acid (AA) according to the reaction:[34]

$$2HAuCl_4+3C_6H_8O_6 \rightarrow 2Au^0+3C_6H_6O_6+8HCl$$

Induction time as a function of pH for gold particles reduced by ascorbic acid with different component concentrations is shown in the following figure.

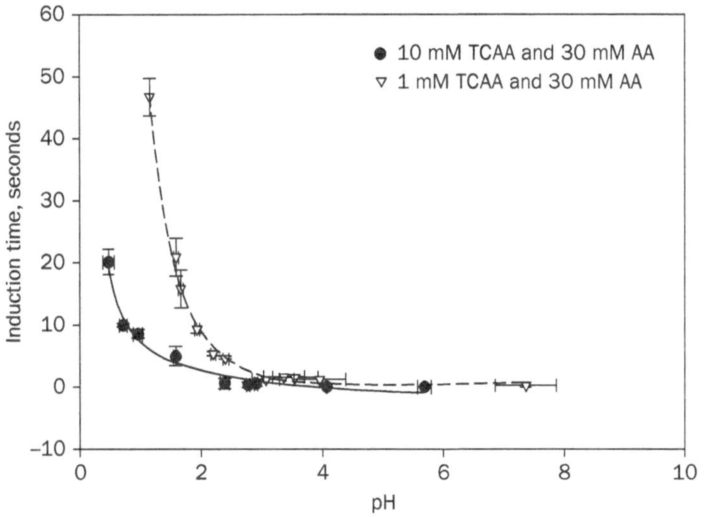

Experiments are carried out at two different initial concentrations of TCAA and AA precursors and inductions times are recorded for the nucleation of gold particles at different pH values as given in the figure above. Explain the changes in induction time with regards to initial precursor concentrations and pH of the reaction solution.

2.14 Magnetic NPs of iron oxides can be used as contrast agents for better imaging in MRI and their efficiency can be further enhanced by controlling the shape and size of final particles. In their study, Mohapatra et al. described the synthesis route for Fe_3O_4 nanospheres by mixing $FeCl_3$ precursor and the reducing agent oleylamine according to the following reactions:[35]

$$Fe^{3+}+e^- \xrightarrow{\text{oleylamine}} Fe^{2+}$$

$$2Fe^{3+}+Fe^{2+}+8OH^- \longrightarrow Fe_3O_4+4H_2O$$

The concentration of initial Fe precursor is 5 mM in each procedure, whereas the concentration of oleylamine was varied, which resulted in final particles with different average size. Suggest a possible explanation on how the concentration of oleylamine affects the final size distribution of particles. The figure below shows TEM micrographs of Fe_3O_4 NPs and their average sizes produced with Fe–oleylamine mole ratios: (A) 1:7 (4 nm), (B) 1:5 (6 nm), (C) 1:3 (9 nm).

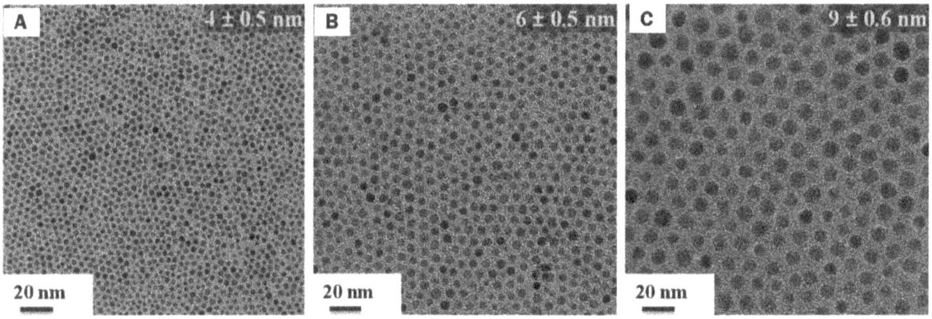

2.5 References

1. Smeets, P. J. M.; Finney, A. R.; Habraken, W. J. E. M.; Nudelman, F.; Friedrich, H.; Laven, J.; De Yoreo, J. J.; et al., A classical view on nonclassical nucleation. *Proceedings of the National Academy of Sciences of the United States of America.* **2017,** *114*(38), E7882–E7890.
2. Kashchiev, D.; van Rosmalen, G. M., Review: nucleation in solutions revisited. *Crystal Research and Technology.* **2003,** *38*(7–8), 555–574.
3. Schüth, F., Nucleation and crystallization of solids from solution. *Current Opinion in Solid State and Materials Science.* **2001,** *5*(5), 389–395.
4. Kalikmanov, V. I., Classical nucleation theory. In: *Nucleation Theory*, Lecture Notes in Physics Series, Bartelmann, M., Hänggi, P., Hjorth-Jensen, M., Jones, R. A. L., Lewenstein, M., von Löhneysen, H., et al., series eds. Dordrecht, the Netherlands: Springer Netherlands, 2013, pp. 17–41.
5. Mullin, J. W., *Crystallization*. Oxford, UK: Butterworth-Heinemann, 1997.
6. Threlfall, T. L.; Coles, S. J., A perspective on the growth-only zone, the secondary nucleation threshold and crystal size distribution in solution crystallisation. *CrystEngComm* **2016,** *18*(3), 369–378.
7. Agrawal, S. G.; Paterson, A. H. J., Secondary nucleation: mechanisms and models. *Chemical Engineering Communications.* **2015,** *202*(5), 698–706.
8. Myerson, A. S., Concluding remarks. *Faraday Discussions.* **2015,** *179*, 543–547.
9. De Yoreo, J. J.; Sommerdijk, N. A. J. M.; Dove, P. M., Nucleation pathways in electrolyte solutions. In: *New Perspectives on Mineral Nucleation and Growth: From Solution Precursors to Solid Materials*, Van Driessche, A. E. S.; Kellermeier, M.; Benning, L. G.; Gebauer, D., eds. Cham, Switzerland: Springer International Publishing, 2017, pp. 1–24.
10. Hamilton, B. D.; Ha, J.-M.; Hillmyer, M. A.; Ward, M. D., Manipulating crystal growth and polymorphism by confinement in nanoscale crystallization chambers. *Accounts of Chemical Research.* **2012,** *45*(3), 414–423.

11. ter Horst, J. H.; Schmidt, C.; Ulrich, J., Fundamentals of industrial crystallization (Chapter 32), In: *Handbook of Crystal Growth,* 2nd edition, Rudolph, P., ed. Boston, MA: Elsevier, 2015, pp. 1317–1349.
12. Roelands, C. P. M.; ter Horst, J. H.; Kramer, H. J. M.; Jansens, P. J., Analysis of nucleation rate measurements in precipitation processes. *Crystal Growth & Design* **2006,** *6*(6), 1380–1392.
13. Shiau, L. D.; Wang, H. P., Simultaneous determination of interfacial energy and growth activation energy from induction time measurements. *Journal of Crystal Growth.* **2016,** *442,* 47–51.
14. Woehl, T. J.; Evans, J. E.; Arslan, I.; Ristenpart, W. D.; Browning, N. D., Direct in situ determination of the mechanisms controlling NP nucleation and growth. *ACS Nano.* **2012,** *6*(10), 8599–8610.
15. Verdoes, D.; Kashchiev, D.; van Rosmalen, G. M., Determination of nucleation and growth rates from induction times in seeded and unseeded precipitation of calcium carbonate. *Journal of Crystal Growth.* **1992,** *118*(3), 401–413.
16. Nývlt, J., The Ostwald rule of stages. *Crystal Research and Technology.* **1995,** *30*(4), 443–449.
17. Herr, U., Metastable phases in interface controlled materials. *Contemporary Physics* **2000,** *41*(2), 93–104.
18. Van Driessche, A. E. S.; Benning, L. G.; Rodriguez-Blanco, J. D.; Ossorio, M.; Bots, P.; García-Ruiz, J. M., The role and implications of bassanite as a stable precursor phase to gypsum precipitation. *Science.* **2012,** *336*(6077), 69–72.
19. Zhao, J.; Liu, Y.; Sun, W.-B.; Yang, X., First detection, characterization, and application of amorphous calcium phosphate in dentistry. *Journal of Dental Sciences.* **2012,** *7*(4), 316–323.
20. Tsuji, T.; Onuma, K.; Yamamoto, A.; Iijima, M.; Shiba, K., Direct transformation from amorphous to crystalline calcium phosphate facilitated by motif-programmed artificial proteins. *Proceedings of the National Academy of Sciences of the United States of America* **2008,** *105*(44), 16866–16870.
21. Jeurgens, L. P. H.; Wang, Z.; Mittemeijer, E. J., Thermodynamics of reactions and phase transformations at interfaces and surfaces. *International Journal of Materials Research* **2009,** *100*(10), 1281–1307.
22. Cardew, P. T.; Davey, R. J., The kinetics of solvent-mediated phase transformations. *Proceedings of the Royal Society of London A: Mathematical, Physical and Engineering Sciences.* **1985,** *398*(1815), 415–428.
23. You, H.; Fang, J., Particle-mediated nucleation and growth of solution-synthesized metal nanocrystals: A new story beyond the LaMer curve. *Nano Today.* **2016,** *11*(2), 145–167.
24. Takesue, M.; Tomura, T.; Yamada, M.; Hata, K.; Kuwamoto, S.; Yonezawa, T., Size of elementary clusters and process period in silver NP formation. *Journal of the American Chemical Society.* **2011,** *133*(36), 14164–14167.
25. Erdemir, D.; Lee, A. Y.; Myerson, A. S., Nucleation of crystals from solution: classical and two-step models. *Accounts of Chemical Research.* **2009,** *42*(5), 621–629.
26. De Yoreo, J. J.; Gilbert, P. U. P. A.; Sommerdijk, N. A. J. M.; Penn, R. L.; Whitelam, S.; Joester, D.; Zhang, H.; et al., Crystallization by particle attachment in synthetic, biogenic, and geologic environments. *Science.* **2015,** *349*(6247), aaa6760. doi: 10.1126/science.aaa6760.

27. Navrotsky, A., Energetic clues to pathways to biomineralization: precursors, clusters, and NPs. *Proceedings of the National Academy of Sciences of the United States of America.* **2004,** *101*(33), 12096–12101.
28. Baumgartner, J.; Dey, A.; Bomans, P. H. H.; Le Coadou, C.; Fratzl, P.; Sommerdijk, N. A. J. M.; Faivre, D., Nucleation and growth of magnetite from solution. *Nature Materials.* **2013,** *12,* 310.
29. Hu, Q.; Nielsen, M. H.; Freeman, C. L.; Hamm, L. M.; Tao, J.; Lee, J. R. I.; Han, T. Y. J.; et al., The thermodynamics of calcite nucleation at organic interfaces: classical vs. non-classical pathways. *Faraday Discussions.* **2012,** *159,* 509–523.
30. Zahn, D., Thermodynamics and kinetics of prenucleation clusters, classical and non-classical nucleation. *ChemPhysChem.* **2015,** *16*(10), 2069–2075.
31. Habraken, W. J. E. M.; Tao, J.; Brylka, L. J.; Friedrich, H.; Bertinetti, L.; Schenk, A. S.; Verch, A.; et al., Ion-association complexes unite classical and non-classical theories for the biomimetic nucleation of calcium phosphate. *Nature Communications.* **2013,** *4,* 1507.
32. Gebauer, D.; Cölfen, H., Prenucleation clusters and non-classical nucleation. *Nano Today.* **2011,** *6*(6), 564–584.
33. Lu, L.; Sevonkaev, I.; Kumar, A.; Goia, D. V., Strategies for tailoring the properties of chemically precipitated metal powders. *Powder Technology.* **2014,** *261,* 87–97.
34. Wang, L.; Andreassen, J.-P., Spherulitic growth of gold particles precipitated from aqueous solution. *Chemical Engineering & Technology.* **2014,** *37*(8), 1399–1407.
35. Mohapatra, J.; Mitra, A.; Tyagi, H.; Bahadur, D.; Aslam, M., Iron oxide nanorods as high-performance magnetic resonance imaging contrast agents. *Nanoscale.* **2015,** *7*(20), 9174–9184.

CHAPTER 3

Crystal Growth

When a driving force causes a system to depart from its equilibrium condition, a nucleus of a crystal can form as the first step to minimize the excess free energy, and it is simultaneously followed by size enlargement.[1] The morphology (external shape) of crystals is a direct outcome of their enlargement process, and the variety of forms crystals can take have been at the center of comprehensive research both due to interest in their fundamental formation mechanism and the practical implications of crystal shape for different applications. Particle size also depends to a great extent on its growth stage. Consequently, understanding how crystals grow enables us to have a deeper understanding and a higher ability to exert control on the important properties of crystals as final size and morphology.

In solution, enlargement of crystalline particles has traditionally been construed within the framework of crystal growth and agglomeration processes.[2] In this classical concept, crystal growth denotes monomeric attachment of building units on the nuclei that follows the lattice structure and stands as the primary process for size enlargement. Agglomeration, however, is considered as a rare event where individual crystals grow together via a crystalline bridge between them when in contact and form a new stable particle. In this context, investigations of agglomeration kinetics have focused on contact caused by shear forces, which implies particles should be in micron range.[2] According to the classical crystal growth theories, the final morphology of a crystal is predicted from an energy minimization viewpoint, and initial implications of this theory have focused on emergence of polyhedral morphologies following the lattice structure of crystals. However, revealing of complex morphologies and surface features with nanosized subunits with advanced characterization techniques have evoked alternative hypotheses to classical growth mechanisms that employ particle aggregation models such as oriented attachment and mesocrystallization for the formation of monocrystalline particles (Fig. 3.1). According to oriented attachment mechanism, supersaturation is eliminated during initial formation of primary nanoparticles (NPs), and these NPs then unite to form single crystals by spontaneous self-organization.[3] The crystallographic orientation of the particles with respect to each other is proposed to be determined by the minimization of the free energy. Oriented attachment has been demonstrated for some systems involving the assembly of up to hundreds of particles.[4] Mesocrystals are also defined as colloidal crystals composed of individual nanocrystals that are aligned via self-assembly. It is proposed that they form in two phase systems, one phase being the NPs (the other can be polymer, etc.). As NPs are aligned, crystal fusion is thermodynamically favored under displacement of the polymers, eventually forming a single crystal.[5]

Due to their small size, NPs have a high surface area to volume ratio and consequently a high surface energy, which makes them inherently unstable. In general,

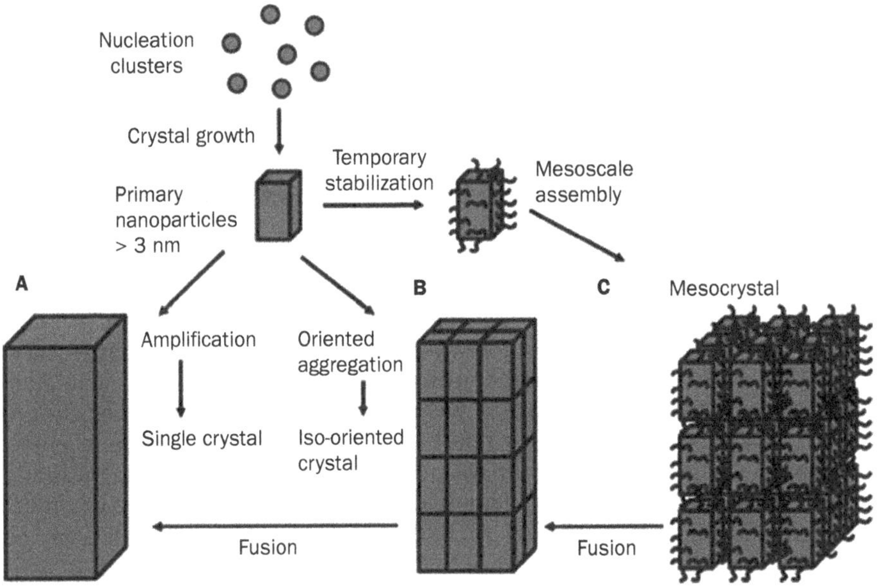

Figure 3.1 Illustration of alternative pathways for crystal growth: (A) classical growth pathway via monomer attachment, (B) oriented attachment of primary NPs forming a single crystal upon fusing, (C) mesocrystal formation via self-assembly of primary NPs that are temporarily stabilized by the polymer phase.[6]

particles at nanoscale are prone to aggregation in an effort to stabilize their high-energy surfaces. Thus, colloidal stability becomes important where the probability of aggregation between two particles increases with their decreasing size as a result of the higher interaction energy in between. Comparing the fundamental idea behind the classical and aggregation-based pathways for size enlargement of NPs shows the contrary assumptions they are based on. Clusters/NPs should overcome an energy barrier and reach a critical size to be able to grow according to the classical theory of crystallization, whereas growth controlled by colloidal stability states that clusters/NPs should continue to grow until they reach a certain size at which they cannot overcome the energy barrier for aggregation.[7]

The non-classical hypotheses based on particle aggregation models define effective parameters of growth related to colloidal stability and assembly mechanisms. However, they still lack a unified theoretical framework complemented with expressions for quantification of reaction kinetics.

Meanwhile, the same observations that led to the non-classical hypotheses are elucidated within the classical models by extending the fundamental approach of classical theory to explain the formation of dendritic and spherulitic morphologies along with polyhedral crystals, which had been treated entirely independent in the early stages of morphology investigations.[1] The classical theory of crystal growth offers a thermodynamic background and associated expressions for interpretation of experimental data. This chapter focuses on the classical framework for crystal growth and discusses growth mechanisms, kinetics, and final particle properties within that framework.

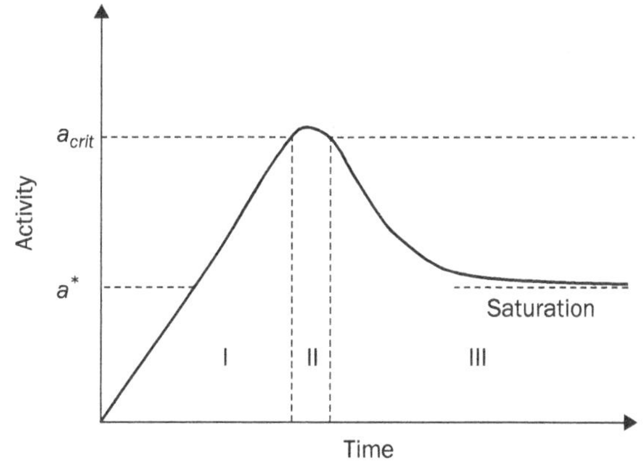

FIGURE 3.2 LaMer diagram illustrating the nucleation and growth stages of crystallization with regions representing (I) prenucleation, (II) nucleation, and (III) growth.[8]

3.1 Classical Theory of Crystal Growth

According to the classical view, as the stable nuclei form in a supersaturated solution, subsequent crystal growth occurs via continual addition of building units on the crystal surfaces and their incorporation into the lattice sites. The monomeric growth units can be atoms, ions, or molecules that are the primary constituents of the crystal lattice. The driving force for this process is the lower chemical potential of the growth units as constituents of the solid, in equilibrium with its solution, than as the corresponding solution species. Thus, the activity-based supersaturation can be defined in the same way for both processes and used as a measure of the driving force [Eqs. (2.1)–(2.7)]. Although the fundamental driving force for nucleation and growth processes is the same, they follow different energy landscapes. Since the solid component is already formed prior to crystal growth, further incorporation of the solute species into the structure proceeds at the expense of a much smaller energy than that required for nucleation. The LaMer diagram explains this phenomenon by conceptually separating the nucleation and growth stages of a crystallization reaction with regards to supersaturation (Fig. 3.2). As illustrated, when solute activity is increased as a function of time, levels above the solubility limit can be reached until the supersaturation reaches a critical value that corresponds to the energy barrier for nucleation as defined in Eq. (2.12). Once this critical value is reached, nucleation takes place and reduces the supersaturation in solution. When the supersaturation falls again below the critical value for nucleation, no more nuclei would form but supersaturation will continue to be consumed by growth until the solution reaches the equilibrium conditions, i.e., the saturation.

Crystal growth in solution is a dynamic process of continuous attachment and detachment of the monomeric units to and from the crystal surface where the attachment rate of monomers exceeds the detachment rate.[9] Monomer attachment scales with the flux of monomers to the surface and thus depends on the solute concentration in the growth medium. In contrast, detachment rate corresponds to the flux of monomers from the surface and is determined by the strength of the bonding between neighboring monomers, which is a function of temperature and nearly independent of solution concentration. This basic concept of growth already defines two important parameters as supersaturation and temperature.

During the growth of crystals from solution, two successive processes occur and exert a decisive influence on the progression of growth: transport of growth units from the solution to the crystal surface by diffusion and incorporation of material into the crystal lattice by surface reactions.[10] The rate and mechanism of growth is determined by the relative rates of these two processes and the characteristics of the solid–liquid interface at the microscopic level.

The Kossel model presented in Fig. 3.3 illustrates the structure of a crystal surface at the atomic level. The surface consists of flat regions named terraces, which contain moving layers, and these steps can contain one or more kinks. The three sites have different binding energies in accordance with the number of nearest neighbors they present. When a growth unit attaches at a terrace, step, or kink, it binds to one, two, or three nearest neighbors, respectively. The corresponding binding energy for each site is defined as the multiplication of the binding energy per pair with the number of binding neighbors. Since the probability of capturing a growth unit at any site scales exponentially with its binding energy, kinks are the most favorable sites for monomer incorporation and the main contributor to the growth.[11] The mechanism of growth changes with the density of kink sites, also termed as the degree of roughness of a crystal surface, and growth rate scales with kink density. Thermally activated attachment and detachment events are the source of kinks that scales with supersaturation and temperature.[12]

The classical model of crystal growth assumes that ions enter kink sites either directly from the solution or after adsorbing and diffusing through the terraces.[10] As the ions incorporate on the kink sites, steps propagate and this process continues until reaching crystal edges and completing the surface layer. At this point, new steps should be generated for growth to continue and growth mechanism defines how that process occurs.

One way of creating new steps is to generate two-dimensional islands of molecules by nucleation on a flat terrace, which requires a critical size and overcoming

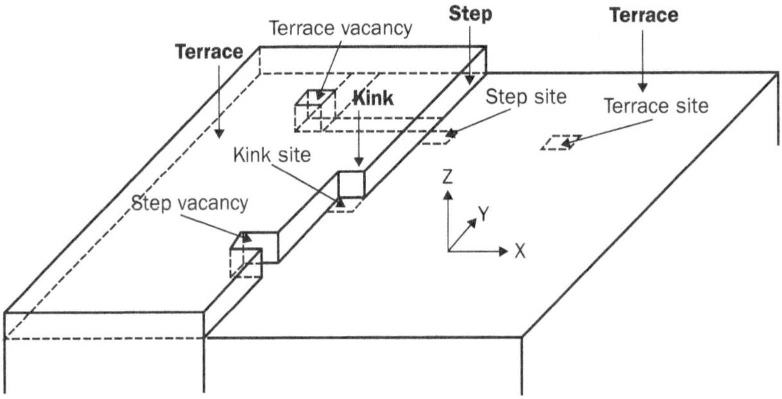

Figure 3.3 Graphical representation of the crystal surface according to the Kossel model.[13]

of the associated energy barrier (Fig. 3.4A). Therefore, sufficiently high supersaturation levels are required for a 2D nucleation mechanism to be active.[9] The combination of 2D nucleation and lateral growth of the islands on the surface is termed as the "birth and spread mechanism" and it shows strong dependence on the supersaturation level.[14] At sufficiently high supersaturation, multiple 2D surface nuclei can form on a crystal surface; moreover, polynucleation of new islands on existing ones can result in high numbers of active sites for growth. Under low supersaturation, critical size for 2D nucleation is large and the probability of reaching it becomes low. However, experimental observations have shown, although slow, crystals continue to grow even at low supersaturation where 2D nucleation could not be accounted for growth. The screw dislocation (spiral growth) theory accounts for these observations by considering the fact that crystals are not perfect.[15] They contain dislocations in the crystal lattice that generate continuous source of steps (Fig. 3.4B).[16] In both 2D nucleation and spiral growth mechanisms, the crystal–solution interface contains a limited number of active sites for growth (smooth surface); thus, incorporation of growth units into the lattice structure by the surface reactions becomes the rate-limiting factor. Therefore, 2D nucleation and spiral growth are denoted as reaction-controlled growth mechanisms. Under very high supersaturation levels, numerous active sites can be generated on the crystal surface, which leads to rapid ion incorporation. With the transition of interface from a smooth to a rough surface, the transport of ions from the bulk solution to the diffusion layer becomes the limiting and controlling factor for the growth rate, and the growth mechanism is called *diffusion-controlled.*

The growth rate and mechanism of a crystal is governed by the roughness of the interface, which varies with crystallographic directions, crystal species, temperature and driving force. It is possible to obtain different and multiple mechanisms dominating at different regions of the supersaturation range as discussed above. In order to explain the correlation of different growth mechanisms with relative

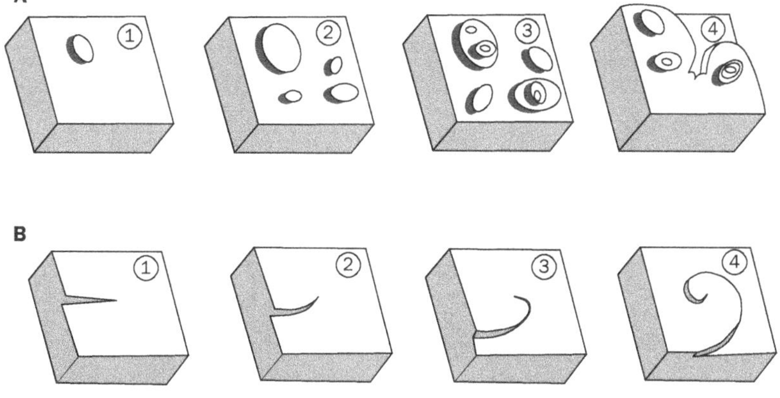

FIGURE 3.4 Reaction-controlled growth mechanisms of (A) 2D nucleation and lateral growth and (B) spiral growth with a continuous step via a screw dislocation.[17]

supersaturation, σ, concentration gradients of mass transfer can be used by applying the assumption[10]

$$\sigma = S - 1 = \left(\frac{a - a^*}{a^*}\right) \sim \left(\frac{c - c^*}{c^*}\right) \tag{3.1}$$

Figure 3.5 shows the mass transfer from the bulk solution with concentration, c, to the crystal surface with equilibrium concentration, c^*. Surface reactions take place in the crystal–solution interface, denoted as the diffusion layer with concentration c_i that is determined by the competition between the transport of monomers from the bulk and their incorporation on the surface. If the processes on the crystal surface are rapid ($c_i \sim c^*$), the transport of ions from the bulk solution to the diffusion layer controls the growth process (diffusion-controlled growth). The rate of this diffusion step is proportional to the concentration difference ($c - c_i$) as given by Fick's law [Eq. (3.2)], which consequently relates the growth rate dr/dt to the relative supersaturation with a linear dependence:[10]

$$\frac{dr}{dt} = DV_m \frac{(c - c^*)}{r} \tag{3.2}$$

where D is the diffusion coefficient, V_m is the molecular volume of the crystallite, and r is the radius of the spherical particle. As a function of their sufficiently small size, NPs introduce important outcomes acting on the mass transfer and concentration gradients that in turn become influential on growth kinetics. Owing to the size effect, the nature of the crystal–solution interface becomes significant in defining the controlling factors for NP growth. Since convection becomes operative for particles above micron size, NPs are surrounded by a stagnant diffusion layer, the thickness of which depends on the temperature and the size of the particle. Due to the presence of this stagnant layer, the concentration at the particle interface, c_i, can be lower than the bulk concentration in solution, which typically induces a diffusion-limited process for NP growth. Yet, c_i should be larger than c^* for growth to occur, where another size-dependent factor comes into play. According to Gibbs–Thomson relation given in Eq. (3.3), at sufficiently small size, the solubility of a particle becomes a function of its size. For a spherical particle with radius r, the solubility, $c(r)$, increases with decreasing particle size:

$$c(r) = c^* \exp\left(\frac{2\gamma V_m}{RTr}\right) \approx c^*\left(1 + \frac{2\gamma V_m}{RTr}\right) \tag{3.3}$$

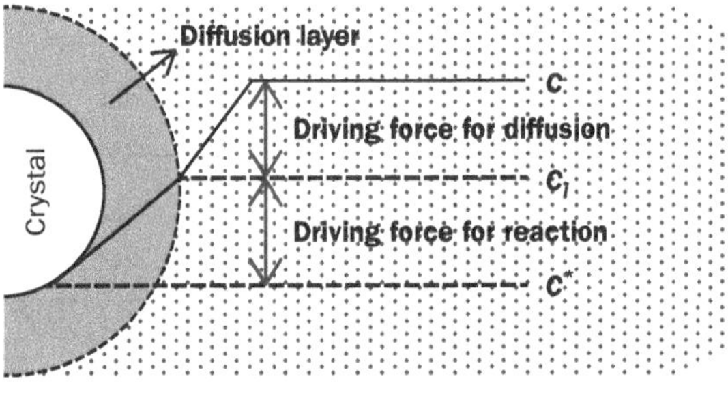

Figure 3.5 Concentration gradients of mass transfer between the bulk solution and crystal surface.

where c^* is the equilibrium solubility, γ is the surface free energy of the solid–liquid interface, R is the gas constant, and T is the absolute temperature. This increase in chemical potential of a particle with decreasing size results from the high surface area to volume ratio and gives rise to a predominant growth mechanism for nanocrystals known as *Ostwald ripening*. The higher equilibrium concentration for the smaller particles results in a concentration gradient that leads to transport of solute from small to large particles and provides the driving force for the growth of larger particles at the expense of smaller ones.[18] In this way, Ostwald ripening allows for reducing the surface area, i.e., the surface free energy, which constitutes a non-negligible percentage of the total energy in the case of small particles and thus plays a key role in the growth of NPs.

When the transport from bulk to crystal surface is rapid ($c_i \sim c^*$), the growth rate is determined by the surface reactions at the interface, including adsorption of growth units to the crystal surface and their incorporation into the lattice sites (reaction-controlled growth). Then, the rate equation for the reaction-limited growth can be expressed by applying the rate constant for surface reaction, k_r, and the concentration gradient at the interface between the diffusion layer and the crystal surface:

$$\frac{dr}{dt} = k_r V_m (c - c^*) \tag{3.4}$$

Spiral growth appears when the growth units integrate in the kinks with a rate proportional to the driving force, as shown in Eq. (3.4). The kink density also varies proportionally with relative supersaturation which results in a parabolic dependence of growth rate on it. The assumptions leading to growth rates are similar for spiral growth and polynuclear growth with the difference in the source of steps.[19] Compared to spiral growth, kink density increases more strongly with increasing supersaturation for polynuclear growth and results in an exponential dependence in between. The general growth rate law summarizes the dependencies of different growth mechanisms on the supersaturation and expresses the growth rate, R, as a function of relative supersaturation and kinetic components:

$$R = k\sigma^g \tag{3.5}$$

where k is the growth rate constant and g is the apparent growth order, the value of which depends on the rate-determining growth mechanism and how it correlates with the supersaturation value. As a result, in general the values of g and the corresponding growth mechanisms can be given as follows:

- $g = 1$; mass transport and diffusion-limited growth
- $g = 2$; screw dislocation (spiral) growth
- $g > 2$; polynuclear (2D surface nucleation) growth

The rate law gives a simplified yet useful representation of the variance of growth mechanism with the driving force and the dependence of growth rate on the relative supersaturation. In practice, rather than clear distinctions between mechanisms attributed to different levels of driving force, it is common that the rate of a crystal growth would be controlled by multiple mechanisms acting simultaneously or consecutively, which would also be reflected on the rate equation. Yet, the growth rate law shows successful implementations in industrial crystallization processes, and interpretation of experimental data in terms of possible mechanisms and reaction kinetics.

3.2 Experimental Determination of Growth Rates

The growth rates of crystalline particles can be quantified in terms of face growth rates or overall growth rates with different experimental techniques. The single crystal growth techniques with microscopic measurements can be used for determination of growth rates of individual faces of a crystal for the fundamental assessment of the growth kinetics and mechanism. However, for more practical purposes overall growth rates can be measured in terms of the total mass deposited per unit time per unit area of the crystal surface regardless of the contributions from individual faces.[20]

Quantification of overall growth rates can be made by monitoring either the size development of solid phase or changes in solution supersaturation. For particles with simple geometric morphologies such as spheres, measuring the particle size can be a convenient method. Then, the overall growth rate is expressed as the change of a typical dimension, L, of the particle with time, t:

$$R = \frac{dL}{dt} \tag{3.6}$$

Alternatively, growth rate can be calculated by monitoring the consumption of precursor ions from the solution due to particle growth with time. Depending on the rate of growth, concentration range of the experimental conditions and experimental design, batch (desupersaturation), or constant composition experiments can be used for growth rate determinations.

3.2.1 Seeded Batch Experiments

In the batch experiments supersaturation is allowed to be continuously consumed by the growth of crystalline particles. For that purpose, a sufficient amount of seed crystals of the particle of interest should be introduced into the supersaturated solution. Then, the growth rate, R, can be determined by monitoring the consumption of supersaturation with respect to time in units of deposited mass (dm/dt) or concentration change of precursor ions (dC/dt) as given:

$$R = \frac{1}{A}\frac{dC}{dt} \tag{3.7}$$

where A is the surface area of the seeds. Since the total surface area in the reaction medium changes with growth, it should be corrected in accordance with the growth mode of particles. For example, assuming three-dimensional growth with a constant shape factor, the mass change of crystals can be related to the change in surface area by a factor of $\left[(w_i/w)^{2/3}\right]$, where w_i is the weight of seed crystals added initially and w is the weight of crystals present at time t. In order to obtain reliable data during desupersaturation, the changes in solution composition due to growth should be accurately detectable. Thus, batch experiments are not efficiently applicable for the kinetic studies of slow growing particles at low supersaturation range. Data obtained by batch experiments cover a range of supersaturation that can evoke nucleation and growth of additional crystal phases or polymorphs, and different growth mechanisms with changing driving force. Therefore, the experimental conditions should be well-defined to prevent supersaturation consumption by any other process than the growth of the crystal of interest, and interpretation of the results to define the growth mechanism should be made with caution.

3.2.2 Seeded Constant Composition Experiments

In the constant composition experiments, initial conditions are maintained by the addition of titrant solutions to compensate for consumption of precursor ions, pH, and ionic strength so that the supersaturation is kept constant at its initial value. The working principle of constant composition method lies in observing the concentration fluctuations of a variable due to growth, upon which addition of titrants is triggered by use of automated titration systems with high sensitivity. Thus, this method provides better accuracy, reproducibility, and control over growth rate determinations at well-defined conditions and enhanced applicability for growth studies of slow growing crystals. A drawback of constant composition method over batch experiments is the surplus experimental work it requires.

After preparation of a metastable supersaturated solution, sufficient amount of seed material is introduced, which initiates the growth process and thus the titrant addition. Growth rates are calculated by relation to the rate of titrant addition:

$$R = \frac{1}{A}\frac{dV}{dt}C_{eff} \tag{3.8}$$

where the effective concentration (C_{eff}) denotes number of moles of solid precipitated per liter of titrant addition and dV/dt is the titrant curve gradient. When it is corrected for the changing surface area with growth, titrant addition curve should give a linear equation for dV/dt since constant composition method suggests the growth reaction to proceed under constant driving force and monomer flux.

3.3 Crystal Size and Size Distribution

The final size and size distribution of NPs are important characteristics regarding their application fields, and for crystalline particles formed via a crystallization process, they are determined by the consumption of total supersaturation in the system during the nucleation and growth processes. The effects of nucleation stage on particle size are discussed in Chap. 2 where influential control can be exercised on size and size distribution. However, the size distribution of particles in a reaction medium greatly improves also through the growth process. Primarily, different growth mechanisms exert dissimilar outcomes on the size distribution in accordance with the correlation between the growth rate and the particle size.

When the growth is diffusion-controlled, the linear growth rate can be expressed as a function of the particle radius by combining Eqs. (3.2) and (3.3):

$$\frac{dr}{dt} = \frac{K_D}{r}\left(\frac{1}{r^*} - \frac{1}{r}\right) \tag{3.9}$$

where K_D is defined as follows:

$$K_D = \frac{2\gamma D V_m^2 c^*}{RT} \tag{3.10}$$

The rate of change in the width of the size distribution (standard deviation of size), $d(\Delta r)/dt$, can be estimated from this expression by considering Eq. (3.9) close to the mean size, $\bar{r}$, and taking the derivative of the rate with respect to $\bar{r}$:[21]

$$\frac{d(\Delta r)}{dt} = \frac{d\left(\frac{dr}{dt}\right)}{d\bar{r}} = \frac{K_D \Delta r}{\bar{r}^2}\left(\frac{2}{\bar{r}} - \frac{1}{r^*}\right) \tag{3.11}$$

Equation (3.11) shows that for $\bar{r} \geq 2r^*$, the change in the width of the size distribution is negative ($d(\Delta r)/dt \leq 0$), which means growth leads to narrowing of the size distribution and for $\bar{r} < 2r^*$, size distribution tends to broaden with further growth (Fig. 3.6). Since the critical radius, r^*, is inversely proportional to supersaturation, we can predicate the boundary for the change of behavior in the size distribution of a system to the supersaturation level of the growth medium. In brief, it can be stated that diffusion-controlled growth can lead to monodisperse particle size distribution under sufficiently high supersaturation.

Similar expressions can be constructed for reaction-controlled growth rate by combining Eqs. (3.3) and (3.4):

$$\frac{dr}{dt} = K_R\left(\frac{1}{r^*} - \frac{1}{r}\right) \tag{3.12}$$

$$K_R = \frac{2\gamma k_r V_m^2 c}{RT} \tag{3.13}$$

The standard deviation in size with respect to time, $d(\Delta r)/dt$, can be derived from Eq. (3.12) in the same manner:[21]

$$\frac{d(\Delta r)}{dt} = \frac{K_R \Delta r}{\bar{r}^2} \tag{3.14}$$

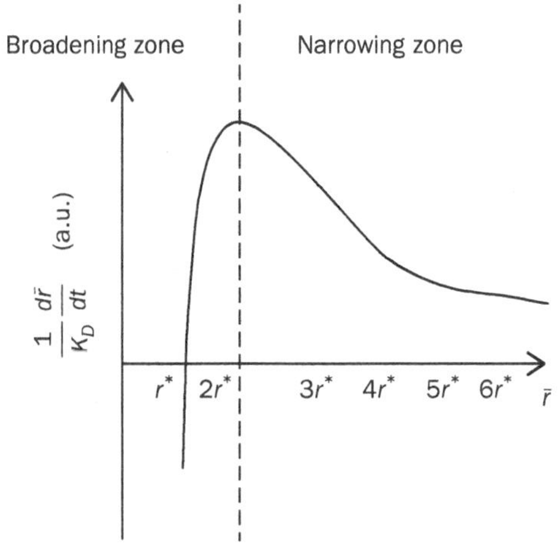

FIGURE 3.6 Two regions of changing particle size distribution with the boundary condition at $2r^*$ for diffusion-controlled growth.[22]

In this case $d(\Delta r)/dt$ is always positive as seen from Eq. (3.14), which indicates that for the reaction-controlled growth mechanism, the particle size distribution will broaden with further growth under all conditions of supersaturation.

The thermodynamic driving force related to solution supersaturation can display certain effects on the size and size distribution of NPs via the different growth mechanisms evoked as summarized above. Yet, effective kinetic control can also be established on growth in a system. Common methods employ the use of additives for this purpose, such as surfactants, ligands, polymers, or dendrimers, to confine the growth in the nanometer regime.[23] A way of kinetic control is to spatially confine the growth so that the growth stops when the limited amount of growth units is consumed or the available space is filled.[24] Fabrication of NPs inside micelles or in microemulsions are examples for confined reactions in restricted space that is generated with the presence of additional molecules in a commonly aqueous precipitation system.[25] In addition, additives that show high affinity for the NP surface can be used to influence the growth kinetics efficiently. Termed as capping agents, these additives can block the active growth sites on the particle surface either temporarily or permanently, and result in retardation or even complete inhibition of growth. Some of these methods will be discussed in Chap. 4.

3.4 Crystal Morphology

The morphology (external shape) of a crystalline particle is determined by two factors: one is the intrinsic characteristics of the crystal structure, which are the crystal lattice and the anisotropic bonding that constitutes the crystal structure, and the other is the external factors of growth conditions such as supersaturation and temperature. The final crystal morphology arises as a result of the relative growth rates of each of its faces that are affected by both internal and external elements of growth. In the end, the slow growing faces become the largest, and the fast growing faces become either small or disappear, as illustrated in Fig. 3.7.

Internal elements introduce faces with varying degrees of roughness due to the differences in bond strength and anisotropy at individual faces of a crystal.[9] Thus, certain faces would have higher kink density, and consequently, higher growth rates compared to smooth faces. If controlled solely by internal factors, the equilibrium morphology of a crystal at a specific temperature and pressure would be the one that minimizes the total surface free energy, and slow growing faces tend to be the low energy faces in a crystal. Thus, as growth proceeds in such a case, rough surfaces would eventually disappear and the crystal would attain a polyhedral morphology dominated by its smooth faces. These equilibrium morphologies driven by minimization of surface

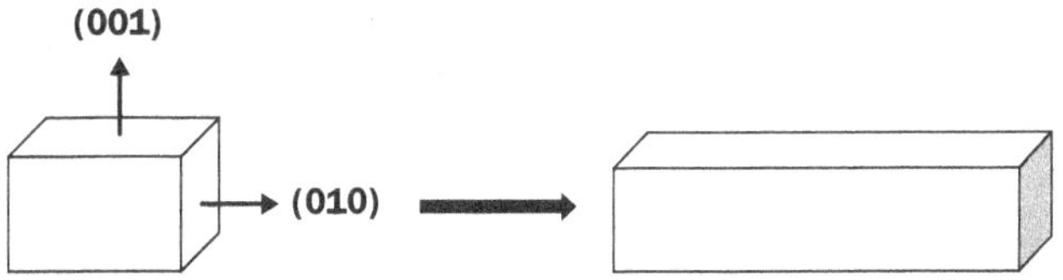

FIGURE 3.7 For a crystal where (010) face grows faster compared to (001) phase, the relative surface area of the slow growing (010) face will be larger with further growth.

energy are called Wulff constructions. However, crystal shape is rather achieved during exposure to growth conditions than by equilibration in a solution at equilibrium conditions. External factors are dominantly influential at determining the final shape of particles by effecting on both kinetics and mechanism of growth.

As discussed in the previous sections, driving force, i.e., the supersaturation of a system, is a decisive variable on the growth mechanism and kinetics, which also reflects upon the morphology. Figure 3.8 shows the coherent logic established with the classical crystal growth theory for the morphology predictions of a crystal in regards to changing driving force. Under low-to-moderate supersaturation levels where growth is surface reaction controlled, monocrystalline particles with polyhedral morphologies are expected outcomes of the advancement of smooth faces along the crystal lattice. With increasing driving force, surface roughness increases due to extensive surface nucleation and stimulates a change in growth mechanism from reaction-controlled to diffusion-controlled growth. The diffusion field around crystals becomes prominent on their development under these conditions and concentration gradients along the diffusion field induce morphological instabilities at the crystal interface. Depending on the diffusion field, certain edges and corners of crystals can have access to higher supersaturation, and consumption of supersaturation by growth of these locations prevents growth of the central part, which results in *hopper crystals*.[2] Further elevation of the characteristics observed in hopper crystals results in a more efficacious advancement in the direction of high supersaturation along the same lattice and gives rise to *dendritic growth*. Snow crystals exhibit well-known examples of dendritic morphologies that are characterized by crystallographic branching, i.e., all the branches of a dendritic crystal follow the same crystallographic orientation.[26] At yet higher supersaturation levels, high interface instability can even lead to non-crystallographic branching and formation of polycrystalline particles. This type of branching often results in spherical, space-filling structures and thus denoted as spherulitic growth.[2] There are yet unresolved questions on the non-crystallographic surface nucleation mechanism, but the general features show crystal inhomogeneity, dislocations, and internal stress underlying this phenomenon.[26]

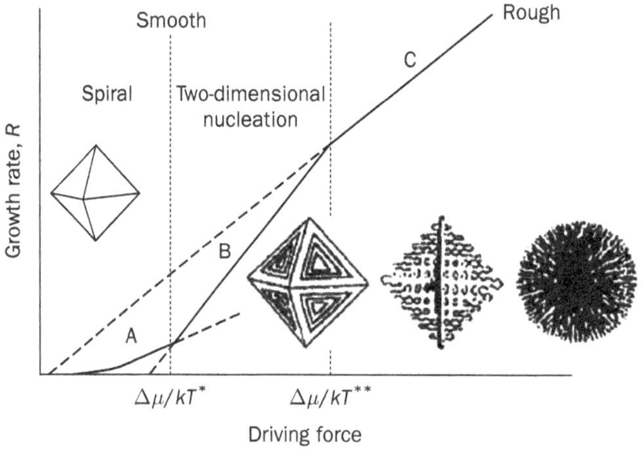

FIGURE 3.8 The change of growth rate, mechanism, and corresponding morphologies of a crystal as a function of the driving force (A: spiral growth, B: 2D nucleation, C: rough growth).[1]

Other factors of the growth environment than supersaturation can also be highly operative on the final crystal morphology. Presence of impurities, or if deliberately added in the growth medium—additives, is another important external factor that modifies morphology via effecting relative growth rates of individual faces of a crystal. Additives have been shown to inhibit or promote crystal growth processes in accordance with the varying mechanisms they evoke. In general, evaluated within the framework of classical theory, inhibition of crystal growth is often a result of the permanent or temporary adsorption of additives to the active growth sites. Promotion of growth in the presence of additives is either a result of formation of a less soluble compound than that formed in the absence of additive, or the lowering of edge free energy that promotes crystal growth, in particular at low supersaturation. When additives show selective interaction to certain faces of a crystal depending on the molecular packing, their effect on crystal morphology becomes prominent. Exploitation of this phenomenon has paved the way for the use of tailored additives to modify crystal shape. The work of Arndt et al. on iron oxide NPs stands as a good example where different size and morphologies are induced by using amine additives with varying chain lengths that preferentially adsorb on the {111} facets of magnetite particles, varying solvents, heating temperature, and aging time.[27] The addition of amines in the reaction medium with the chosen solvents allowed for efficient particle tuning to enhance the applications of magnetite as a contrast agent in MRI imaging (Fig. 3.9).

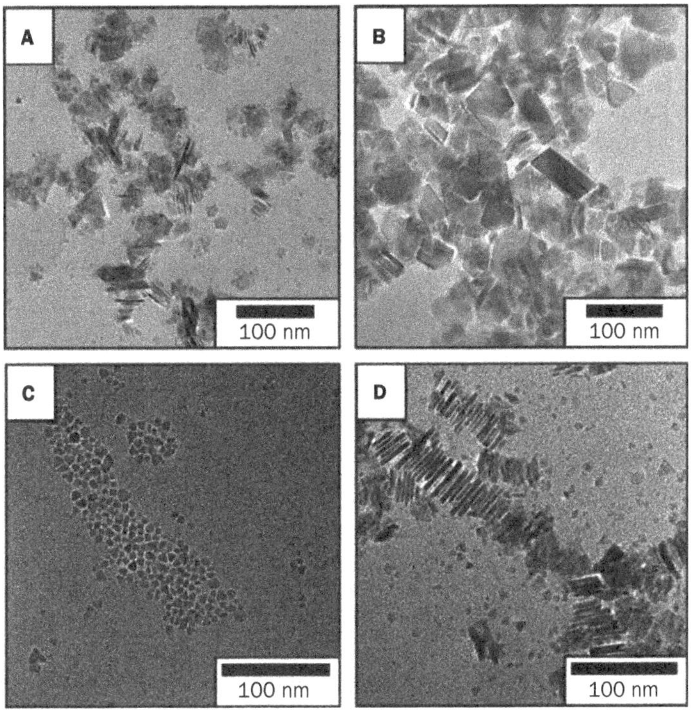

FIGURE 3.9 TEM pictures of iron oxide NPs precipitated in (A, C) triethylene glycol and (B, D) tetraethylene glycol solvent and in the presence of (A, B) diethylene triamine and (C, D) ethylene diamine additives.[27]

It is common to use the same strategy to produce NPs with high aspect ratios, which can be an important parameter for some applications. Aspect ratio can be tuned via inducing an asymmetric growth with additives while they act as directional agents. Molecules with specific affinity to certain crystalline faces of the particle of interest result in stupendous change in the relative growth rates of different faces and orient the growth through a certain direction. For example, in the production of gold nanorods the most common synthetic approach is the growth of gold seeds in the presence of cetyltrimethylammonium bromide (CTAB) as the surface-directing agent. CTAB preferentially binds to {110} facets of gold seeds and hinders the growth along these facets, which initiates an asymmetric one-directional growth in the {100} direction.[28] The role of CTAB in promoting one-directional growth is discussed in the next chapter.

3.5 Concluding Remarks

NPs can take various forms depending on the fabrication conditions, and the significant particle characteristics such as particle size, size distribution, and morphology are finalized during the size enlargement process. Since many properties of NPs are size- and shape-dependent for applications in electronics, optics, catalysis, diagnostics, or therapy, it has been of utmost importance to understand the governing mechanisms on this process and identify the parameters acting on it to further improve particle tuning. The intense research activity in this field has brought up two main growth theories: classical growth that proceeds by monomeric attachment of growth units to the crystal surface and particle-based pathways that rely on aggregation and self-assembly. In both theories, there are yet aspects of growth, which lacks a deeper understanding. Particle-based hypotheses have not yet provided satisfactory evidence for the bottom-up formation route they propose and lack in a theoretical framework complemented with quantitative expressions. Classical growth theory stands as a strong tool in this aspect where the well-established theoretical background allows us to both qualitatively and quantitatively interpret the growth phenomena. However, future investigations are still needed to clarify how classical mechanisms lead to complex morphologies and what are the roles of parameters such as driving force, temperature, and additives.

3.6 Study Questions

3.6.1 Concept-Based Questions

3.1 What are the two successive processes that occur during the growth of a crystal in its supersaturated solution?

3.2 Name two important parameters that are effective on the crystal growth mechanisms.

3.3 How is the solubility of a solid related to its size?

3.4 Discuss how growth mechanisms change with the supersaturation level.

3.5 Explain the quantitative dependence of growth rate on supersaturation.

3.6 How can particle size distribution be controlled during growth stage?

3.7 What does the Wulff construction denote?

3.8 Explain the differences and similarities between dendritic and spherulitic growth.

3.9 Seeded batch experiments were conducted to calculate the growth rate of brushite ($CaHPO_4 \cdot 2H_2O$). For this purpose, 100 mg of seeds with a surface area of 0.64 m^2/g were added to a metastable supersaturated solution and the solution was analyzed at determined time points to determine the momentary calcium ion concentration and supersaturation. The table below shows measured values of calcium ion concentration in the solution and momentary supersaturation values at determined time points. By using the data given in the table below, calculate the growth rate constant, k, and the apparent growth order, g. What is the effective growth mechanism of brushite crystals under corresponding conditions?

Time (min)	Ca concentration (mM)	Supersaturation
6	14.20	1.51
10	14.05	1.34
18	13.89	1.21
25	13.85	1.14
30	13.78	1.11
35	13.77	1.09

3.10 Seeded constant composition experiments were conducted to quantify the effects of three different additives on the growth rate of nanocrystalline hydroxyapatite (HA) particles. The figure below shows titrant addition curves for the seeded constant composition experiments. The slope of each curve is given in the table. The 30 mg of HA seeds with a surface area of 62.52 m^2/g were used in each experiment and the effective concentration of titrant solutions, C_{eff}, was 0.5 mM. The slopes of the titrant addition curves are given below. Calculate the growth rate for each experimental condition and compare the effect of additives on the growth kinetics of HA.

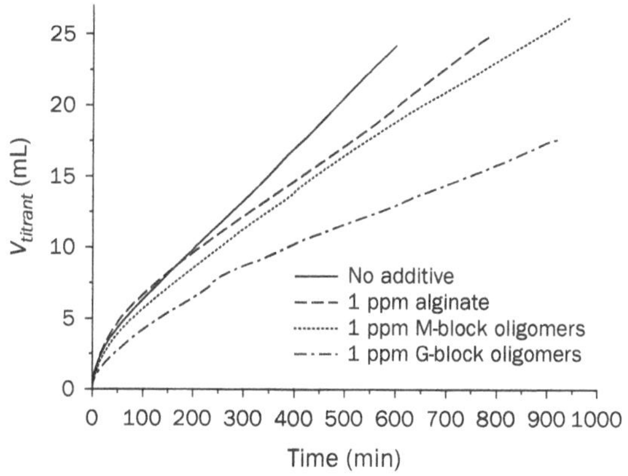

Experiment	Slope
No additive	0.036
Alginate	0.027
M-block	0.025
G-block	0.017

3.6.2 Research-Based Questions

3.11 Halaciuga and Goia produced silver particles by reduction of a silver amine complex in aqueous medium with ascorbic acid. The figure below shows final crystals with different morphology and crystallinity when the rate of reduction was altered. The figure also shows dispersed silver particles with different internal structure: (A) single crystalline polyhedral particles formed by slow reduction; (B) polycrystalline spherical particles formed by rapid reduction of the silver amine complex with ascorbic acid. Suggest a possible explanation for this outcome.[29]

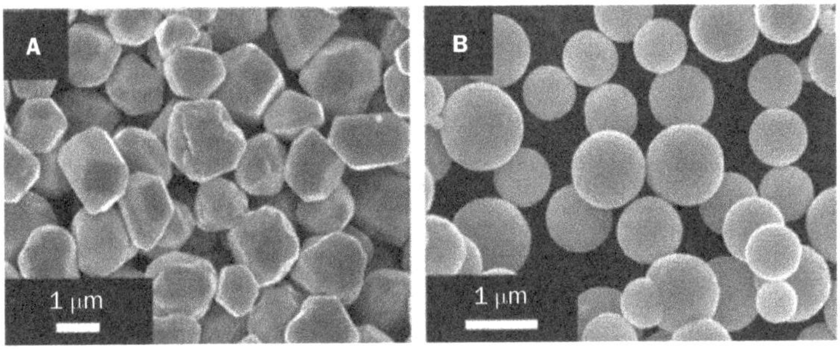

3.12 In their study, Dakhlaoui et al. investigated the morphology of zinc oxide (ZnO) crystals when they were grown in the presence of sodium hydroxide (NaOH), which was introduced to the reaction medium as an additive as shown in the figure below.[30] The basic ratio, *b*, was defined as the molar ratio of NaOH to zinc precursor and was varied from 0 to 4. The final morphology of ZnO crystals showed differences with changing *b* value. What can be a possible reason for this behavior?

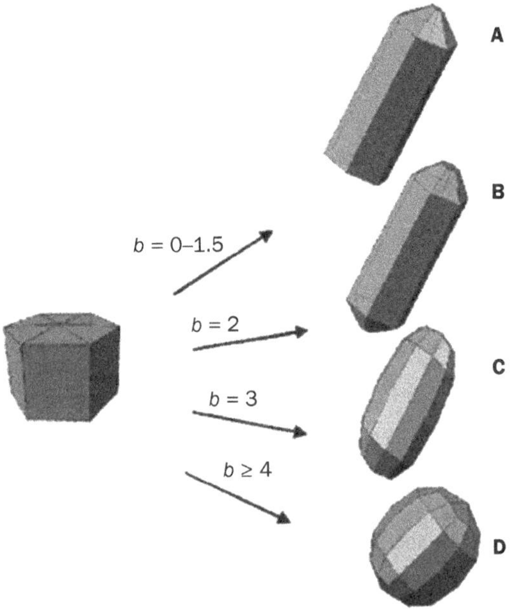

3.7 References

1. Crystal growth. In: *Crystals: Growth, Morphology, & Perfection*, Sunagawa, I., ed. Cambridge, UK: Cambridge University Press, 2005, pp. 20–59.
2. Andreassen, J.-P.; Lewis, A. E., Classical and nonclassical theories of crystal growth. In: *New Perspectives on Mineral Nucleation and Growth: From Solution Precursors to Solid Materials*, Van Driessche, A. E. S.; Kellermeier, M.; Benning, L. G.; Gebauer, D., eds. Cham, Switzerland: Springer, 2017, pp. 137–154.
3. Ivanov, V. K.; Fedorov, P. P.; Baranchikov, A. Y.; Osiko, V. V., Oriented attachment of particles: 100 years of investigations of non-classical crystal growth. *Russian Chemical Reviews.* **2014,** *83*(12), 1204–1222.
4. Yuwono, V. M.; Burrows, N. D.; Soltis, J. A.; Penn, R. L., Oriented aggregation: formation and transformation of mesocrystal intermediates revealed. *Journal of the American Chemical Society.* **2010,** *132*(7), 2163–2165.
5. Niederberger, M.; Cölfen, H., Oriented attachment and mesocrystals: non-classical crystallization mechanisms based on nanoparticle assembly. *Physical Chemistry Chemical Physics.* **2006,** *8*(28), 3271–3287.
6. Sebastian, W.; Nicola, P.; Markus, A.; Helmut, C., Polymer-induced alignment of DL-alanine nanocrystals to crystalline mesostructures. *Chemistry—A European Journal.* **2005,** *11*(10), 2903–2913.
7. Polte, J., Fundamental growth principles of colloidal metal nanoparticles—a new perspective. *CrystEngComm.* **2015,** *17*(36), 6809–6830.
8. Sugimoto, T., Underlying mechanisms in size control of uniform nanoparticles. *Journal of Colloid and Interface Science.* **2007,** *309*(1), 106–118.
9. De Yoreo, J. J.; Vekilov, P. G., Principles of crystal nucleation and growth. *Reviews in Mineralogy and Geochemistry.* **2003,** *54*(1), 57–93.
10. Nielsen, A. E., Electrolyte crystal growth mechanisms. *Journal of Crystal Growth* **1984,** *67*(2), 289–310.
11. Sangwal, K., ed. Kinetics and mechanism of crystal growth: an overview. In: *Additives and Crystallization Processes*, West Sussex, England: Wiley, 2007, pp. 65–107.
12. De Yoreo, J. J.; Zepeda-Ruiz, L. A.; Friddle, R. W.; Qiu, S. R.; Wasylenki, L. E.; Chernov, A. A.; Gilmer, G. H.; et al., Rethinking classical crystal growth models through molecular scale insights: consequences of kink-limited kinetics. *Crystal Growth and Design.* **2009,** *9*(12), 5135–5144.
13. Zhang, J.-H.; Zhang, Y.; Wen, Y.-H.; Zhu, Z.-Z., Comparative study of Cu13 and Co13 clusters deposition and diffusion on the Cu(001) surface. *Computational Materials Science.* **2010,** *48*(2), 250–257.
14. ter Horst, J. H.; Schmidt, C.; Ulrich, J., Fundamentals of industrial crystallization (Chapter 32), In: *Handbook of Crystal Growth*, 2nd edition, Rudolph, P., ed. Boston, MA: Elsevier, 2015, pp. 1317–1349.
15. Frank, F., The influence of dislocations on crystal growth. *Faraday Discussions.* **1949,** *5*, 48–54.
16. Uwaha, M., Introduction to the BCF theory. *Progress in Crystal Growth and Characterization of Materials.* **2016,** *62*(2), 58–68.
17. Li, J.; Tilbury, C. J.; Kim, S. H.; Doherty, M. F., A design aid for crystal growth engineering. *Progress in Materials Science.* **2016,** *82*, 1–38.

18. Viswanath, R.; Sarma, D. D., Growth of nanocrystals in solution. In: *Nanomaterials Chemistry*, Rao, C. N. R., Müller, A., Cheetham, A. K., eds., Weinheim, Germany: Wiley-VCH Verlag GmbH, 2007, pp. 139–170.
19. Teng, H. H.; Dove, P. M.; De Yoreo, J. J., Kinetics of calcite growth: surface processes and relationships to macroscopic rate laws. *Geochimica et Cosmochimica Acta* **2000,** *64*(13), 2255–2266.
20. Mullin, J. W., *Crystallization*. Oxford, MA: Butterworth-Heinemann, 1997.
21. Sugimoto, T., –Growth (Chapter 2). In: *Monodispersed Particles*, Sugimoto, T., ed. Amsterdam: Elsevier, 2001, pp. 86–117.
22. Thanh, N. T. K.; Maclean, N.; Mahiddine, S., Mechanisms of nucleation and growth of nanoparticles in solution. *Chemical Reviews*. **2014,** *114*(15), 7610–7630.
23. Jana, N. R.; Gearherat, L., Murphy, C. J., Seed-mediated growth approach for shape-controlled synthesis of spheroidal and rod-like gold nanoparticles using a surfactant template. *Advanced Materials*. **2001,** *13*(18), 1389–1393.
24. Guozhong, C., *Nanostructures and Nanomaterials: Synthesis, Properties and Applications*. Singapore: World Scientific Publishing, 2004.
25. Eastoe, J.; Hollamby, M. J.; Hudson, L., Recent advances in nanoparticle synthesis with reversed micelles. *Advances in Colloid and Interface Science*. **2006,** *128–130*, 5–15.
26. Shtukenberg, A. G.; Punin, Y. O.; Gunn, E.; Kahr, B., Spherulites. *Chemical Reviews*. **2012,** *112*(3), 1805–1838.
27. Arndt, D.; Zielasek, V.; Dreher, W.; Bäumer, M., Ethylene diamine-assisted synthesis of iron oxide nanoparticles in high-boiling polyolys. *Journal of Colloid and Interface Science* **2014,** *417*, 188–198.
28. Sarah, J.; Moritz, T.; Luca, G.; Andreas, F.; A., A. P. R.; Nicolas, P. P., Silver-assisted synthesis of gold nanorods: the relation between silver additive and iodide impurities. *Small*. **2018,** *14*(20), e1703879.
29. Halaciuga, I.; Goia, D. V., Preparation of silver spheres by aggregation of nanosize subunits. *Journal of Materials Research*. **2008,** *23*(6), 1776–1784.
30. Dakhlaoui, A.; Jendoubi, M.; Smiri, L. S.; Kanaev, A.; Jouini, N., Synthesis, characterization and optical properties of ZnO nanoparticles with controlled size and morphology. *Journal of Crystal Growth*. **2009,** *311*(16), 3989–3996.

CHAPTER 4

Synthesis of Metal Nanoparticles

Metal nanoparticles (NPs) form an important class of nanomaterials that possess remarkable physico-chemical properties such as magnetic, optical, electronic, and so on. Synthesis plays a major role in controlling size, shape, morphology, and surface functionality of such NPs and hence their final applications.

This chapter builds upon the fundamental principles of nucleation and growth explained in Chaps. 2 and 3. Thereon, it introduces commonly used top-down and bottom-up methods for the synthesis of metal NPs. Although the majority of the methods are discussed with respect to gold NPs as an example, the general principle can be extended for noble metals and several other metals. Starting from a homogeneous nucleation, this chapter goes on to show examples where small seeds are used to synthesize anisotropic structures, with gold as an example. Thus, this chapter will equip you to synthesize a wide array of metallic NPs using the fundamental principles of nucleation and growth.

4.1 Reduction of Metal Precursors in Solution

One of the most common methods to synthesize NPs using a bottom-up approach is by reduction of metallic precursors in solution. To form NPs of a certain metal, the corresponding metallic precursor is reduced in the presence or absence of a passivating ligand. Besides providing stabilization to the NPs in solution, the passivating ligand also provides functionality. However, the size of the NPs is controlled by the strength of the reducing agent. A rule of thumb is, the stronger the reducing agent, the smaller the NPs. This happens because a strong reducing agent provides higher or quicker supersaturation, leading to a large number of nuclei and hence smaller size for a constant mass of available metallic precursor.

4.1.1 Mechanism[1]

Reduction

The first step to the formation of metallic NPs is the reduction of oxidized metallic species leading to neutral atoms, as shown below. `M` refers to the metal, `Red` is the reducing agent, and `Ox` stands for the oxidized product.

$$a\mathrm{M}^{n^+} + b\mathrm{Red} \rightarrow a\mathrm{M}^0 + b\mathrm{Ox} \tag{4.1}$$

Nucleation

The formed metal atoms unite to form dynamic clusters in a continuous dissociation–condensation process, since they are insoluble in the liquid. Only those embryos that contain a magic number of metal atoms are stable and the other intermediate-sized embryos dissociate or keep growing to reach the stable state.

$$xM^0 \rightleftharpoons (M_x^0)_{em} \tag{4.2}$$

Once the embryos reach a critical size, they separate from solution as solid particles, the nuclei.

$$(M_x^0)_{em} + yM^0 \rightarrow (M_{x+y}^0)_{nucl} \tag{4.3}$$

As discussed in Chap. 2, the number and the size of the nuclei depend on the solute concentration, temperature, solvent viscosity, and surface tension. In addition to the above parameters, the redox potential of the reduction reaction is an important parameter.

Growth

The nuclei continue to grow by further addition of metal atoms to form primary particles, which are unstable owing to large free energy.

$$(M_n^0)_{nucl} + mM^0 \rightarrow (M_{n+m}^0)_{pp} \tag{4.4}$$

The primary particles further grow slowly via a diffusion mechanism or rather fast via an aggregation mechanism, yielding stable NPs.

$$(M^0)_{pp} + nM^0 \rightarrow (M^0)_p \tag{4.5}$$

$$(M^0)_{pp} + nM^0 \rightarrow (M^0)_p \tag{4.6}$$

4.1.2 Parameters Influencing Nucleation and Growth

The final size of the NP depends on three important parameters: the supersaturation, the fraction of solute species involved in the nucleation step relative to the total amount of metal in the system, and the extent of aggregation. Among these, supersaturation is the easiest to control. A high supersaturation will generate a large number of nuclei consuming a major fraction of the metal species. However, the smaller the particle, the higher is the surface energy. This means that aggregation of the NPs will almost be inevitable. However, if aggregation can be inhibited by the use of stabilizing ligands, subsequent increase in size will reflect only growth by exhaustion of the remaining metallic species in solution.

The role of a passivating ligand comes into play when the aggregation process needs to be arrested in the early stages of particle formation. Stabilization can happen through electrostatic, steric, electrosteric, or hydration mechanisms. Out of these modes, steric and electrosteric stabilization, comprising adsorption of surfactants, polymers, or polyelectrolyte molecules on metal NPs, are the most common mechanisms for concentrated dispersions of larger metal particles. This happens via screening of unbalanced van der Waal's forces. Electrostatic stabilization, based on relative balance between attractive and repulsive Coulombic forces, is more common for dilute systems, where charge plays a major role. However, the hydration stabilization mechanism is important for stabilizing hydrophilic NPs.

Thus, a rule of thumb to generate stable nanosized particles in solution will almost always require three important ingredients: metallic precursor, reducing agent, and stabilizing ligand. This opens up a world of possibilities to cherry-pick several combinations to alter particle size, shape, morphology, and their physico-chemical properties.

4.1.3 Examples of Metallic Precursor Reduction in Solution

Metallic precursor reduction in solution has been widely employed for the synthesis of noble metal NPs including Au, Pt, Co, among others. Some commonly studied examples are highlighted in this section.

Turkevich Method

One of the foremost examples employing this method is the Turkevich method[2-4] developed in the 1950s. It is still one of the widely used methods for the synthesis of monodisperse gold NPs in the size range of 2–200 nm. The method employs citrate as a reducing agent as well as passivating ligand, with the synthesis temperature being typically 100°C with refluxing conditions. One of the major benefits of this method is the ease with which a wide array of surface functionalities can be ascribed to the spherical gold NPs through simple ligand exchange reactions. The citrate on the surface of the NPs can be easily displaced by thiols, isothiocyanates, and phosphines.

Several researchers have looked into the sequence of reactions in the multiple-step process leading to the synthesis of final NPs.[5,6]

The initial step is the oxidation of citrate yielding dicarboxy acetone.

$$\underset{(Ctr^{3-})}{(^{-}O\overset{O}{\overset{\|}{C}}CH_2)_2C(OH)\overset{O}{\overset{\|}{C}}O^{-}} \longrightarrow \underset{(ACDC^{2-})}{(^{-}O\overset{O}{\overset{\|}{C}}CH_2)_2C{=}O} + CO_2 + H^+ + 2\,e^- \tag{4.7}$$

The dicarboxy acetone is rapidly degraded to acetone at the reaction temperature.

$$2H_2O + ACDC^{2-} \rightarrow (CH_3)_2C = O + 2CO_2 + 2\,(OH)^- \tag{4.8}$$

Both dicarboxy acetone and the degradation products assist in additional redox reactions when the Na citrate/$HAuCl_4$ ratio is less than 1.5. Further, the dicarboxy acetone plays the role of organizer molecule—three Au+ can be tethered by a minimum of two dicarboxy acetone molecules, leading to the formation of complex.

Reduction of auric salt to aurous salt is the next step.

$$AuCl_4^- + 2e^- \rightarrow AuCl + 3Cl^- \tag{4.9}$$

The next step, disproportionation of aurous species, leads to the formation of gold atoms.

$$3AuCl \rightarrow 2Au^0 + AuCl_3 \tag{4.10}$$

Once gold particles are formed, disproportionation can occur on the particle surface, too. The overall stoichiometry of the reaction boils down to the following equation. However, this equation does not take into account the degradation of dicarboxy acetone [Eq. (4.8)].

$$2AuCl_4^- + 3Ctr^{3-} \rightarrow 2Au^0 + 3ACDC^{2-} + (CH_3)_2C = O + 3CO_2 + 8Cl^- + 3H^+ \tag{4.11}$$

Considering the fact that some of the degradation products (acetone primarily) also assist in reduction of the auric chloride to form Au^0 particles, the following reaction has been suggested:

$$4AuCl_4^- + 6H_2O + (3CH_3)_2C=O \rightarrow 4Au^0 + 9CH_2O + 12H^+ + 16Cl^- \quad (4.12)$$

Considering this effect and correcting for stoichiometry, the final equation representing the overall process of reduction, degradation, nucleation, and disproportionation is as follows:

$$2AuCl_4^- + Ctr^{3-} + 2H_2O \rightarrow 2Au^0 + 3CH_2O + 3CO_2 + 8Cl^- + 3H^+ \quad (4.13)$$

pH of the solution has been shown to be one of the most important parameters for this reaction. pH is primarily varied by tuning the concentration of sodium citrate in the traditional synthesis method.[5] In the pH range of 6.5–7.7, a traditional nucleation followed by a diffusion-controlled growth is the widely accepted suggested mechanism. However, at pH values below 6.5 down to 3.7, nucleation, aggregation, and intraparticle ripening are found to dominate the reaction pathway. While nucleation and aggregation show fast kinetics, the smoothing process with an aim to minimize the total surface energy of the nanocrystal happens to be very slow. Figure 4.1 shows the different pH-dependent mechanistic pathways for the Turkevich reaction.

Thus, in a traditional Turkevich method, sodium citrate not only acts as the reducing agent and passivating ligand, but also mediates the pH of the reaction. The mole ratio between the reductant and metallic precursor leads to control of Au NPs of different sizes, size/shape distribution, and the size/shape evolution pattern of the resulting gold NPs.

Although an extensive research has focused on gold NPs using this method, other NPs have also been synthesized including Ag, Pt, and other noble metals.[7–10]

Brust Reaction[11,12]

In this method, stable functionalized gold NPs can be prepared by simultaneous reduction of tetrachloroaurate ions and attachment of bifunctional organic thiol molecules. First, chloroauric acid (Au^{3+}) is phase transferred from an aqueous phase into toluene by employing a phase transfer catalyst. This is followed by phase separation and addition of a desired amount of dodecanethiol to the separated organic phase. The concentration and type of dodecanethiol is responsible for controlling the reduction of Au^{3+} to Au^{1+} in this step. Thereafter, the aqueous solution containing a strong

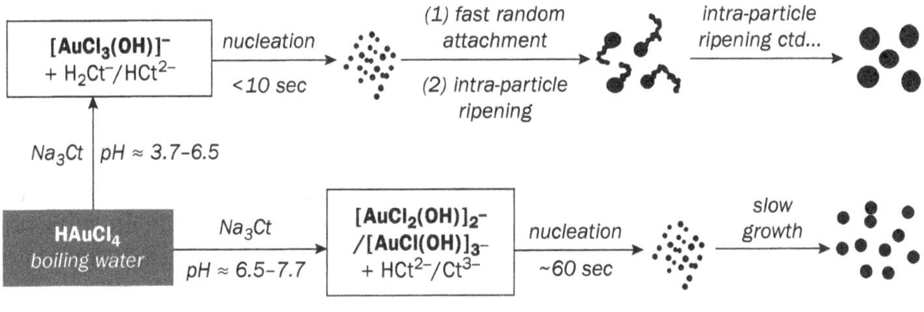

Figure 4.1 Schematic showing the different mechanistic pathways for the Turkevich reaction operating at different reaction pHs.[6]

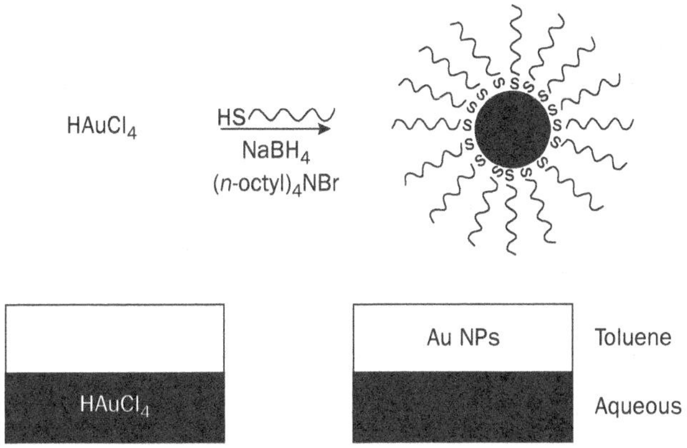

FIGURE 4.2 Schematic showing the synthesis and phase transfer of thiol-coated Au NPs in Brust reaction.

reducing agent, sodium borohydride, is contacted with the organic phase. A change in color of the organic phase indicates particle formation. A simple schematic of the overall process is shown in Fig. 4.2.

Unlike in the Turkevich method, where the reducing agent (sodium citrate) plays a major role in controlling the final size of the particle, the particle size in the Brust reaction is controlled by addition of different amounts of alkanethiol. This method produces thiol-functionalized stable NPs at room temperature in the size range of 1.5–8 nm. The mechanism relies on thiol-based reduction of Au^{3+} to Au^{1+} ions partially or completely (depending on concentration and type of thiol) followed by strong reduction initiated by sodium borohydride forming Au atoms. These Au atoms nucleate to form primary particles that undergo growth and capping either sequentially or simultaneously. For further understanding of the mechanism, the reader is referred to the excellent work by Perala and Kumar,[13] whereby using a detailed population balance model, the authors have concluded that particle formation is controlled by a new synthesis route: continuous nucleation, growth, and capping of particles throughout the synthesis process unlike the classical LaMer mechanism.

Goia Reaction

Another convenient and reproducible method widely used for the synthesis of stable Au NP dispersion in water in a range of sizes (from nanoparticles to microparticles) is by using the Goia reaction developed in the late 1990s. The method employs reduction of tetrachloroauric acid in the presence of iso-ascorbic acid at room temperature. An overall equation representing the process can be summarized as follows:

$$2HAuCl_4 + 3C_6H_8O_6 \rightarrow 2Au^0 + 3C_6H_6O_6 + 8HCl \tag{4.14}$$

The size of the resulting NPs is affected by the concentration and pH of the gold solution, while the stability of the electrostatically stabilized Au NPs is strongly dependent on the excess of reductant in the system, the ionic strength, and the temperature of precipitation. A wide range of sizes can be synthesized ranging from 30 to 100 nm (stabilizer free)[14] or 80 nm to 5 μm (in the presence of gum Arabic).[15] The ascorbic acid can be easily displaced from the surface using thiols, making way for a wide range of surface functionalizations.

This method has also been extended for synthesizing Ag NPs by reducing silver nitrate in the presence of iso-ascorbic acid at room temperature. However, the particle sizes in this case vary between 40 and 70 nm and the synthesis reaction needs to be carried out at pH values above 10. For control of particle sizes and shapes, the reader is referred to a series of papers by Goia and his group.[16,17]

Hydrogen Reduction of Metallic Precursor

Rampino and Nord employed a hydrogen-based reduction of Pt metallic precursor to form Pt NPs as early as 1941.[18] The method relied on boiling a $PtCl_4^{2-}$ solution in the presence of sodium hydroxide to form hydroxide precursor, followed by the reduction using hydrogen gas. The initial method employed has been refined over time, leading to finer control over particle size, shape, and stability.

Henglein et al. have shown that by aging the $PtCl_4^{2-}$ solution for a day, it is possible to increase the rate of reduction, which otherwise takes several days.[19] This happens because of the formation of aquated complexes within a few hours at ambient temperatures, as shown by the following equations:

$$PtCl_4^{2-} + H_2O \rightarrow PtCl_3(H_2O)^- + Cl^- \tag{4.15}$$

$$PtCl_3(H_2O)^- + H_2O \rightarrow PtCl_2(H_2O)_2 + Cl^- \tag{4.16}$$

However, the rate of reduction also depends on the use of polymeric stabilizers and the type of stabilizer used. For instance, using polyphosphate increases the rate in comparison to polyacrylate. The stabilizer also plays a dominant role in preventing aggregation of the particles, as can be seen in Fig. 4.3.

To further improve the stability of the Pt NPs and investigate into the mechanism of growth, Henglein and Giersig conducted several experiments with and without stabilizers.[20] Addition of sodium hydroxide reduced the rate of reaction by at least one order of magnitude. It is also believed to affect the structure of the starting material and produce a narrow particle size distribution. When sodium citrate, a stabilizer for colloidal particles, was further added, the reaction occurred slowly. Increasing the citrate concentration led to decrease in particle sizes; however, there was noticeable change in the particle morphology as well. Without the presence of citrate, cubes and truncated octahedra-shaped particles were primarily observed. Increasing the concentration of citrate led to the frequent appearance of tetrahedra-shaped particles. The authors

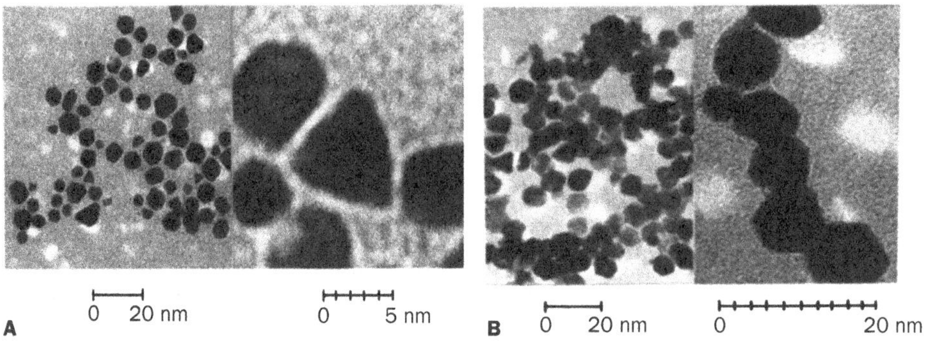

FIGURE 4.3 Electron micrographs of Pt particles synthesized (A) in the presence of polyacrylate stabilizer and (B) without any stabilizer.[20]

hypothesize that the tetrahedra are first formed and then grow into cubes or truncated octahedra, unless their growth is hampered by adsorbed stabilizer.

In this section, we have presented a few examples in which the metallic precursor is reduced in solution to yield NPs. In the next section, we will discuss what happens when the same reduction takes place in confined spaces.

4.2 Colloidal Templating: Reverse Micelles as Spherical Nanoreactors

Colloidal templating is one of the most important synthesis methods owing to ease of size control of the final NP simply by manipulating the drop size and control of their coalescence-exchange dynamics.

At and above the critical micelle concentration (CMC), surfactant micelles are formed. The CMC is affected by various parameters, which include solvent interactions between the surfactant tails and bulk organic solvent, the geometry of the surfactant, the addition of co-solutes or salts, and temperature. For critical packing parameter (CPP) values greater than one, reverse micelles are formed. Reverse micelles are characterized by an aqueous core dispersed within an organic bulk phase. Water with a pre-dissolved reactant when added into such a solution goes only inside the templates forming nanometer-sized spherical water droplets. These nanosized droplets act as the nanoreactors for the synthesis of desired NPs of various shapes and sizes. As the reactants are confined within the droplets, the size and shape control of the synthesized NPs are easily achieved by controlling the size of these droplets. An overall schematic of the reverse micelle–based synthesis of metallic NPs is shown in Fig. 4.4.

The water droplet sizes can be increased by increasing water-to-surfactant molar ratio. However, the coalescence-exchange of water drops can be varied over two orders of magnitude by changing the organic solvent (oil phase), nature of surfactant, or temperature.[21] The other advantage of using this method is that the morphology of the product can also be controlled as the surfactant aggregates form a variety of structures depending on the constituents and the location within the phase diagram.[22] Additional control parameters for this synthetic route arise from the fact that the rate of a chemical reaction can be reduced when it is conducted in a water in oil (w/o) micro-emulsion rather than in bulk aqueous medium. This happens as the reactants are partitioned into different water drops and the time scale of coalescence-exchange of these drops are few orders less than the diffusion time scale of the reactants in bulk.

The method employs the following three steps, schematically shown in Fig. 4.4.

1. Mixing reverse micellar solutions: One reverse micellar solution containing the metal precursor and the other reverse micellar solution containing the reducing agent are mixed. Passivating ligands may be added to the solutions to render additional stability to the final NPs to be formed.
2. Fusion of cores: The drops collide owing to Brownian motion, occasionally coalesce, and exchange their contents.
3. Formation of NPs: Upon fusion of the cores, the metallic precursor comes in contact with the reducing agent resulting in the formation of the NPs following classical nucleation and growth.

For detailed understanding of the modeling of such systems, the reader is referred to the works of Li and Park [23] and Bandyopadhyay et al.[24] This method of synthesis has been used extensively in the preparation of metal NPs,[25,26] core-shell metal NPs,[27,28] and

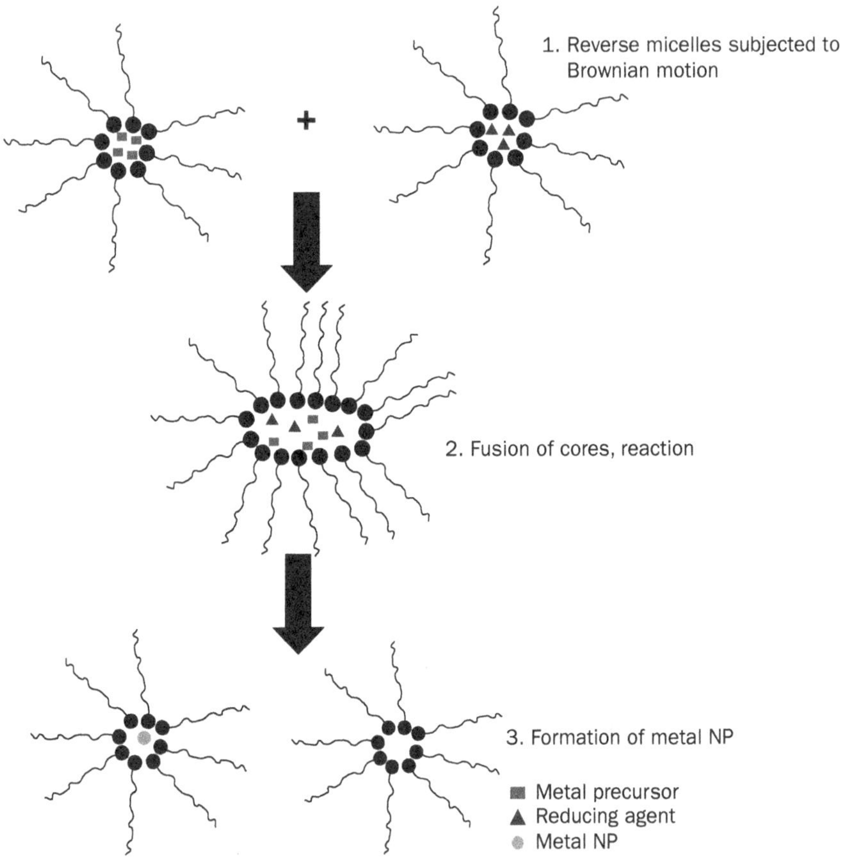

FIGURE 4.4 Schematic showing reverse micelle–based synthetic route for metal NPs.

various other combinations such as CdS-Ag_2S,[29] Ag-Silica,[30] and so on. Although the method of the reverse micelle–based synthesis of metallic NPs is relatively fast and precise in controlling particle size, additional steps are required to get rid of the surfactants.

Core-shell iron–gold (Fe@Au) NPs have been synthesized by Jafari et al. using a reverse micelle procedure.[31] The Fe core was first synthesized by mixing two reverse micellar solutions, one containing $FeCl_3$ and the other containing reducing agent $NaBH_4$. Cetyltrimethylammonium bromide (CTAB) and 1-butanol were utilized as surfactant and co-surfactant, respectively. The oil phase was *n*-octane. Thereafter, the Au shell was formed by adding two reverse micellar solutions, one containing $HAuCl_4$ and the other containing $NaBH_4$.

4.3 Sol-Gel Method

The sol-gel process is a versatile solution process for making NPs. Sol-gel methods generally refer to the hydrolysis and condensation of metal alkoxides or alkoxide precursors, leading to dispersions of oxide particles in a 'sol'. The 'sol' is then dried or 'gelled' by solvent removal or by chemical reaction. The solvent used is generally water, but the precursors can also be hydrolyzed by an acid or base.

In a typical sol-gel process, the precursor is subjected to a series of hydrolysis and polycondensation reactions to form a colloidal suspension, or a "sol," as presented in the following equations:

Hydrolysis:

$$\mathrm{M(OEt)_4} + x\mathrm{H_2O} \rightarrow \mathrm{M(OEt)_{4-x}(OH)_x} + x\mathrm{EtOH} \tag{4.17}$$

Polycondensation:

$$\mathrm{M(OEt)_{4-x}(OH)_x} + \mathrm{M(OEt)_{4-x}(OH)_x} \rightarrow \mathrm{(OEt)_{4-x}(OH)_{x-1}MOM(OEt)_{4-x}(OH)_{x-1}} + \mathrm{H_2O} \tag{4.18}$$

The rates of hydrolysis and condensation are important parameters that affect the properties of the final products. Smaller particle sizes are obtained at slower and more controlled hydrolysis rates. The particle size also depends on the solution composition, pH, and temperature. However, for nano-composites derived from gels, structural parameters and material porosity are determined by the rate of hydrolysis and condensation of the gel precursors and also by other oxidation–reduction reactions that occur during the gelling and subsequent heat treatment stages.[32] Iron-oxide supported on silica gel,[33] magnetite, and maghemite thin films[34] has been prepared extensively using this method.

The low processing temperature and molecular level homogeneity are advantages of this method, but the contamination from byproducts of reactions and the need for post-treatment of the products often limit their applicability.

4.4 One-Pot Synthesis Methods: Globular Proteins, Viruses, and Microorganisms

The concept of one-pot synthesis methods relies on the utilization of different templates (that perform the role of reducing agent and/or passivating ligand) onto which the metal NPs need to be fabricated or encapsulated within. Researchers have invested interests in searching for such templates, which can range from proteins to microorganisms.

Proteins are the basis for cellular functions, which in turn arise from the structure of the proteins. Some of these proteins have affinity for metals and ions, while some others can perform biomineralization of metals. Bovine serum albumin (BSA) was one of the first proteins employed to synthesize gold NPs.[35] When incubated with gold ions under basic conditions (pH above 10), an increasing number of amino acids comprising BSAs such as tyrosine, histidine, and cysteine become negatively charged. This promotes the adsorption of gold ions electrostatically to the regions rich in these amino acids. Once on the protein backbone, there occurs the reduction of gold ions to atomically sized gold clusters, known as *gold nanoclusters*. Depending on the application at hand, external reducing agents such as sodium borohydride may be simultaneously used to increase the rate of reduction or change the particle properties. This method of applying additional external reducing agents is termed as *extrinsic method* in comparison to the intrinsic method which solely depends on the reducing capacity of the constituent amino acids. With increase in time and concentration of gold, the nanoclusters develop into NPs.[36,37] Although this is an effective method to load different structures of gold

NPs on protein backbones, their incorporation might change the structure of the protein leading to loss of protein function.

Viruses represent a confined nano-environment and a unique protein surface topology for the synthesis of NPs. In addition, they possess high symmetry, robust functional protein capsids, and unique structural architectures among other properties. Virus-based NPs are highly homogeneous because they are synthesized in living cells, and can be modified to gain new functionalities through genetic engineering, in addition to chemical reactions.[38] Further, their preparation and purification can be low-cost and environment-friendly. Some of the most commonly used virus platforms include wild-type tobacco mosaic virus (TMV), cowpea chlorotic mottle virus (CCMV). The viral-mediated reduction of $AuCl_4^-$ has been shown on the CCMV template. The viral capsid actively potentiated the reduction of $AuCl_4^-$ by electron transfer from surface tyrosine residues at pH values less than 10. This further led to the reduction of Au^{3+} to Au^0 NPs.

Microorganisms employ metal-binding proteins and peptides to cause heavy metal detoxification. This is also the principle underlying the synthesis of metal NPs using microorganisms. These NPs can be synthesized at mild temperatures without the use of toxic chemicals and thus are environmentally friendly. These also open up a wide range of applications ranging from removal or recovery of metal ions to cancer treatment, biosensing, drug delivery, among others. Several different approaches for the synthesis of NPs based on different systems, such as wild-type versus recombinant microorganism strains, whole live cells versus cell extracts, and bulk phase versus micro-droplet, can be carried out in order to tailor the NPs for specific applications. The reader is referred to a recent review by Park et al. to examine the various kinds of microorganisms that have already been employed to synthesize a wide range of metal NPs.

4.5 Nucleotide-Mediated Synthesis of Metal Nanoparticles[39]

Nucleic acids consist of chains of monomeric units called nucleotides that are linked together by phosphodiester bonds. The building blocks of nucleic acids are the nucleosides, which consist of a nucleobase—adenine, thymine, uracyl, guanine, and a ribose or 2′-deoxyribose sugar ring. Depending on the type of constituting nucleotides, their sequence, and the type of linkage holding them together, nucleic acids can form excellent templates for biomineralization. The different steps leading to the formation of NPs can be outlined as follows:

1. *Initiation:* The first event is the formation of a precursor complex via specific interactions between a metal ion and the nucleic acid/nucleotide. Both metal–phosphate and metal–nucleotide interactions are crucial in this initial step. The type of metal–nucleic acid binding is determined by the nature of the metal ion, but the strength of this interaction is dependent on other factors such as the solvent or ionic strength.
2. *Growth:* Small clusters formed during initiation are coated with negatively charged layer of nucleic acid/nucleotide molecules. These tend to lose capping through thermodynamically favorable aggregation forming larger clusters carrying a larger number of capping molecules. The tendency toward aggregation is counterbalanced by the electrostatic barrier formed by the nucleotide/nucleic acid cap.

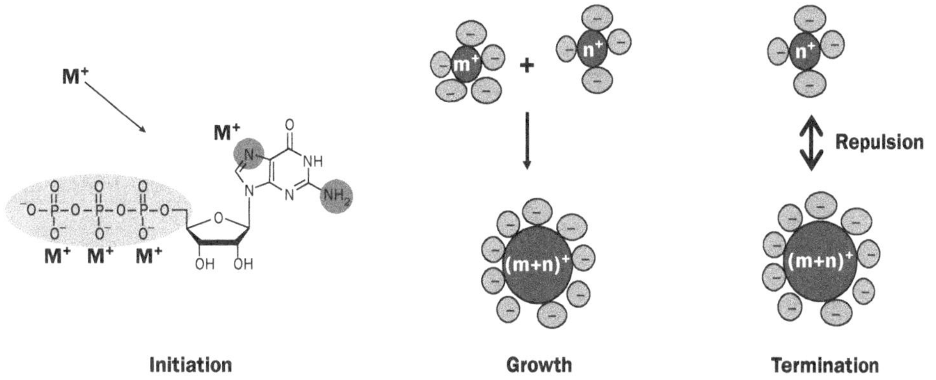

Figure 4.5 Schematic showing the three important steps of nucleotide-mediated synthesis of metal NPs: initiation, growth, and termination. The nucleic acid/nucleotide molecules also provide passivation and solubilization to the metal NPs.

3. *Termination:* The nucleic acid/nucleotide capping acts as a stabilizing electrostatic shell and provides a termination point in growth. Upon reaching a critical size, further aggregation is prevented since the electrostatic barrier required to overcome in order to bring clusters into close proximity is too large.
4. *Passivation:* Nucleic acid/nucleotide molecules serve to saturate empty orbitals and dangling bonds resulting in improved physico-chemical characteristics such as narrower emission bands, high quantum yields, and so on.
5. *Solubilization:* Nucleic acid/nucleotide ligands ensure that NPs remain dispersed in aqueous solution.

A schematic of steps 1 to 3 is shown in Fig. 4.5.

The ability to synthesize and screen libraries of modified nucleotides is promising for the synthesis and selection of non-natural sequences that might lead to the synthesis of NPs with unprecedented properties.

4.6 Natural Selection of Biomolecules Capable of Nanoparticle Formation[40]

A substantial amount of research has focused on *in vitro* selection schemes in order to explore possible biomolecule–inorganic materials interaction. This is an attempt to mimic natural evolutionary processes. The concept relies on generating a huge library of random sequence biomolecules such as RNA, DNA, and peptides. These are then incubated with precursor ions or pre-synthesized nanocrystal to isolate sequences capable of assembling monomeric metallic precursors into nanocrystals synthesized using high-temperature methods (Fig. 4.6A) or isolate sequences that bind to nanocrystals (Fig. 4.6B). In the latter case, these sequences can subsequently be tested for their ability to direct the synthesis of the identical nanocrystal at lower temperatures. The experimental process by which large libraries (about 10^{14}) of unique single-stranded or double-stranded RNA or DNA sequences are subjected to iterative cycles of screening, selection, and amplification resulting in the isolation of sequences capable of performing a task such as high-affinity binding to a certain protein or formation of a

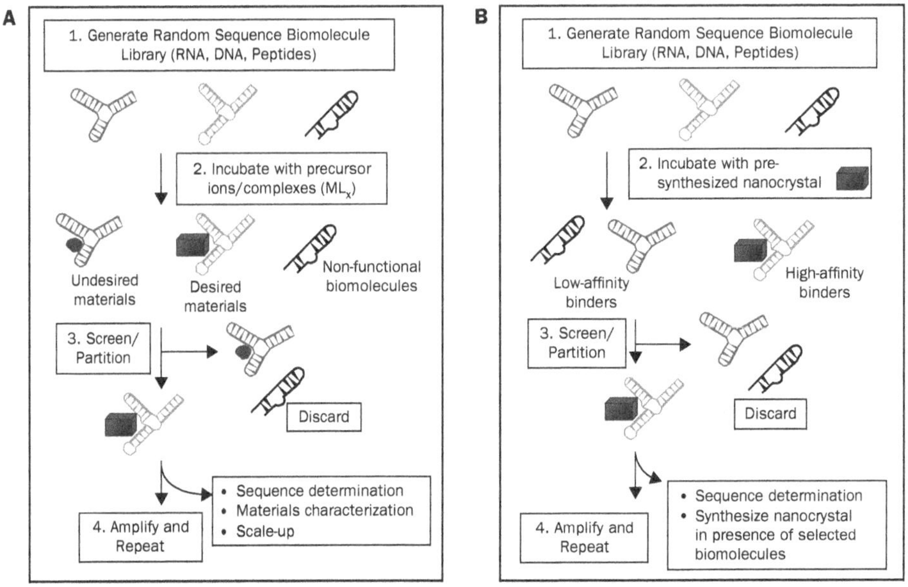

Figure 4.6 Generic *in vitro* selection schemes for exploring sequences (A) capable of forming NPs or (B) binding to specific nanocrystals.

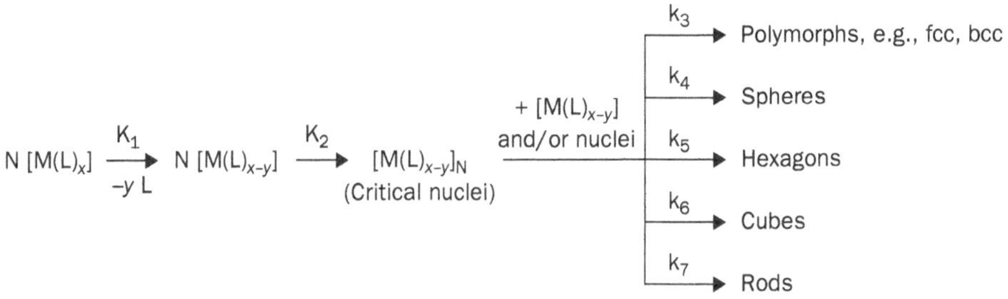

Figure 4.7 Schematic of the fundamental steps required to generate metal crystals from the generic complex.

characteristic nanocrystal is known as SELEX method. SELEX stands for Systematic Enrichment of Ligands by Exponential enrichment.

Using such methods, it has been possible to isolate biomolecules with high binding affinities and specificities for certain inorganic faces. Several biomolecules possessing catalytic activities toward the synthesis of materials have also been isolated. Another expected outcome is to isolate biomolecules capable of altering the formation of a spontaneously growing crystal. This would affect the size, size distribution, shape selectivity, growth rate, and polymorph selectivity of the resulting crystals. This phenomenon of crystal growth mediation by employing a biomolecule as a template or mediator is schematically shown in Fig. 4.7.

For crystal growth to happen, monomer units $M(L)_{x-y}$ must be generated from the metal complex $M(L)_x$ by loss of y number of ligands. When $x = y$, free metal atom is generated which then becomes the monomer. Crystal nuclei grow by monomer addition or nuclei aggregation. Once a crystal nucleus has formed, the resulting crystal shape is determined by a growth phase either controlled thermodynamically or kinetically. In a thermodynamic growth regime (k_3), monomers will add to the highest energy facets of the growing crystal until they merge into more stable crystalline facets. In a kinetically controlled growth regime (k_4–k_7), the relative rates of monomer addition to different facets dictates the resulting shape. This is where the biomolecule can bind selectively to certain facets, speeding or slowing monomer addition to that facet relative to others resulting in non-thermodynamic and often highly anisotropic crystal shapes. In addition, it can speed up or slow the rate at which a monomer is generated, thereby changing its concentration and solubility equilibria or crystal nucleation and growth rates. A biomolecule that coordinates strongly to a metal center could prevent crystallization altogether.

4.7 Seeded Growth of Gold Nanostructures[41]

Another way to control the shape of nanostructures is by using seeded growth. This section is devoted to understanding how to grow anisotropic gold nanostructures and understand the mechanisms of growth.

Fine-controlling the aspect ratio (defined as the ratio of length to diameter) of Au nanostructures has become important, as this enables wide range of optical properties that can be explored by changing the aspect ratio. Anisotropy splits up the single localized surface plasmon resonance (LSPR) band, generally observed for spherical Au NPs, into at least two distinct peaks, representative of the transverse and longitudinal axes. The larger aspect ratio structures have a larger extinction coefficient, whereby they scatter more light at longitudinal plasmon wavelength, thereby rendering improved applications in optical imaging. However, the smaller aspect ratio structures absorb more light at the longitudinal plasmon wavelength, paving the way for photothermal applications (Fig. 4.8).

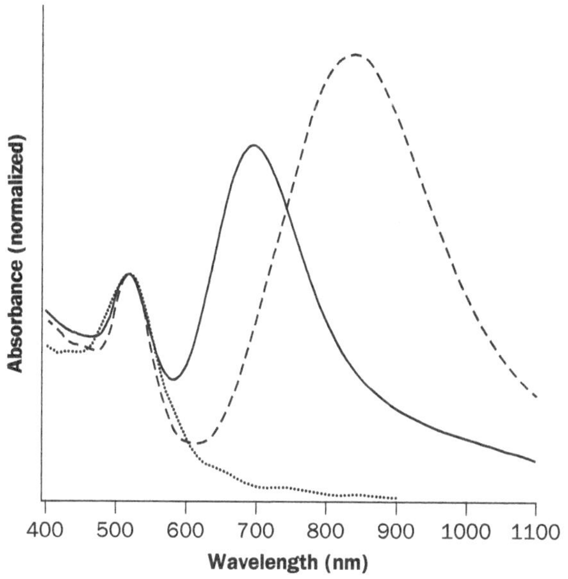

FIGURE 4.8 UV-Vis spectra of colloidal water suspensions of Au nanostructures: dotted curve: nanospheres (diameter 15–25 nm); solid curve: low aspect ratio nanorods; dashed curve: high aspect ratio nanorods. Extinction is normalized at about 520 nm.

The most common synthetic methods for Au nanorods can be broadly divided into (a) electro-templated synthesis,[42] (b) photochemical reduction technique,[43] (c) three-step seeded growth,[44] (d) one-step Ag-assisted seeded growth,[45] and (e) "seedless" growth approach.

In recent years, the seeded growth synthesis has gained immense popularity owing to the fact that a wide array of shapes and sizes can be synthesized using modifications that include small organic additives, binary surfactant mixtures, and other synergistic strategies to help control growth. The seeded growth procedure utilizes small Au NP seeds (about 1.5 nm) stabilized with a cationic surfactant, CTAB (cetyltrimethylammonium bromide). The growth solution is made of chloroauric acid ($HAuCl_4$), CTAB, and a small amount of silver nitrate ($AgNO_3$) wherein ascorbic acid that reduces Au^{3+} to Au^{1+} is added. Thereafter, the seed solution is added to the growth solution and the nanostructures are allowed to grow at 35°C. The resulting solutions are purified using dialysis or centrifugation. This method always results in lower aspect ratio (1.5–5) rods as compared to the seeded growth without Ag.[46] However, modifications in this method using binary surfactant mixtures have also made it possible to synthesize high aspect ratio structures.[47] A schematic of the overall process is depicted in Fig. 4.9.

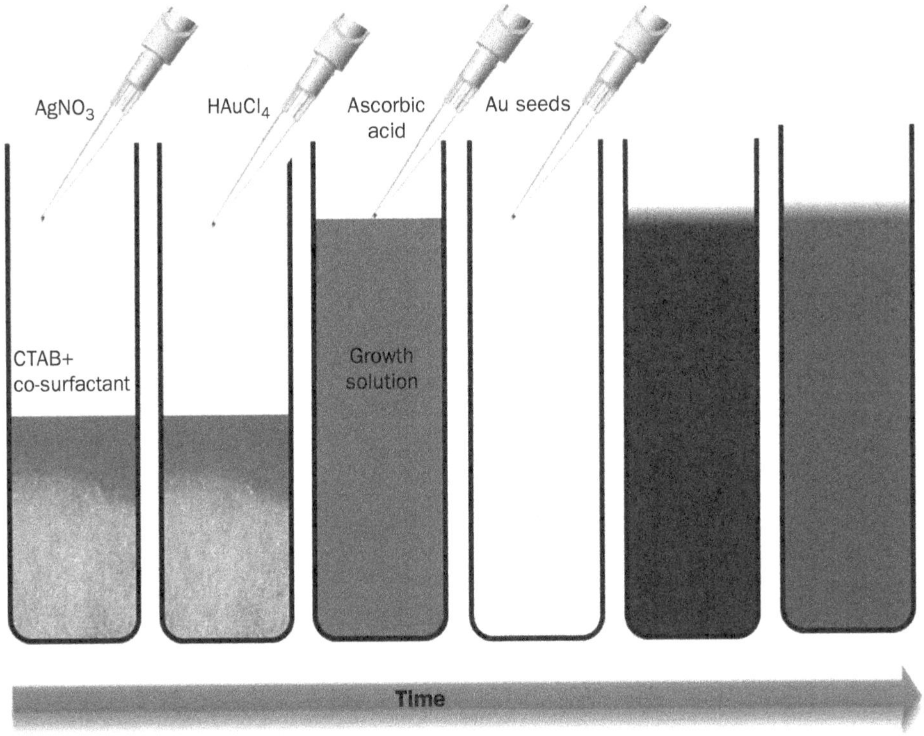

FIGURE 4.9 Schematic showing an Ag-assisted seeded growth process for synthesis of anisotropic gold nanostructures.

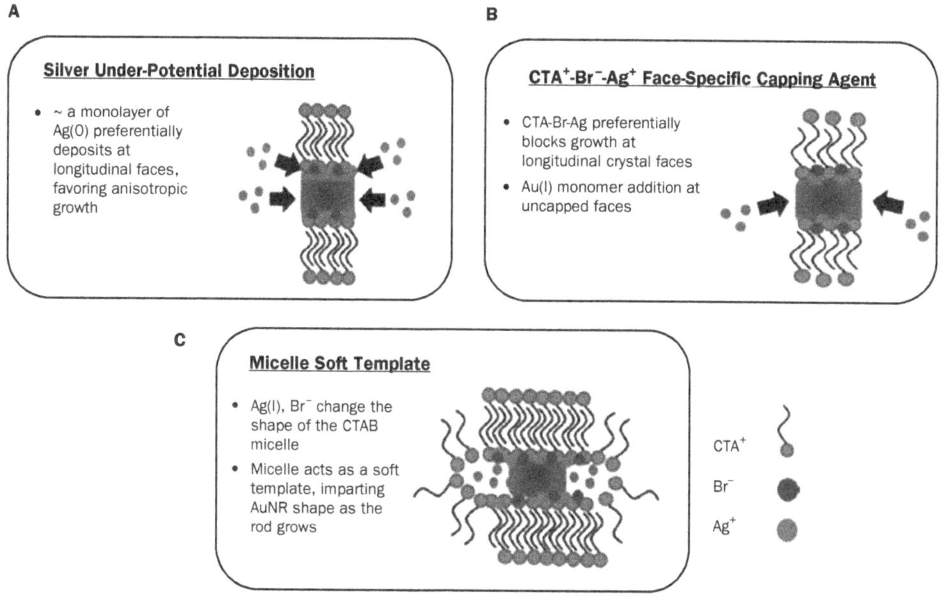

FIGURE 4.10 Simplistic overview of (A) Ag under potential deposition, (B) face-specific capping, and (C) surfactant templating as mechanism for growth of Au nanorods.

The underlying principle for seeded growth method comprises two fundamental steps: (i) generation of Au seed particles at high values of supersaturation, (ii) followed by directed growth of these seed particles in the presence of surfactants, organic additives, or binary surfactant mixtures at low values of supersaturation. In the first step, high supersaturation is used to generate very small and uniform, spherical seed particles. Thereafter, the reaction conditions are altered by using a milder reducing agent than used in the first step. The growth of the nanostructures is affected at a much slower rate in this step by using a templating surfactant or molecule that directs the seeds to grow into larger particles of defined morphologies.[48] In order to explain the differences in structures obtained using the Ag-assisted seeded growth synthesis, three different mechanisms have been proposed.

These include metallic silver serving as capping agent at certain faces of the nanostructures, the Ag-surfactant complex serving as a face-specific capping agent, and silver ions changing the shape of surfactant micelles (micellar template growth) (Fig. 4.10).[49] These mechanisms are further detailed in the subsequent sections.

4.7.1 Under-Potential Deposition

Under-potential deposition (UPD) refers to the electrodeposition of metal monolayer(s) on a foreign metal substrate at potentials significantly less negative than that for deposition on the same metal surface as the adsorbate. The under-potential shift is proportional to the work function difference of the two metals. The work function of Ag is lower than that of Au by more than 0.5 eV; therefore, the UPD of Ag over Au is expected.

A compact silver monolayer can be formed preferably on the {110} times facets of the gold nanocrystals. The {110} facet is not a densely packed surface, and it has higher energy than the {100} and {111} facets.[50] The silver monolayer over Au {110} acts as a strongly binding surfactant to protect the facet from further growth. The top faces of the growing nanorods (i.e., Au {100} facets) are only partially covered by Ag, and therefore grow faster, leading to a one-dimensional growth along the {100} direction. The orientation of the nanorods on the carbon grid indicates that the {110} facets are in fact dominant during the growth process. Although advanced characterization techniques should be able to locate the presence of Ag on the Au nanorod surfaces, it is very difficult to ascertain the state of silver (metallic or ionic), since these exist in a dynamic equilibrium under atmospheric conditions.[51] This has led researchers to believe that instead of elemental Ag, face-specific capping could be due to other Ag-based structures.

4.7.2 Face-Specific Capping

This mechanism of Au nanorod growth relies on the fact that an Ag(I) bromide complex adsorbs to specific faces, promoting anisotropic growth. A complex between CTAB and Ag bromide has been recognized by nuclear magnetic resonance (NMR) and x-ray photoelectron spectroscopy (XPS) to be the entity at the surface of Au nanorods, resulting from an *in situ* formation during the synthesis protocol.[52] The basic difference with the previously mentioned mechanism is what initially binds to the surface with minor difference in how the reaction proceeds.

The reduction of silver ions under synthesis conditions can be neglected since the reducing power of ascorbic acid is too positive at low pH.[53] However, the concentrations are above the solubility product of bulk AgBr ($K_{sp} = 5.35 \times 10^{-13}$), thereby facilitating the formation of AgBr on the gold nanorods. This AgBr is presumed to deposit on rod surfaces at the Au-CTAB interface. This helps in stabilizing the rods and directing growth by hindering growth from a specific facet. The less densely covered facets are thereby promoted to grow in this approach. The evidence that points to the presence of AgBr on the nanorods includes chemical shifts in ^{1}H NMR spectra for capped Au nanorods being identical to that of a AgBr-CTAB preformed complex and XPS data that suggest that the detectable silver is present as Ag(I).

4.7.3 Surfactant Templating

One of the most widely accepted mechanisms in connection to growth of Au nanorods by the seed-mediated approach is surfactant templating. Surfactants are well-known to act as soft templates or nanoreactors in determining the shape and size of the formed NPs. The micelles containing reactants collide owing to Brownian motion occasionally coalesce and exchange their contents leading to reaction, as discussed in Sec. 4.2. The critical packing parameter (CPP) of a surfactant that takes into account the volume of the hydrophobic chain (V_C), the cross-sectional area of the hydrophobic core of the aggregate expressed per molecule in the aggregate (a), and the length of the hydrophobic chain (L_C) gives an overall behavior of the surfactants in solution.

$$CPP = \frac{V_C}{L_C a} \tag{4.19}$$

Among the major properties that can be exploited from the knowledge of CPP, the most important for the ongoing discussion is the different shapes of micelles that stem from varying CPP. The micelles can range from spherical, cylindrical, bilayers, or even inverted structures when the CPP varies.[54] The mechanism employing surfactant templating emphasizes that CTAB micelles are inherently rod-shaped under Au nanorod synthesis conditions. This happens due to a simultaneous action of the silver and bromide ions, which alter the shape of the CTAB micelle. Thus, the micelles do act as soft templates to direct Au nanorod growth.

Although the above-mentioned hypotheses are often used to explain the growth of Au nanostructures in case of seed-mediated method, it is very difficult to account for various shapes with absolute certainty. Therefore, several combinations are often used to interpret the growth processes. An alternative to this growth limit has been proposed as a two-step process. Initially, the Au NPs grow anisotropically up to a maximum aspect ratio. Thereafter, the micelles burst out releasing the nanorods, which finish their growth isotropically until the growth reagents are exhausted.[55] With the advent of new *in situ* characterization techniques such as liquid-cell TEM, small-angle x-ray scattering (SAXS), and already existing robust techniques such as XPS, UV-Vis, and NMR, it might be possible to address some of the key issues toward identifying which mechanistic route the synthesis is following. Although atomistic description of the role of different compounds in the synthesis of nanorods has been discussed elsewhere,[56] it might be possible to do molecular simulations to scan through energy configurations of intermediate entities expected to be formed. It might also be possible to do a detailed energy configuration analysis of the various facets to better appreciate the growth parameters.

Thus, you have seen how by employing different metallic precursors, reducing agents, and passivating or stabilizing ligands, many synthetic routes can be controlled to yield NPs using a bottom-up approach. Depending on the desired particle size, shape, surface functionality, and degree of surface ligand substitution, specific methods can be cherry-picked. The preferred synthesis procedure is heavily dependent on intended application.

4.8 Study Questions

4.8.1 Concept-Based Questions

4.1 Enlist the generic components and conditions necessary for synthesizing metal NPs and stabilizing them in a solvent.

4.2 Name an example of a synthesis procedure where the reducing agent also acts as a passivating ligand.

4.3 Enumerate the role played by passivating ligand in solution based synthesis of metal NPs.

4.4 Being seriously fascinated by the discussion in this chapter, some of you decide to investigate more with bottom-up synthesis of core-shell NPs suitable for targeted drug delivery. Your supervisor has always been working with lithography and similar techniques for synthesis. You need to convince him why bottom-up strategies are favored.

a. List some of the reasons why using bottom-up strategy is an advantage in this case.

Fully convinced by your reasoning, your supervisor gives the green light to the project! All excited, you start to search the state of the art in this field, in order to select the method of synthesis. Taking a quick look at the notes from this chapter, you decide to try hands on with reverse micelle-based synthetic route.

b. What is the general principle underlying this synthetic protocol?

The oil phase you select to use is decane. 1-Butanol is added as a co-surfactant. Copious amounts of CTAB are used to form reverse micelles. Aqueous solution of $FeCl_2$ is reduced with $NaBH_4$ in these reverse micelles to form Fe NPs. Literature suggests that you can expect a S(T)EM size of around 7 nm.

c. State a rule of thumb by which you can control the size of Fe NPs in this route. Suggest a simple composition change in this method with which you can achieve this.

You are all set to use them for growing a shell of gold around them. Believing in old school chemistry, you choose a modified Turkevich method to achieve your goal.

d. What kind of nucleation would you expect in this case? List at least two ways how you can control the Au shell thickness.

4.5 a. What are the different mechanisms that can explain how spherical Au NPs grow into anisotropic structures?

b. In one specific case, you increase the concentration of ascorbic acid and observe that the nanorods get shorter (lower aspect ratio). Explain your observation.

4.6 In another synthesis method, you use dendrimers to form Au nanoparticles. The reagents you use are $HAuCl_4$ and ascorbic acid. Identify the role of each reagent. What will be the role of pH in this case?

4.8.2 Research-Based Questions

4.7 In a seeded growth method for synthesis of anisotropic Au NPs, didodecyldimethylammonium bromide (DDAB) is used as a co-surfactant in one case, while tridodecylmethylammonium iodide (TDAI) is used in the other case. All other reaction parameters are kept constant with both cases using Ag-assisted seeded growth method. The shapes obtained are shown in the figure below. (Images have been obtained using NTNU Nanolab Electron Microscope.) Explain by using nucleation and growth concepts learned in Chaps. 2 and 3 why you get different shapes.

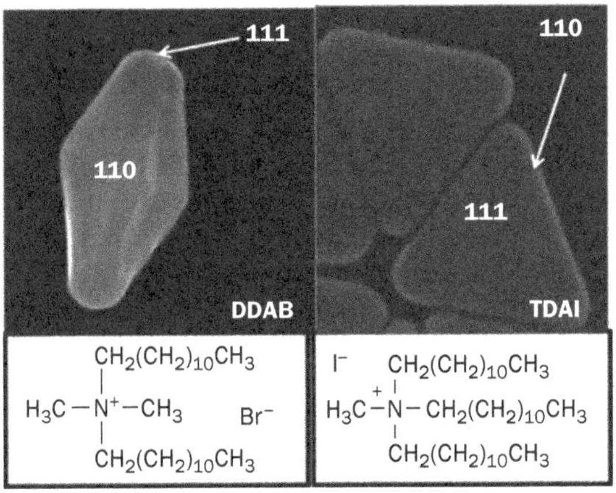

4.8 Sivaraman et al.[57] synthesized Au NPs using the Turkevich method. To study the influence of reactants in the order of addition to the reaction mixture, two sets of experiments were conducted. In the first case 0.254 mM of 24.75 mL $HAuCl_4$ was heated until it started boiling and then 0.25 mL of sodium citrate was added to it. The concentration of sodium citrate was varied to obtain the desired mole ratio (ratio of $HAuCl_4$ to citrate). In the second case 5.2 mM sodium citrate solution was heated until the solution started boiling, to which 25.4 mM $HAuCl_4$ solution was added. The volume of citrate and $HAuCl_4$ solution was adjusted to the desired mole ratio. In both cases the total volume of the reaction mixture was fixed at 25mL.

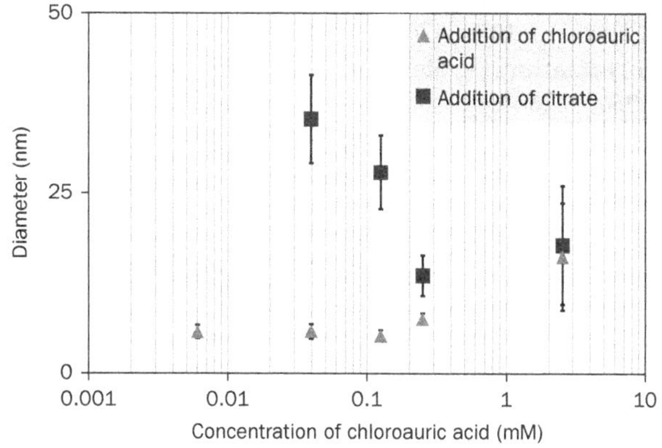

The above figure shows the plot of size of the obtained Au nanoparticles measured from Field Emission Scanning Electron Microscope (FESEM) images to the concentration of $HAuCl_4$ for both cases. Comment on the observed trends on the sizes against gold concentration for both cases.

4.9 Kimling et al.[58] synthesized Au nanoparticles with the Turkevich method by bringing 95 mL of gold solution to boil, and then 5 mL of pre-heated sodium citrate was added to it quickly and stirred vigorously. The reaction time was defined to be 15 minutes. The reactions were conducted in three different gold concentrations of 0.8, 1.0, and 1.2 mM and the concentration of citrate was increased for the desired mole ratio. The figure below shows a plot of gold to citrate mole ratio against surface plasmon resonance (SPR) peak wavelength and average particle size measured using SEM images.

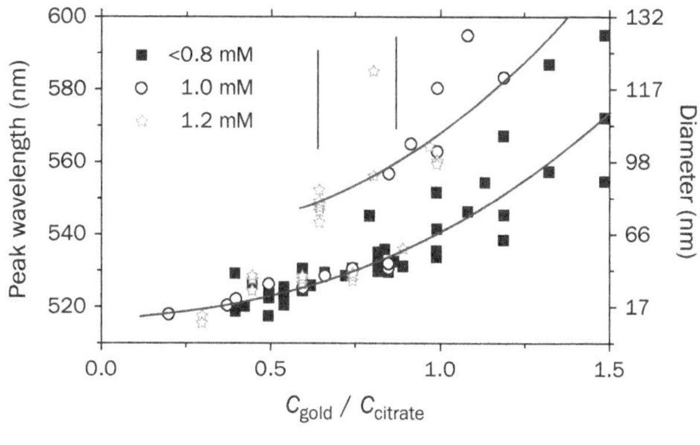

Comment on the influence of increase in citrate concentration on the observed sizes and the shift in peak wavelength.

4.9 References

1. Goia, D. V.; Matijevic, E., Preparation of monodispersed metal particles. *New Journal of Chemistry.* **1998,** *22*(11), 1203–1215.
2. Turkevich, J.; Stevenson, P. C.; Hillier, J., A study of the nucleation and growth processes in the synthesis of colloidal gold. *Discussions of the Faraday Society.* **1951,** *11,* 55–75.
3. Turkevich, J.; Stevenson, P. C.; Hillier, J., The formation of colloidal gold. *Journal of Physical Chemistry.* **1953,** *57*(7), 670–673.
4. Turkevich, J.; Garton, G.; Stevenson, P. C., The color of colloidal gold. *Journal of Colloid Science.* **1954,** *9,* 26–35.
5. Ji, X.; Song, X.; Li, J.; Bai, Y.; Yang, W.; Peng, X., Size control of gold nanocrystals in citrate reduction: the third role of citrate. *Journal of American Chemical Society.* **2007,** *129*(45), 13939–13948.
6. Ji, X.; Song, X.; Li, J.; Bai, Y.; Yang, W.; Peng, X., Size control of gold nanocrystals in citrate reduction: the third role of citrate. *Journal of the American Chemical Society.* **2007,** *129*(45), 13939–13948.
7. Lee, P. C.; Meisel, D., Adsorption and surface-enhanced Raman of dyes on silver and gold sols. *Journal of Physical Chemistry.* **1982,** *86*(17), 3391–3395.
8. Dong, X.; Ji, X.; Wu, H.; Zhao, L.; Li, J.; Yang, W., Shape control of silver nanoparticles by stepwise citrate reduction. *Journal of Physical Chemistry C* **2009,** *113*(16), 6573–6576.
9. Pillai, Z. S.; Kamat, P. V., What factors control the size and shape of silver nanoparticles in the citrate ion reduction method? *Journal of Physical Chemistry B.* **2004,** *108*(3), 945–951.
10. Furlong, D. N.; Launikonis, A.; Sasse, W. H. F.; Sanders, J. V., Colloidal platinum sols. Preparation, characterization and stability towards salt. *Journal of the Chemical Society, Faraday Transactions 1: Physical Chemistry in Condensed Phases.* **1984,** *80*(3), 571–588.
11. Brust, M.; Walker, M.; Bethell, D.; Schiffrin, D. J.; Whyman, R., Synthesis of thiol-derivatised gold nanoparticles in a two-phase Liquid-Liquid system. *Journal of the Chemical Society, Chemical Communications.* **1994,** 7, 801–802.
12. Brust, M.; Fink, J.; Bethell, D.; Schiffrin, D. J.; Kiely, C., Synthesis and reactions of functionalised gold nanoparticles. *Journal of the Chemical Society, Chemical Communications.* **1995,***16,* 1655–1656.
13. Perala, S. R. K.; Kumar, S., On the mechanism of metal nanoparticle synthesis in the Brust–Schiffrin method. *Langmuir.* **2013,** *29*(31), 9863–9873.
14. Andreescu, D.; Sau, T. K.; Goia, D. V., Stabilizer-free nanosized gold sols. *Journal of Colloid and Interface Science.* **2006,** *298*(2), 742–751.
15. Goia, D.; Matijević, E., Tailoring the particle size of monodispersed colloidal gold. *Colloids and Surfaces A: Physicochemical and Engineering Aspects.* **1999,** *146*(1–3), 139–152.
16. Sondi, I.; Goia, D. V.; Matijević, E., Preparation of highly concentrated stable dispersions of uniform silver nanoparticles. *Journal of Colloid and Interface Science.* **2003,** *260*(1), 75–81.
17. Suber, L.; Sondi, I.; Matijević, E.; Goia, D. V., Preparation and the mechanisms of formation of silver particles of different morphologies in homogeneous solutions. *Journal of Colloid and Interface Science.* **2005,** *288*(2), 489–495.

18. Rampino, L. D.; Nord, F. F., Preparation of palladium and platinum synthetic high polymer catalysts and the relationship between particle size and rate of hydrogenation. *Journal of American Chemical Society.* **1941,** *63*(10), 2745–2749.
19. Henglein, A.; Ershov, B. G.; Malow, M., Absorption-spectrum and some chemical-reactions of colloidal platinum in aqueous solution. *Journal of Physical Chemistry.* **1995,** *99*(38), 14129–14136.
20. Henglein, A.; Giersig, M., Reduction of Pt (II) by H_2: effects of citrate and NaOH and reaction mechanism. *Journal of Physical Chemistry B.* **2000,** *104*(29), 6767–6772.
21. Friberg, S. E.; Bothorel, P., *Microemulsions: Structure and Dynamics*, United States: N. p., 1987.
22. Ganguli, A. K.; Ahmad, T.; Vaidya, S.; Ahmed, J., Microemulsion route to the synthesis of nanoparticles. *Pure and Applied Chemistry.* **2008,** *80*(11), 2451–2477.
23. Li, Y.; Park, C.-W., Particle size distribution in the synthesis of nanoparticles using microemulsions. *Langmuir.* **1999,** *15*(4), 952–956.
24. Bandyopadhyaya, R.; Kumar, R.; Gandhi, K. S.; Ramkrishna, D., Modeling of precipitation in reverse micellar systems. *Langmuir.* **1997,** *13*(14), 3610–3620.
25. Egorova, E. M.; Revina, A. A., Synthesis of metallic nanoparticles in reverse micelles in the presence of quercetin. *Colloids and Surfaces A: Physicochemical and Engineering Aspects.* **2000,** *168*(1), 87–96.
26. Bagwe, R. P.; Khilar, K. C., Effects of intermicellar exchange rate on the formation of silver nanoparticles in reverse microemulsions of AOT. *Langmuir.* **2000,** *16*(3), 905–910.
27. Liu, C. H.; Zhou, Z. D.; Yu, X.; Lv, B. Q.; Mao, J. F.; Xiao, D., Preparation and characterization of Fe3O4/Ag composite magnetic nanoparticles. *Inorganic Materials.* **2008,** *44* (3), 291–295.
28. Chao, S. M.; Meen, T. H.; Chen, W. R.; Wu, K. H.; Liu, Y. S.; Tzou, W. C.; Huang, C. J. Synthesis of Fe-core/Au-shell nanoparticles under ambient pressure. *Key Engineering Materials*, **2010**, *434–435*, 799–802.
29. Hota, G.; Jain, S.; Khilar, K. C., Synthesis of CdS–Ag 2 S core-shell/composite nanoparticles using AOT/n-heptane/water microemulsions. *Colloids and Surfaces A: Physicochemical and Engineering Aspects.* **2004,** *232*(2), 119–127.
30. Zhang, D.; Cheng, H.; Ma, J.; Wang, Y.; Gai, X., Synthesis of silver-coated silica nanoparticles in nonionic reverse micelles. *Journal of Materials Science Letters.* **2001,** *20*(5), 439–440.
31. Jafari, T.; Simchi, A.; Khakpash, N., Synthesis and cytotoxicity assessment of superparamagnetic iron–gold core–shell nanoparticles coated with polyglycerol. *Journal of Colloid and Interface Science.* **2010,** *345*(1), 64–71.
32. Tavakoli, A.; Sohrabi, M.; Kargari, A., A review of methods for synthesis of nanostructured metals with emphasis on iron compounds. *Chemical Papers.* **2007,** *61*(3), 151–170.
33. Tadić, M.; Marković, D.; Spasojević, V.; Kusigerski, V.; Remškar, M.; Pirnat, J.; Jagličić, Z., Synthesis and magnetic properties of concentrated c′-Fe_2O_3 nanoparticles in a silica matrix. *Journal of Alloys and Compounds.* **2007,** *441*(1–2), 291–296.
34. Sedlar, M., The preparation and magnetic properties of sodium-modified iron oxide thin films by a sol· gel method. *Ceramics International.* **1994,** *20*(1), 73–78.
35. Xie, J.; Zheng, Y.; Ying, J. Y., Protein-directed synthesis of highly fluorescent gold nanoclusters. *Journal of American Chemical Society.* **2009,** *131*(3), 888–889.

36. McDonagh, B. H.; Singh, G.; Bandyopadhyay, S.; Lystvet, S. M.; Ryan, J. A.; Volden, S.; . . .Glomm, W. R., Controlling the self-assembly and optical properties of gold nanoclusters and gold nanoparticles biomineralized with bovine serum albumin. *RSC Advances.* **2015,** *5*(122), 101101–101109.
37. McDonagh, B. H.; Volden, S.; Lystvet, S. M.; Singh, G.; Ese, M.-H. G.; Ryan, J. A.; Lindgren, M.; et al., Self-assembly and characterization of transferrin-gold nanoconstructs and their interaction with bio-interfaces. *Nanoscale.* **2015,** *7*(17), 8062–8070.
38. Li, F.; Wang, Q., Fabrication of nanoarchitectures templated by virus-based nanoparticles: strategies and applications. *Small.* **2014,** *10*(2), 230–245.
39. Berti, L.; Burley, G. A., Nucleic acid and nucleotide-mediated synthesis of inorganic nanoparticles. *Nature Nanotechnology.* **2008,** *3*(2), 81–87.
40. Feldheim, D. L.; Eaton, B. E., Selection of biomolecules capable of mediating the formation of nanocrystals. *ACS Nano.* **2007,** *1*(3), 154–159.
41. Bandyopadhyay, S., Smart and Multifunctional Core-Shell Nanoparticles (NPs) for Drug Delivery. 2016. Available at: https://brage.bibsys.no/xmlui/handle/11250/2423731.
42. Hulteen, J. C.; Martin, C. R., A general template-based method for the preparation of nanomaterials. *Journal of Materials Chemistry.* **1997,** *7*(7), 1075–1087.
43. Kim, F.; Song, J. H.; Yang, P., Photochemical synthesis of gold nanorods. *Journal of American Chemical Society.* **2002,** *124*(48), 14316–14317.
44. Gole, A.; Murphy, C. J., Seed-mediated synthesis of gold nanorods: role of the size and nature of the seed. *Chemistry of Materials.* **2004,** *16*(19), 3633–3640.
45. Sau, T. K.; Murphy, C. J., Room temperature, high-yield synthesis of multiple shapes of gold nanoparticles in aqueous solution. *Journal of American Chemical Society.* **2004,** *126*(28), 8648–8649.
46. Sau, T. K.; Murphy, C. J., Seeded high yield synthesis of short Au nanorods in aqueous solution. *Langmuir* **2004,** *20*(15), 6414–6420.
47. Nikoobakht, B.; El-Sayed, M. A., Preparation and growth mechanism of gold nanorods (NRs) using seed-mediated growth method. *Chemistry of Materials.* **2003,** *15*(10), 1957–1962.
48. Grzelczak, M.; Pérez-Juste, J.; Mulvaney, P.; Liz-Marzán, L. M., Shape control in gold nanoparticle synthesis. *Chemical Society Review.* **2008,** *37*(9), 1783–1791.
49. Lohse, S. E.; Murphy, C. J., The quest for shape control: a history of gold nanorod synthesis. *Chemistry of Materials.* **2013,** *25*(8), 1250–1261.
50. Wang, Z.; Gao, R.; Nikoobakht, B.; El-Sayed, M., Surface reconstruction of the unstable {110} surface in gold nanorods. *Journal of Physical Chemistry B.* **2000,** *104*(23), 5417–5420.
51. Glover, R. D.; Miller, J. M.; Hutchison, J. E., Generation of metal nanoparticles from silver and copper objects: nanoparticle dynamics on surfaces and potential sources of nanoparticles in the environment. *ACS Nano.* **2011,** *5*(11), 8950–8957.
52. Hubert, F.; Testard, F.; Spalla, O., Cetyltrimethylammonium bromide silver bromide complex as the capping agent of gold nanorods. *Langmuir.* **2008,** *24*(17), 9219–9222.
53. Pal, T.; De; Jana, N. R.; Pradhan, N.; Mandal, R.; Pal, A.; Beezer, A. E.; Mitchell, J. C., Organized media as redox catalysts. *Langmuir.* **1998,** *14*(17), 4724–4730.
54. Hiemenz, P. C.; Rajagopalan, R., *Principles of Colloid and Surface Chemistry*. Boca Raton, FL: CRC Press, 1997.

55. Hubert, F.; Testard, F.; Thill, A.; Kong, Q.; Tache, O.; Spalla, O., Growth and overgrowth of concentrated gold nanorods: time resolved SAXS and XANES. *Crystal Growth & Design*. **2012,** *12*(3), 1548–1555.
56. Almora-Barrios, N.; Novell-Leruth, G.; Whiting, P.; Liz-Marzán, L. M.; López, N. R., Theoretical description of the role of halides, silver, and surfactants on the structure of gold nanorods. *Nano Letters*. **2014,** *14*(2), 871–875.
57. Sivaraman, S. K.; Kumar, S.; Santhanam, V., Monodisperse sub-10nm gold nanoparticles by reversing the order of addition in Turkevich method: The role of chloroauric acid. *Journal of Colloid and Interface Science*. **2011,** *361*(2), 543–547.
58. Kimling, J.; Maier, M.; Okenve, B.; Kotaidis, V.; Ballot, H.; Plech, A., Turkevich method for gold nanoparticle synthesis revisited. *Journal of Physical Chemistry B*. **2006,** *110*(32), 15700–15707.

CHAPTER 5

Functionalization of Metal Nanoparticles

By now you would have painstakingly controlled nucleation rates and growth solutions, and you would have succeeded in making metal nanoparticles (NPs) with the intended structure, of the right metal, with planned size distribution and with a high yield. If so, congratulations! You should take a moment to bask in the glory of a job well done. If you didn't do so, go ahead—we will wait.

Now that you are back, you may be surprised that the gorgeous NPs you've made are unlikely to be suitable for the application you had in mind. It is because though the synthesis procedure you followed gives you the right type of NP, it does not take the physicochemical environment of your intended application into consideration. For example, the requirements for successful incorporation of your NPs in a polymer nanocomposite for the automobile industry are very different from the requirements for a drug delivery vector, a catalyst carrier, or a home pregnancy kit. This is where tuning the surface chemistry—also called functionalization—of your NPs becomes crucial, as the compatibility between the NPs and the surroundings is predominantly controlled by the surface properties.

Generally, there are two pathways toward surface functionalization: *in situ* and *post situ* (also known as post-synthesis) functionalization, respectively. *In situ* functionalization—sometimes referred to as *direct functionalization*—implies that the passivating or capping ligand you use during synthesis perfectly suits your application, which is arguably the preferred scenario, as it reduces the number of processing steps. Post-synthesis functionalization implies that you modify the NP surface after synthesis, which allows for a great deal of flexibility with respect to what surface groups you introduce and how.

5.1 *In Situ* Functionalization

As luck would have it, you have already been through a number of synthesis methods in Sec. 4.1.3 that yield *in situ* functionalization:

1. *The Turkevich method:* In this method, the citrate acts both as a reducing agent and passivating ligand, with the citrate surface layer providing electrostatic stability in an aqueous environment. Despite the vast literature on gold NP synthesis using the Turkevich method or adaptations thereof, the exact structure of the citrate layer remains to be fully elucidated. This is partly because both hydroxyl groups and carboxylic acid moieties can bind to gold, and that once a

monolayer of citrate is formed, it can form multilayers via hydrogen bonding. However, what is known is that the citrate layer can be displaced by stronger nucleophiles (indicating weak interactions with the gold surface), by dilution (further indicating physisorption to the surface), by increasing the ionic strength in the bulk phase, or by reducing pH to or below the lowest pK_a of citrate (pH ~4).

2. *The Brust reaction:* In this method, a strong reducing agent is added to the gold salt in the presence of a bifunctional thiol-containing molecule. The thiol groups bind covalently to the surface of the gold NPs, resulting in stable suspensions. As thiol-containing bifunctional molecules are commercially available with a large selection of end groups and molecular weights, this provides great design flexibility in surface chemistry for a range of applications. However, the need for a strong reducing agent limits the upper size of the NPs produced, which could further limit the application range.
3. *The Goia reaction:* It is also called the ascorbic (or iso-ascorbic) acid method, as it relies on the use of—you guessed it—ascorbic acid as a combined reducing agent and passivating ligand. As such, it strongly resembles the Turkevich method with one notable exception: Iso-ascorbic acid likely forms a resonant complex with the gold surface by sharing of electron pairs, resulting in a significantly stronger surface binding than what is observed for citrate. This confers a higher pH stability onto gold NPs made by the Goia reaction, with NPs being stable even below pK_{a1} of ascorbic acid.[1]
4. *Thermal decomposition* (also called *hot injection*): This method is often used for synthesizing iron oxide NPs by injecting an iron oleate complex into a heated organic solvent such as octadecene in the presence of a surfactant such as oleic acid or hexadecylamine (refer to research work by Singh et al.[2] and references therein). The surfactant can be selected according to the thermal stability and intended application, and forms strong interactions with the NP surface. An illustration of the hot injection method is presented in Fig. 5.1.

5.2 Post-Synthesis Functionalization

The other—and most common—general approach to functionalization of metal NPs is to modify the surface post-synthesis. There are a number of possibilities as to what type of synthesis method was used, and what type of surface ligand is in place. Your variables can often come down to the following:

1. Binding to or replacing the existing surface ligand
2. Intermolecular forces, such as hydrophobic, van der Waals, electrostatic, or covalent interactions with the particle
3. Number of attachment sites, thickness, and density of surface layer
4. Number of different surface groups needed to perform the intended function

Some of the most common examples are as follows:

1. *Non-specific binding:* This is typically done by introducing surface active polymers such as polyvinylpyrrolidone (PVP) which bind to the particle

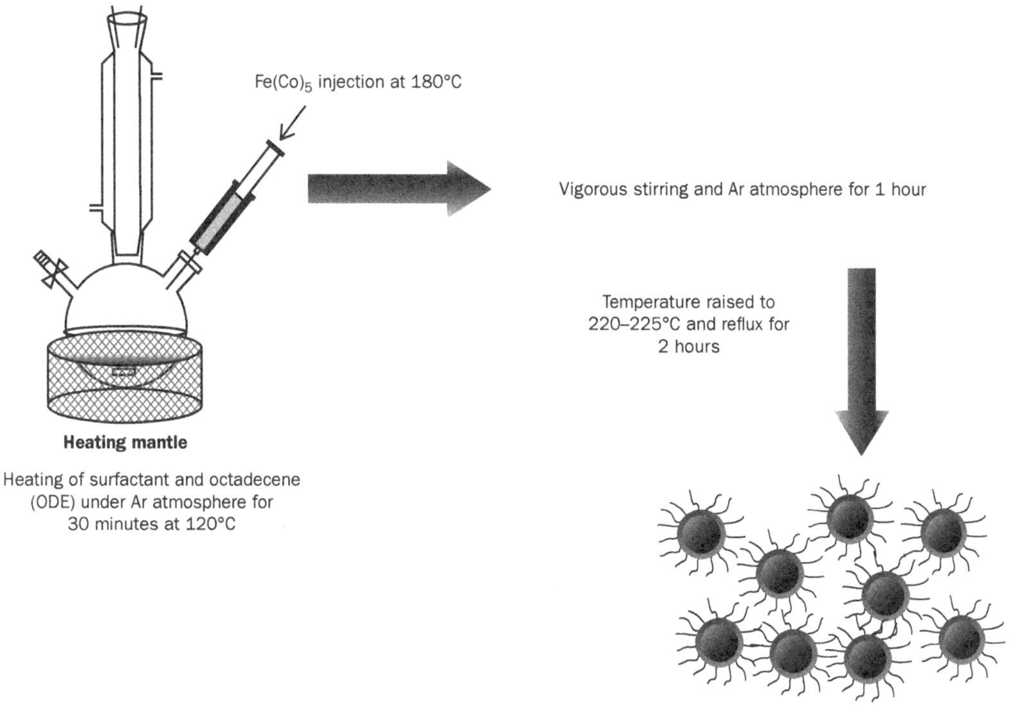

Figure 5.1 Illustration of the hot injection method for *in situ* functionalization of iron oxide NPs.

surface. This type of functionalization generally results in steric stabilization of the NPs, without imparting any specific functionality.

2. *Displacement:* Weakly to moderately bound passivating ligands such as citrate or iso-ascorbic acid are easily displaced from the surface by stronger nucleophiles such as thiols or phosphines. For example, by adding thiol-containing bifunctional molecules (such as in the Brust reaction) to gold NPs made by the Turkevich method, you can replace the citrate. This has dual benefits of yielding much more stable surface layers and having a range of functionalities to select from.
3. *Layer-by-layer (LbL):* Layer-by-layer is a charge-based surface functionalization method used to form thick, dense surface layers. Starting with a charged surface (e.g., a gold NP produced by the Turkevich method), you sequentially add oppositely charged polymers, as shown in Fig. 5.2. The charge of the final NP construct and the thickness/density of the surface composite layer are controlled by varying the number of layers and the molecular weight of the polyelectrolytes used. LbL can also be used to protect unstable biological molecules such as RNA or DNA from degradation by sandwiching between cationic layers.
4. *Protein binding:* Binding of proteins to the surface of metal NPs is typically done either to camouflage the NP from the immune system (when used in nanomedicine) or to impart a specific functionality such as enzymatic catalysis.

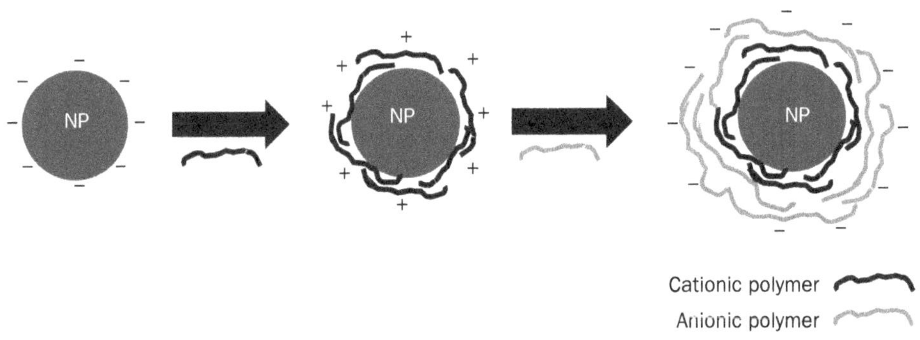

FIGURE 5.2 Schematic representation of layer-by-layer (LbL) method for functionalization of NPs.

Proteins can either be adsorbed non-specifically onto the NPs (typically via electrostatic or hydrophobic interactions) or be covalently linked to surface groups. Non-specific adsorption of proteins usually forms strongly bound surface layers, as each protein has multiple attachment points along the NP surface, resulting in an effect similar to superglue. Covalent binding of proteins can be done via a number of approaches, including reaction between epoxy groups on the surface and primary amines on the protein (requiring initial surface functionalization with an epoxy-containing passivating ligand). Another widely used method for protein binding is to use coupling agents which bind to both the NP and the protein (such as carbodiimide chemistry, which is beyond the scope of this chapter).

In addition to the four methods mentioned above, it is possible to graft polymer chains onto (i.e., covalently bind to) the NPs or to grow polymer chains extending from the surface *in situ*. For an excellent overview of surface modification via grafting, see the review article by Christian von der Ehe and coworkers.[3]

For functionalization of NPs, there are fundamental differences between using small molecules (such as organothiols or ascorbic acid) and large molecules (such as polymers and proteins). To discuss the differences, we need to establish a more formal framework for adsorption.

5.3 Langmuir Adsorption

The most common way of describing the adsorption process to a particle is to model or fit the relationship between the surface-bound and the free (solution) concentration of an adsorbate at a fixed temperature. This type of graphic model of surface-bound versus total (or free) concentration of an adsorbate is called an adsorption isotherm, and can be empirically or theoretically derived (see Fig. 5.3 for an example of an adsorption isotherm). While an empirical adsorption isotherm simply constitutes a best fit to the surface concentration versus bulk concentration of the adsorbate, a theoretical model comprises several assumptions regarding how the adsorption occurs as well as what is the maximum theoretical surface coverage.

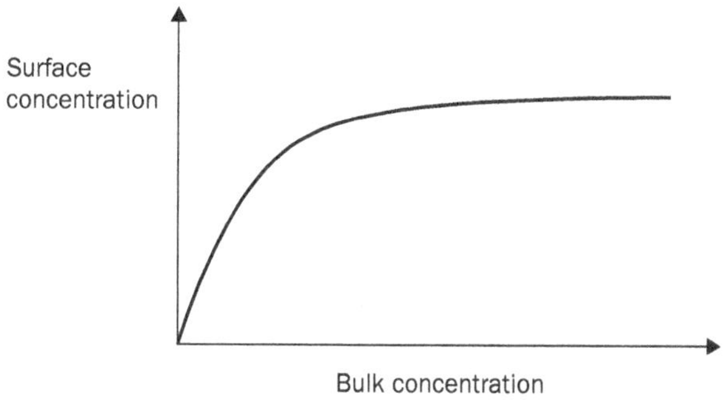

Figure 5.3 Generic depiction of an adsorption isotherm, showing the relationship between the surface concentration and bulk concentration of an adsorbate at a fixed temperature.

The Langmuir isotherm is one of the most widely applicable (and utilized) models for adsorption onto metal surfaces and particles, and also forms a good basis for theoretical understanding of adsorption. The model relates fractional surface coverage (θ) with the bulk concentration of the adsorbate (C).

$$\theta = \frac{KC}{1 + KC} \tag{5.1}$$

where K is the equilibrium constant.

5.3.1 Theoretical Basis

Let us now take a quick look into the theoretical basis for Eq. (5.1). For a thorough understanding of the concepts, you may take a closer look at Chap. 7 of the excellent text book by Hiemenz and Rajagopalan.[4]

Adsorption can be viewed as an equilibrium process depicted by the following equation:

$$A + S \underset{k_d}{\overset{k_a}{\rightleftharpoons}} AS \tag{5.2}$$

where A is the adsorbate, S is the adsorbent or substrate, and AS is A adsorbed on S. k_a and k_d are the adsorption and desorption rate constants, respectively.

For simplicity, let us define the surface fractional coverage θ as the ratio of the number of molecules adsorbed (N) onto surface to the number of molecules adsorbed at monolayer coverage (N_0):

$$\theta = \frac{N}{N_0} \tag{5.3}$$

To better understand the adsorption process, let us take a real-life example. Imagine a classroom with empty chairs. Each chair is to accommodate one student. The rate at which the seats are occupied by students would then vary with the number of students, number of empty seats, and whether or not the students are motivated for the class.

Now, if you substitute a student with an adsorbate molecule and a chair with an active site of adsorption on the surface, the rate of adsorption (ν_a) can be written as follows:

$$\nu_a = \frac{\partial \theta}{\partial t} = k_a CJ(1-\theta) \tag{5.4}$$

where $(1-\theta)$ represents the number of free active sites on the surface, J is a proportionality constant, and k_a is the adsorption rate constant.

Now, imagine that students are allowed to leave the classroom anytime during the lecture hours. The rate at which the students will leave the classroom will depend on the number of occupied chairs (since students can only leave from seated places) and the level of motivation/proximity to exam.

From a desorption point of view, the rate of desorption (ν_d) can be written as follows:

$$\nu_d = -\frac{\partial \theta}{\partial t} = k_d J\theta \tag{5.5}$$

The next question is, what happens at equilibrium? Hypothetically, it is the condition when the rate at which the students enter the classroom is the same as the rate at which they leave.

Therefore at equilibrium,

$$\nu_a = \nu_d \tag{5.6}$$

Yielding

$$k_a CJ(1-\theta) = k_d J\theta \tag{5.7}$$

If we define the equilibrium constant, $K = \frac{k_a}{k_d}$ and solve the above expression, we get

$$\theta = \frac{KC}{1+KC} \tag{5.8}$$

One step ahead, we can substitute the definition of θ to get

$$\frac{N}{N_{mono}} = \frac{KC}{1+KC} \tag{5.9}$$

Linearization yields

$$\frac{C}{N} = \frac{C}{N_{mono}} + \frac{1}{KN_{mono}} \tag{5.10}$$

Equation (5.10) enables you to utilize linear regression to a set of experimental data and use the slope and intercept to infer information about the adsorption process (Fig. 5.4). The slope can be directly used to calculate the number of molecules adsorbed at monolayer coverage, N_{mono}. This information can be further used to evaluate the equilibrium constant K.

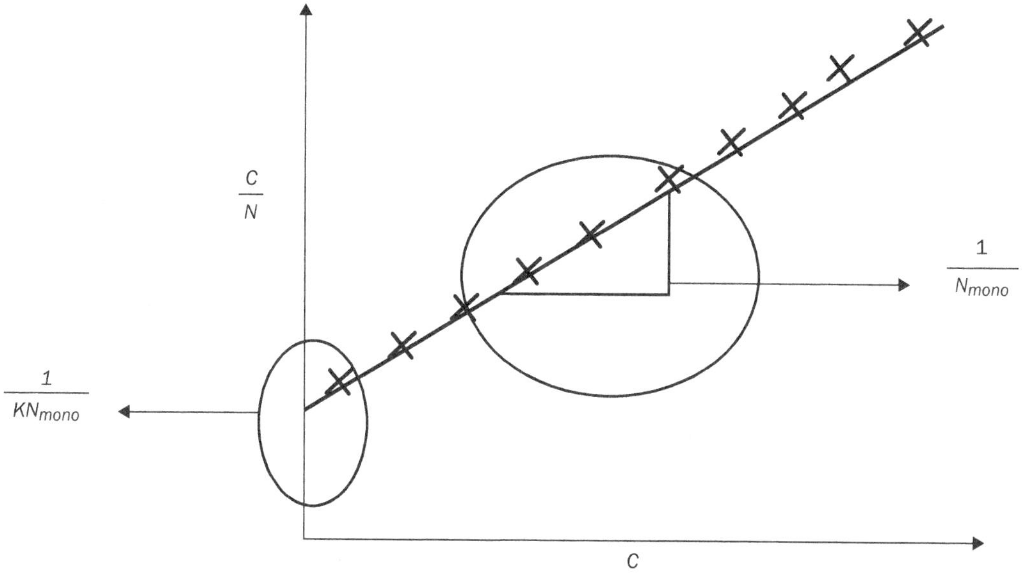

FIGURE 5.4 Schematic plot of linearized Langmuir isotherm showing how the slope and intercept of the regression line may be used to obtain parameters—equilibrium constant (K) and number of molecules adsorbed at monolayer coverage (N_{mono}).

5.3.2 Assumptions, Caveats, and Validity

From the derivation in the previous section, it follows that the Langmuir adsorption isotherm is based on the following assumptions:

1. Adsorption does not proceed beyond monolayer coverage.
2. All adsorption sites are equivalent, and the surface is completely uniform.
3. Adsorption to a given adsorption site is independent of the surface coverage (no lateral interactions between adsorbate molecules).
4. There is no surface diffusion among bound (localized) adsorbate molecules.
5. The adsorption is completely reversible.

Despite its theoretical foundation, the Langmuir isotherm often reverts to an empirical adsorption isotherm, as real systems typically deviate from the underlying assumptions. Deviation from the theoretical model can emanate from a number of causes, including the following:

1. The assumption of surface homogeneity is central to the Langmuir isotherm, as it allows for use of a single equilibrium constant K. For heterogeneous surfaces, the value of K would vary across the surface, leading to increased mathematical complexity or use of an average or median value. In real systems, very few surfaces are completely uniform or flat at a molecular level, with a few exceptions such as mica. In other words, surface roughness

is likely to affect the organization of binding sites and thus both the adsorption process and the resulting surface structures. For adsorption to particles, curvature is a separate parameter related to surface roughness, and will be discussed later.

2. For all binding sites to be equivalent, the size of the adsorbate (i.e., the molecule binding to the surface, or adsorbent) must be smaller than the dimensions of the binding/adsorption site, or that the spacing between the binding sites significantly exceeds the molecular dimensions of the adsorbate. For situations where the adsorbate is capable of binding to or at least stretching across multiple adsorption sites, other incoming molecules will at least be sterically hindered from binding, also violating the assumption of adsorption being independent of surface coverage.
3. Another assumption which excludes interaction between adsorbate molecules—both lateral and otherwise—is the assumption of adsorption not proceeding beyond monolayer coverage. Bilayer or even multilayers are not uncommon results of adsorption for cases where the adsorbates can lower the free energy of the system via favorable intermolecular interactions. These include hydrogen bonding, electrostatic interactions between oppositely charged species or moieties, van der Waals forces, hydrophobic interactions (such as the formation of a surfactant bilayer on a hydrophilic surface), or limited solubility of the adsorbate (common for adsorption of polymers).
4. Finally, for adsorption to be completely reversible, the interactions between the adsorbate and the adsorbent should be weak, or at least desorption should depend on a single event. For chemical adsorption, such as via binding a thiol moiety to a gold surface, the adsorption can be considered as irreversible in the absence of energy input. Polymer adsorption is also often considered as irreversible (depending on the size and composition of the polymer), as the macromolecule typically binds to multiple sites on the adsorbent. While each of these binding events could comprise weak interactions and as such be fully reversible, desorption of the entire polymer chain depends on desorption at each binding site occurring simultaneously, with a similar time constant. Given the low odds of all these events being synchronized, polymer adsorption is for practical purposes considered to be irreversible.

Did we mention that the Langmuir model originally was derived for adsorption of gas molecules onto solids? Despite all its flaws, the Langmuir isotherm or one of its derivations is still one of the most applied models for adsorption for the simple reason: It works. Specifically, it provides an empirical model which fits a wide range of data sets, allowing for estimation of useful parameters such as binding affinity and maximum surface coverage from available experimental data.

5.3.3 Macromolecular Adsorption

In the previous section, Langmuir-type adsorption is discussed in terms of a single event, with the resulting isotherm only depending on the rate constant K and the bulk concentration c. However, when a bare surface is exposed to a solution of adsorbates, each adsorbate molecule must pass through three steps in the formation of an adsorbed layer:[5]

1. Transport toward the surface via diffusion and convection
2. Attachment or binding to the surface
3. Relaxation or spreading into the preferred (lowest energy) surface conformation

These three steps are illustrated for adsorption of macromolecules in Fig. 5.5. While all adsorbate molecules pass through all these three steps, spreading or relaxation is typically only considered for macromolecules such as polymers and proteins for two reasons: (i) they typically occupy more than a single adsorption or binding site, and (ii) for smaller molecules such as surfactants, the relaxation or spreading step occurs very fast compared to the experimental time scale. In other words, for smaller adsorbates, spreading is not experimentally observed, and consequently no information about the spreading process is obtained.

However, for larger molecules, the time constants involved are large enough that spreading could have an impact on the experimental results. If the molecules require a time τ_s for spreading and a time τ_d for deposition onto the surface, we can define two distinct cases:

If deposition is slow ($\tau_s < \tau_d$), each adsorbing molecule has time to unfold and spread before it is enclosed by other adsorbates. Each adsorbate molecule will occupy a large area, resulting in a thin surface layer and a low adsorbed amount.

Conversely, if deposition is fast ($\tau_s > \tau_d$), each adsorbate molecule is enclosed before it has time to spread, resulting in a thicker surface layer and a high adsorbed amount. These two cases are illustrated in Fig. 5.6.

If one were so inclined—and for the purposes of this chapter we are—this concept can be expressed more quantitatively by defining the occupied surface area (a) as a function of time (t). Assuming that the adsorbing polymers spread on more or less the same time scale τ_s, we can define an expression

$$a = a_0 + \alpha(1 - e^{-t/\tau_s}) \tag{5.11}$$

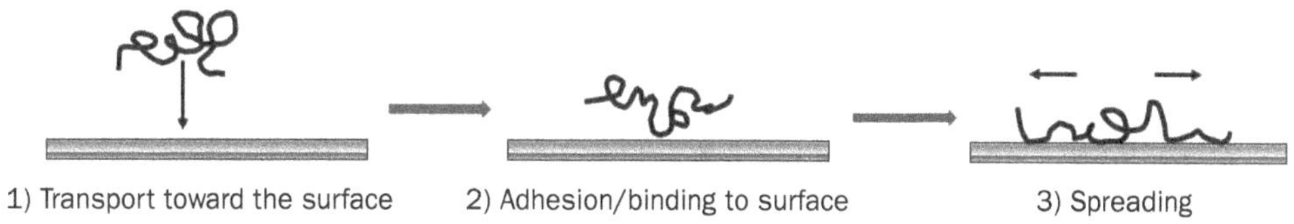

Figure 5.5 The three steps involved in forming an adsorbed layer of macromolecules.

Figure 5.6 Adsorbed layers obtained at fast spreading (left) and slow spreading (right).

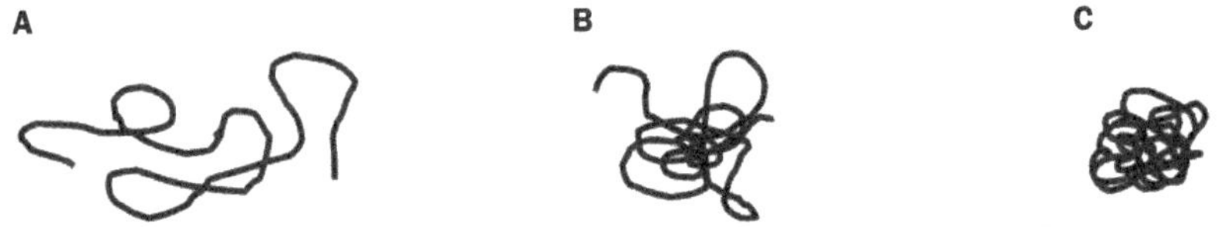

FIGURE 5.7 Polymer molecules in solution under (A) good, (B) borderline, and (C) poor solvent conditions.

where a_0 is the surface area occupied by the polymer upon first attachment, and α denotes the extra surface occupied at full spreading.

While a derivation of differential equations for adsorbed mass versus time is beyond the scope of this chapter, there is a silver lining: The information above can actually be used to control the structure of the adsorbed layer. For a given system consisting of a polymer (adsorbate) and a surface (planar or particle), we can reasonably assume that the rate of spreading will be more or less constant. The interaction between adsorbate and surface is dictated by the solubility and composition of the polymer as well as the roughness and functional groups present on the surface. In other words, the options for manipulating τ_s are limited once adsorbate and surface have been defined. This leaves the rate of deposition τ_d, which is directly proportional to the bulk concentration of polymer. As the forward rate of the adsorption equilibrium—i.e., attachment—only depends on the bulk concentration and the diffusion coefficient/hydrodynamic conditions, the structure of the adsorbed layer can be tuned simply by varying the polymer concentration. At low concentrations, the supply rate will be low, leading to a thinner layer and lower adsorbed amount, whereas increasing the bulk concentration of adsorbate will lead to a thicker layer.

Moreover, as polymers generally are large molecules, they can assume a large number of shapes or conformations. In the absence of externally applied forces, the polymer chain will fluctuate between a number of conformations, assuming an average, approximately spherical coil shape of a characteristic size R. The size of R depends on the number of monomers N per chain—i.e., the molecular weight of the polymer—as well as the polymer structure, and can generally be expressed as a power law where $R \sim N^{\nu}$. The characteristic size of R also depends strongly on the solvent. Specifically, in a good solvent, the monomers in the chain are kept more or less separate, leading to an extended conformation of the polymer chain. In less good solvents, the monomers prefer to interact with each other rather than with the solvent, leading to more compact structures, as illustrated in Fig. 5.7.

The preferred polymer conformation as a function of solvent conditions is also largely retained upon adsorption to a surface; under good solvent conditions, the polymer adopts a spread-out surface structure, maximizing solvent contact. Conversely, under poor solvent conditions, the polymer forms a collapsed, dense surface structure which minimizes contact with the solvent, as illustrated in Fig. 5.8. The configuration of a polymer as a function of solvent quality is the subject of Flory–Huggins theory, which is a topic in polymer chemistry and is beyond the scope of this chapter.

The summary of some other critical parameters for polymer adsorption is as follows:

Polymer structure: The polymer structure determines the conformational flexibility, solubility in the solvent, and surface affinity. The more rigid the polymer structure—either due to branching or bulky side groups on the monomers—the fewer conformations are available both in solution and on the surface. A more rigid

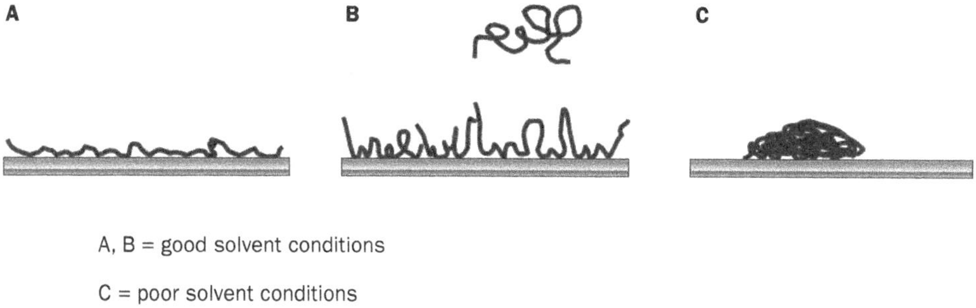

FIGURE 5.8 Structures of adsorbed polymers under good solvent conditions (A and B) and poor solvent conditions (C).

polymer structure also limits the degree of surface spreading following deposition. If the solvent conditions for the polymer are poor, unfavorable contact with the bulk phase can be limited and the free energy can be lowered by adsorption. Polymer structure can also dictate surface affinity, for example, via shielding of hydrophobic domains by way of adsorption of block copolymers.

Molecular weight: Generally, solubility decreases with increasing molecular weight, which again favors adsorption. Thus, for a given polymer structure, increasing the molecular weight would likely promote adsorption. However, as (i) a higher molecular weight indicates longer chains which occupy a larger surface area/more adsorption sites than smaller adsorbates and (ii) longer chains are more likely to sterically hinder incoming polymer molecules from attaching to the surface, a significant increase in molecular weight could result in lower adsorbed amounts.

Charge: For charged polymers (i.e., polyelectrolytes), adsorption depends strongly on the nature of the adsorbent and solvent. In aqueous solution, the charged groups confer significant solubility to the polymer. Consequently, adsorption is preferred for oppositely charged adsorbate–adsorbent pairs, whereas adsorption is electrostatically inhibited if the polymer and surface have the same charge sign—provided the charges are isotropically distributed along the polymer chain.

5.3.4 Protein Adsorption

Proteins—biological macromolecules—comprise an interesting and versatile subclass of polymers. They are copolymers of 22 different amino acids (i.e., monomers) of varying hydrophobicity.[6] To further add to the complexity, a number of the amino acid residues are amphiprotic—they contain pH-dependent positive or negative charges. The amino acids are covalently bound to each other via peptide linkages, which are condensation reactions between the non-side chain carboxylic acid moiety on one amino acid and the non-side chain amine moiety on another. Proteins are a combination of hydrophilic, hydrophobic, anionic, cationic, and neutral amino acids, which means they typically are highly surface active polyelectrolytes. Depending on the spatial organization of the amino acids along the polypeptide chain, proteins can be broadly divided into three categories:

1. Highly solvated and flexible proteins which assume disordered (random) coiled structures in an aqueous phase. The natural function of these proteins is typically nutritional, such as casein in milk or glutelins in wheat.

2. Proteins which have adopted highly regular structures such as helices and pleated sheets—often referred to as *fibrillar proteins*. These are usually found in connective tissue including tendons and ligaments, and are insoluble (or at least sparingly soluble) in water.
3. Proteins which contain a range of structural elements—random coils, helices, and pleated sheets—and are folded up into spherical, dense structures; the so-called *globular proteins*. Globular proteins have evolved to carry out specific biological functions, with examples such as enzymes, transport proteins, and antibodies. A consequence of these highly evolved structures is that their biological function is directly related to their native three-dimensional (tertiary) structure; significant deviation from their native structure results in denaturation and (irreversible) loss of functionality. A tetrameric protein hemoglobin structure is used as an example to illustrate the four levels of organization of a protein structure, as shown in Fig. 5.9.

Given that globular proteins by far comprise the most complex and interesting class of proteins, it should come as no surprise that globular proteins are the most relevant proteins for surface functionalization. Examples of applications where the adsorption of globular proteins are crucial include drug delivery vectors, biomedical and tissue engineering, biosensors, immunological test systems, immobilized enzyme bioreactors, biofouling, and chromatography.

While proteins in many respects behave similarly to copolymers, there are some trends in the adsorption of globular proteins which should be addressed:

1. *Proteins yield high-affinity adsorption isotherms:* Because proteins are surface active, they typically adsorb fast even at low bulk concentrations, resulting in a steep slope of the adsorption isotherm.

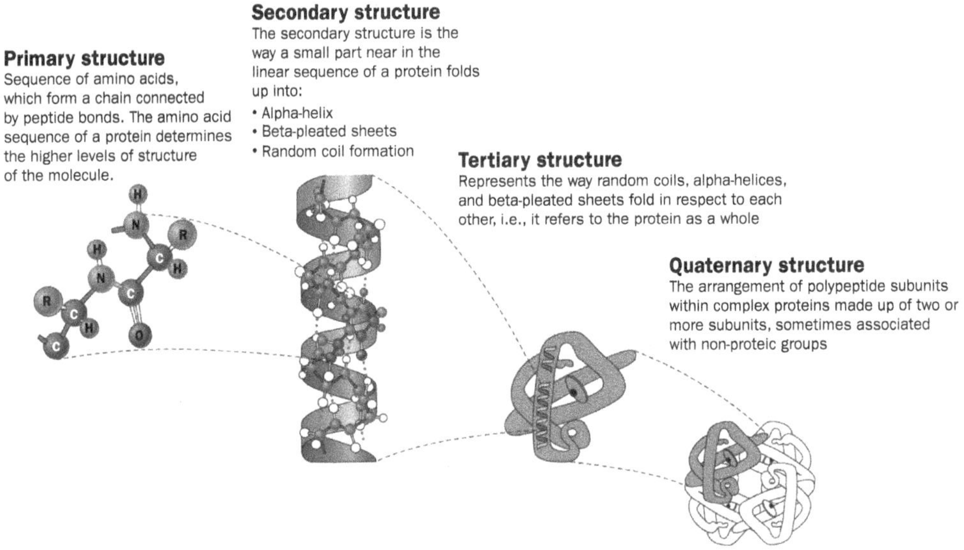

FIGURE 5.9 A tetrameric protein hemoglobin structure used as an example to illustrate the four levels of the organization of a protein structure: primary, secondary, tertiary, and quaternary.[7]

2. *Charge density matching is more important than net charge:* As a protein is comprised of both acidic (Asp and Glu) and basic (Lys and Arg) amino acid residues, the net charge is determined by the isoelectric point (pI) for a given pH. The isoelectric point represents the pH where the protein is electrically neutral, as determined by the statistical mean taken over the pK_a values for the charged amino acids and their relative population along the polypeptide chain. Below the pI, the protein molecule thus carries a net positive charge, and above the pI, it carries a net negative charge. This has a very interesting implication for protein adsorption to charged surfaces. While conventional wisdom would dictate that a charged protein molecule would be repelled by a surface with the same charge sign—for example, a protein interacting with a negatively charged surface at a pH above the pI—this ignores that the negative charges are not necessarily homogeneously distributed along the polypeptide chain. Rather, a globular protein will have cationic and anionic domains scattered along its external surface, with the overall charge being negative. Consequently, the protein can reorient itself to maximize favorable interactions between oppositely charged domains—the so-called *charge density matching*, as illustrated in Fig. 5.10.[8] This means that proteins can—and in many cases do—bind to charged surfaces irrespective of the protein pI.
3. *Surface packing of the protein depends on pH:* When proteins are (more or less) randomly packed on a surface, they generally repel each other for pH values significantly above or below the pI. In other words, highly charged proteins will not pack very tightly on a surface due to electrostatic repulsions, resulting in low adsorbed amounts. However, at or close to the pI, the proteins are electrically neutral, which means that they can pack tighter, yielding higher adsorbed amounts.
4. *Structural rigidity:* As discussed above, proteins are comprised of different structural motifs, ranging from highly flexible (random coils) to highly ordered (pleated sheets). This affects both their solubility (thus their propensity for adsorption) and how they behave once they are adsorbed on the surface. Specifically, spreading is limited by the protein structure as well as bulk concentration and the physicochemical properties of the surface. The higher the content of structured motifs (helices and sheets), the more likely the protein is to retain its native structure on the surface.

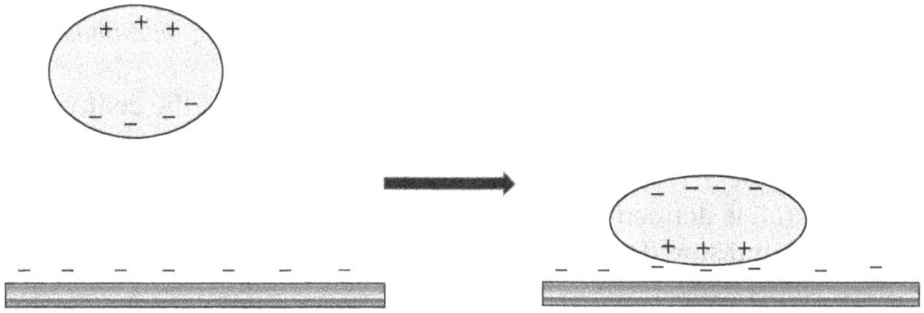

FIGURE 5.10 Illustration of a protein with a net negative charge binding to and spreading on a negatively charged surface by charge density matching between oppositely charged domains on protein and surface.

5. *Curvature:* For adsorption of proteins onto NPs or even rough surfaces, the protein will have to deform or spread more in order to cover the surface area compared to a flat surface (think the arc length compared to the diameter for a semicircle). Thus, spreading out on a curved surface comes at a significant cost in terms of conformational entropy for the protein, which in turn limits the deformation upon binding. In other words, the higher the curvature, the more likely the protein is to retain its native structure (i.e., not spread or deform). For planar surfaces, meaningful measurement of curvature can be difficult to obtain with respect to adsorption, as the degree of roughness is likely to vary significantly. However, for microparticles and NPs, the curvature is directly correlated to the particle size; the smaller the particle, the higher the curvature. Therefore, for a given protein and a given surface, the tendency toward surface-induced conformational changes (i.e., spreading) can be tuned simply by varying the particle size.[9,10]

As a first step to understand more about the interactions between adsorbate and adsorbent, it is useful to look at the interactions with a planar surface.

5.3.5 Quartz Crystal Microbalance for Studying Adsorption

The quartz crystal microbalance (QCM) technique provides an *in situ* measurement of the adsorption of molecules onto solid surfaces. It is based on the piezoelectric effect and is capable of detecting changes down to 0.1 ng by registering the changes in inertia of a vibrating quartz crystal (which might have an additional coating with, for example, gold).[11] By applying an alternating electrical field across the quartz crystal through electrodes on each side of the quartz disk, a mechanical oscillation of the characteristic frequency (f) is generated in the crystal. This oscillation frequency is related to the mass of the oscillating body, and thus a change in f can be directly correlated to a change in bound mass on the crystal via the Sauerbrey equation:[12]

$$\Delta m = -C\frac{\Delta f}{n} \tag{5.12}$$

Here, Δm is the change in mass, C is the sensitivity constant (17.8 ng cm^{-2} Hz^{-1}), Δf is the frequency shift, and n is the number of harmonic or overtone. The Sauerbrey equation is valid for an adsorbed mass smaller than the mass of the quartz crystal, preferentially rigid layers which are evenly distributed along the surface.

Hence, the QCM apparatus largely consists of a crystal which oscillates at a given frequency, and when molecules adsorb and the crystal gets heavier, vibrations are slowed down in proportion to the increased mass.

It is also possible to get additional information about the conformation and rigidity of the adsorbed layers by monitoring the dissipation of the crystal by periodically switching off the driving power and recording the decay of the dampened signal. The dissipation (D) is defined according to the equation below, where E_{diss} and E_{stor} are the energies lost and stored, respectively, during one oscillation cycle:

$$D = \frac{E_{diss}}{2\pi E_{stor}} \tag{5.13}$$

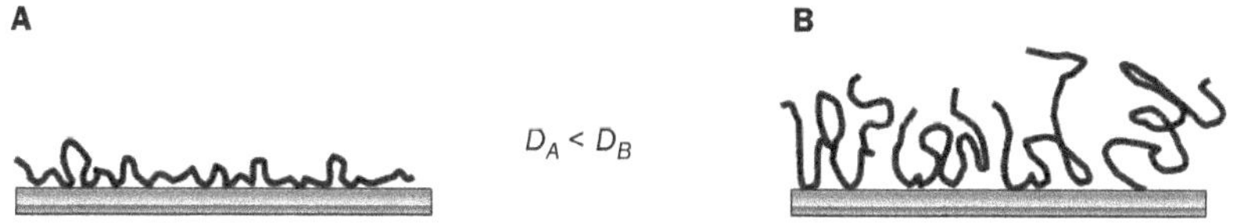

FIGURE 5.11 Illustration of dissipation values for a rigid (A) and a viscoelastic/"floppy" (B) adsorbed layer.

When more energy is lost from the oscillation, the value of dissipation will be higher, indicating a more viscoelastic (as opposed to rigid) layer.

In other words, dissipation is measured by turning the power off and measuring how long it takes for the oscillation to stop. If the layer is "floppy," oscillations will diminish faster than if the layer is rigid; the higher the value of D, the more energy is dissipated, and the faster oscillations are dampened (Fig. 5.11).

Moreover, if we combine the information from the frequency shifts and the changes in dissipation values, we can gather a lot of information about the adsorption process. By plotting ΔD vs Δf (D-f plots) as pioneered by Fredrik Höök and coworkers,[13] concentration (if the adsorbate is added sequentially in increasing concentrations) or time (if a single adsorbate concentration is used) is eliminated as explicit parameter. The value $\left|\frac{\Delta D}{\Delta f}\right|$ equals the induced energy loss per coupled unit mass, which provides a relative measure of the rigidity of the adsorbed layer. Specifically, the smaller the value, the more rigid the layer.

D-f plots enable us to extract a lot of information, both via the absolute values and from the slope, about the following:

- Different kinetic processes during adsorption
- Conformational changes during adsorption
- Formation of multilayers during adsorption
- Strength of interaction between the adsorbate and the surface
- The rigidity of the adsorbed layer

Figure 5.12 shows a possible scenario for a D-f plot where there is an abrupt change in the dissipated value per adsorbed unit mass, as shown by the dotted line for two structural or conformational regimes 1 and 2. If there is no change in the structure of the layer during adsorption, one would expect a first-order linear relationship akin to what is observed in Regime 1. For the D-f plot in Fig. 5.12, we can envision two possible physical interpretations as depicted in Fig. 5.13.

A sudden increase in the flexibility of the adsorbed layer can either be ascribed to a change in conformation or orientation as shown in scenario 1, or by the formation of a multilayer as shown in scenario 2. Determining which of these two scenarios best fits the adsorption process for a given system is not trivial, and requires further knowledge of the surface, the protein (or polymer), and maybe additional experiments—preferably using a complementary technique. For the two cases described above, simple dilution

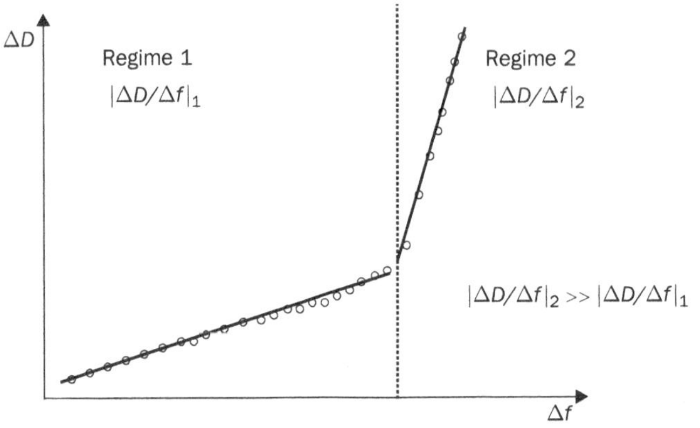

FIGURE 5.12 D-f plot showing an adsorption process with an abrupt change in the rigidity of the adsorbed layer.

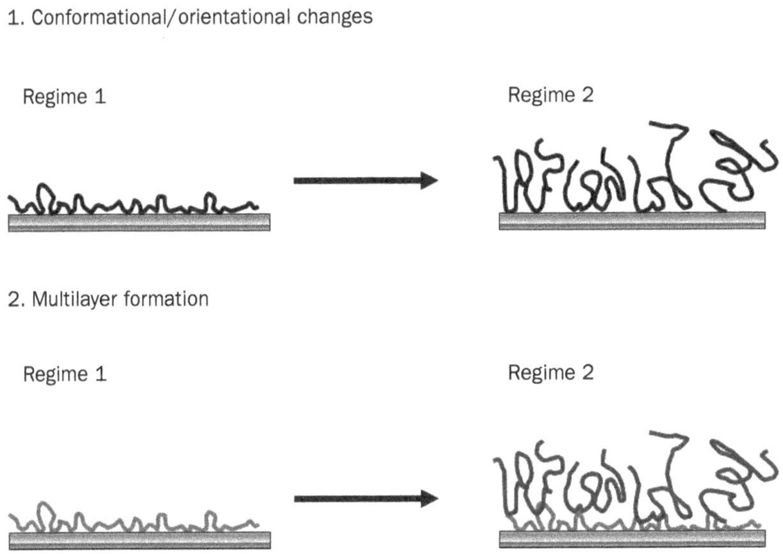

FIGURE 5.13 Possible physical interpretations for the D-f plot shown in Fig. 5.12.

might be sufficient to elucidate the structure of the adsorbed layer, as a loosely bound multilayer is much more likely to be washed off than a reoriented monolayer. Thus, if flushing the samples with water does not result in any changes in either frequency or dissipation, scenario 1 is likely correct.

Utilizing both frequency shifts and dissipation changes as well as D-f plots, several authors have been able to explain adsorption processes for polymers and proteins. For example, one study used QCM to distinguish between five different adsorption behaviors for a small library of proteins: random sequential adsorption, adsorption followed by reorientation (such as scenario 1 in Fig. 5.13), adsorption followed by spreading, multilayer formation (such as scenario 2 in Fig. 5.13), and surface-induced micellization.[8]

Now that you have gone through in detail how metal NPs may be functionalized, you must have noted several caveats in the process. These issues and challenges make research in this field an intriguing adventure. In the next two chapters, you will take a systematic look into polymer-based NPs. Some of the strategies learned in this chapter will come in handy.

5.4 Study Questions

5.4.1 Concept-Based Questions

5.1 Give an example each of *in situ* and *post situ* functionalization of metal NPs.

5.2 Derive and linearize the Langmuir equation for adsorption. State the assumptions under which this equation is valid.

5.3 Calculate N_{mono} and the affinity (K) for the proteins BSA, Cas, and Hb based on the data presented in the table below. Discuss if the Langmuir isotherm provides a good fit and provide possible explanations if this is not the case.

Bovine Serum Albumin (BSA)		α-Casein (Cas)		Hemoglobin (Hb)	
Conc. (M)	Adsorbed Molecules per cm²	Conc. (M)	Adsorbed Molecules per cm²	Conc. (M)	Adsorbed Molecules per cm²
5.15×10^{-7}	9.91×10^{11}	5.09×10^{-7}	1.72×10^{13}	5.03×10^{-7}	4.24×10^{12}
1.03×10^{-6}	1.25×10^{12}	1.02×10^{-6}	2.26×10^{13}	2.52×10^{-6}	5.34×10^{12}
5.15×10^{-6}	2.00×10^{12}	5.09×10^{-6}	2.83×10^{13}	5.03×10^{-6}	6.70×10^{12}
1.03×10^{-5}	2.47×10^{12}	1.02×10^{-5}	2.97×10^{13}	1.01×10^{-5}	1.02×10^{13}
1.54×10^{-5}	2.90×10^{12}	1.53×10^{-5}	3.00×10^{13}	1.51×10^{-5}	1.54×10^{13}
2.06×10^{-5}	3.00×10^{12}	2.04×10^{-5}	3.00×10^{13}	2.01×10^{-5}	3.23×10^{13}
2.57×10^{-5}	3.24×10^{12}	2.54×10^{-5}	3.01×10^{13}	2.52×10^{-5}	3.78×10^{13}
3.09×10^{-5}	3.34×10^{12}	3.05×10^{-5}	3.00×10^{13}	3.02×10^{-5}	4.22×10^{13}
4.12×10^{-5}	3.48×10^{12}	4.07×10^{-5}	3.04×10^{13}	4.03×10^{-5}	4.64×10^{13}
5.15×10^{-5}	3.59×10^{12}			5.03×10^{-5}	5.03×10^{13}

5.4 What is the fundamental working principle for QCM?

5.5 Discuss the caveats of using Langmuir adsorption for studying adsorption of proteins and polymers.

5.6 From a QCM study with casein, you have obtained a D-f plot as shown in the figure below. What does this tell you about the adsorbed layer?

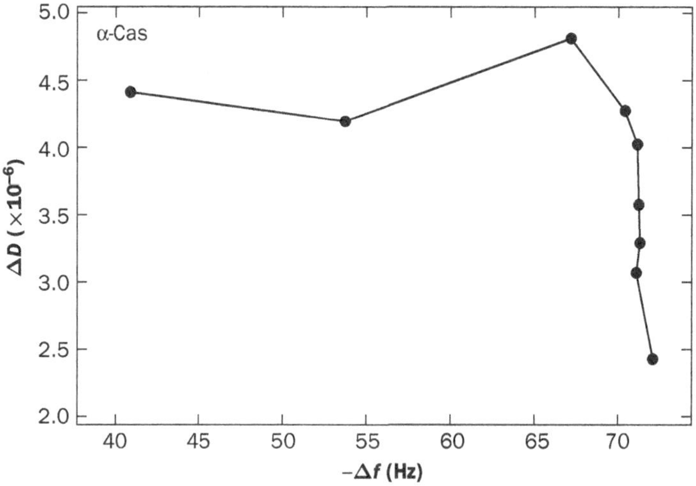

5.7 After synthesizing core-shell NPs, your supervisor wants you to study the adsorption of a biologically relevant drug carrier, an albumin protein, BSA, to which you will finally bind the drug.

a. The technique that you plan to use is the QCM. Through a sequential deposition of BSA on a citrate-covered gold-coated quartz crystal, a plot was made between (C/N) vs C, where C is the concentration in molarity and N is the number of molecules of BSA adsorbed per square centimeter of the crystal surface. The slope of this line was obtained as 2.65×10^{-13} and the intercept as 9.93×10^{-19}. Calculate the maximum number of molecules adsorbed per square centimeter and the equilibrium constant.

b. Constant for the quartz crystal can be assumed to be 17.8 ng/Hz-cm². At 30-μM BSA concentration, the difference between the frequency of oscillation of the crystal when adsorbed with BSA and the baseline is measured as −20.7 Hz. Calculate the mass of BSA adsorbed per unit area of the crystal.

c. Calculate the surface fractional coverage when the concentration of BSA is the same as in (b). Molecular weight of BSA is 67,000 g/mol. The active area of the crystal is estimated to be 3.14 cm².

Your curious supervisor also wants you to be critical on what assumptions of the model you have used that may not be valid for adsorption of BSA.

You have almost done the required characterization of the BSA layer, except that you have not measured the thickness. A simple Google search gives you the specific volume of BSA as 0.77 cm³/g.

d. Calculate the thickness of the adsorbed BSA layer for the conditions mentioned in (b).

5.4.2 Research-Based Questions

5.8 A sequential QCM study was carried out on a gold-coated quartz crystal to understand binding of P-lysine and siRNA that may be sandwiched in a layer-by-layer approach on Au NPs. The QCM plot is illustrated in the following figure. Based on this figure, answer the following questions.

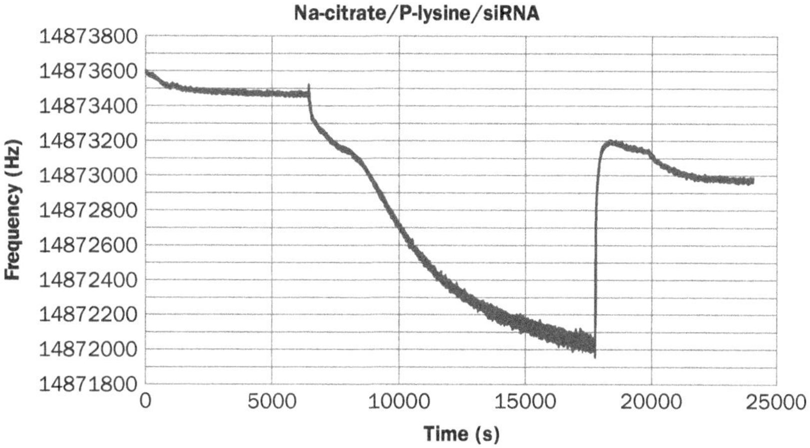

a. For which of the molecules (P-lysine or siRNA) is equilibrium saturation reached faster? Explain.

b. Calculate the amount of P-lysine that has bound per unit area of the gold surface.

c. After addition of the si-RNA, an increase in the frequency is observed. What could be the possible reason for this?

d. Is it possible to calculate the actual amount of bound si-RNA? If not, what changes in the experimental setup may be carried out to obtain this?

5.9 As a potential application in drug delivery, you decide to study the interaction of Cyt C with Au NPs. But, as a startup, you perform a sequential deposition study of Cyt C on a citrate-covered gold-coated quartz crystal using the QCM. The table below shows *N* vs *C*, where *C* is the concentration in M and *N* is the number of molecules of Cyt C adsorbed per square centimeter of the crystal surface.

C (M)	*N* (molecules/cm^2)
5.57×10^{-7}	2.95×10^{12}
1.11×10^{-6}	4.06×10^{12}
5.57×10^{-6}	7.03×10^{12}
1.11×10^{-5}	9.07×10^{12}
1.67×10^{-5}	1.12×10^{13}
2.23×10^{-5}	1.24×10^{13}

2.78×10^{-5}	1.34×10^{13}
3.34×10^{-5}	1.40×10^{13}
4.46×10^{-5}	1.56×10^{13}
5.57×10^{-5}	1.64×10^{13}

a. Starting from equilibrium consideration, derive an expression for surface fractional coverage.

b. Using the surface fractional coverage derived above, comment on the adsorption affinity of Cyt C with gold at 35 μM concentration.

c. How valid are the assumptions underlying this approach?

5.5 References

1. Halskau, O.; Volden, S.; Calvo, A. C.; Martinez, A.; Glomm, W. R., Adsorption and bioactivity of tyrosine hydroxylase on gold surfaces and nanoparticles. *Protein & Peptide Letters.* **2010,** *17*(11), 1376–1382.
2. Singh, G.; Kumar, P. A.; Lundgren, C.; van Helvoort, A. T. J.; Mathieu, R.; Wahlstrom, E.; Glomm, W. R., Tunability in crystallinity and magnetic properties of core-shell Fe nanoparticles. *Particle and Particle Systems Characterization.* **2014,** *31*(10), 1054–1059.
3. von der Ehe, C.; Weber, C.; Gottschaldt, M.; Schubert, U.S., Immobilized glycopolymers: synthesis, methods and applications. *Progress in Polymer Science.* **2016,** *57*, 64–102.
4. Hiemenz, P. C.; Rajagopalan, R., *Principles of Colloid and Surface Chemistry,* revised and expanded. Boca Raton, FL: CRC Press, 1997.
5. Malmsten, M., *Biopolymers at Interfaces.* New York, NY: CRC Press, 1998.
6. Creighton, T. E., *Proteins: Structures and Molecular Properties.* 2nd ed. New York, NY: WH Freeman and Company, 1993.
7. Hardin, J.; Bertoni, G.; Kleinsmith, L. J., The macromolecules of the cell. In: *Becker's World of the Cell,* 8th ed. Boston, MA: Benjamin Cummings, 2012.
8. Glomm, W. R.; Halskau, O.; Hanneseth, A. M. D.; Volden, S., Adsorption behavior of acidic and basic proteins onto citrate-coated Au surfaces correlated to their native fold, stability, and pI. *Journal of Physical Chemistry B.* **2007,** *111*(51), 14329–14345.
9. Lystvet, S. M.; Volden, S.; Yasuda, M.; Halskau, O.; Glomm, W. R., Emergent membrane-affecting properties of BSA-gold nanoparticle constructs. *Nanoscale.* **2011,** *3*(4), 1788–1797.
10. Kane, R. S.; Stroock, A. D., Nanobiotechnology: protein–nanomaterial interactions. *Biotechnol Progress.* **2007,** *23*(2), 316–319.
11. O'Sullivan, C. K.; Guilbault, G. G., Commercial quartz crystal microbalances—theory and applications. *Biosensors and Bioelectronics.* **1999,** *14*(8–9), 663–670.
12. Sauerbrey, G., Verwendung von Schwingquarzen zur Wägung dünner Schichten und zur Mikrowägung. *Zeitschrift für Physik.* **1959,** *155*(2), 206–222.
13. Höök, F.; Rodahl, M.; Brzezinski, P.; Kasemo, B., Energy dissipation kinetics for protein and antibody-antigen adsorption under shear oscillation on a quartz crystal microbalance. *Langmuir.* **1998,** *14*(4), 729–734.

CHAPTER 6

Synthesis of Polymer-Based Nanoparticles

Macromolecules containing repeating units (monomers) constitute an important part of our daily lives—clothes made from synthetic fibers, PET bottles of soda, fiberglass, tennis balls, bandages, plastic bags, polymer-based paints, contact lenses, Teflon-coated cookware, and an endless number of other utilities. Although the definition of polymers as "large molecules" has been used since the 1830s, natural polymeric materials such as cellulose, proteins, and DNA have always existed.

Polymeric materials of small dimensions were already in use in several applications even before the advent of *nano* revolution. However, properties of these materials had never been studied from a nano perspective until the late 1990s when these materials saw an escalated demand in terms of research and applications. Thus, the fabrication and application of polymer-based nanoparticles (NPs) have regained renewed interest among the scientific community to study them as a subgroup of nanomaterials.

Imagine how the scenario of international trade would look like if large ships would not be protected using marine paints. Marine vessels account for almost 90 percent of the global trade and hence their prevention from corrosion as well as fouling is an important challenge for paint industries. The rate of fouling determines the cruising speed, the fuel consumption (and in effect greenhouse gas emissions), and the interval between dry dockings. In marine environments, fouling refers to the growth of slime, seaweeds, algae, mussels, barnacles, and tube worms among others over time when the vessels are immersed in water. One way to prevent the film formation on vessel surfaces is by releasing biocides from the antifouling paint film. One of the high-end technologies controlling the release of biocides employs acrylate-based polymeric nanocapsules, which work on the following principle. The nanocapsules comprise a permeable hydrophobic shell and a hydrophilic core. Upon contact, water permeates through the shell into the core causing it to hydrolyze and subsequently swell. The hydrophilic–hydrophobic balance of the polymeric nanocapsules determines the hydrolysis rate (polishing of the paint film), which in turn controls the release rate of biocides into seawater. This is an example where understanding the fundamentals of polymeric NPs not only protects property but also impacts the consequences of global trade on the environment. Some other practical applications exploiting the material design and characterization of polymeric NPs in catalysis, drug delivery and biomedical applications, water management, and energy harvesting will be explored in Chaps. 9 to 12.

As of now, you may focus on the classification, synthesis, and physicochemical properties of polymer-based NPs, which will help fabricate nanomaterials for tailored end uses.

6.1 Classification of Polymer-Based Nanoparticles

One broad way to classify polymer-based NPs depends on the origin of the polymers. These could be either naturally occurring polymers such as chitosan, alginates, DNA and other proteins, polysaccharides, polypeptides, lipids found in nature, or synthetic polymers. However, the classification can be done based on factors that influence the physicochemical properties of the polymeric NPs and their applications. Classifications based on the morphology of the NPs, surface charge, amphiphilicity, and preparation techniques are presented in this section, as outlined in Fig. 6.1. However, a detailed overview of different classifications is presented throughout this chapter.

6.1.1 Morphology

Two broad categories can be distinguished based on the morphology of the NPs. Figure 6.2 shows different classification of polymer-based NPs ranging from nanospheres to dendrimers. Arrangement of the polymeric chains in the form of a matrix leads to the formation of nanospheres or NPs. In such particles, the entire mass is solid. However, nanocapsules are vesicular systems characterized by a liquid core (either oil or water) surrounded by a polymeric shell. In the latter case, the entrapped substance is confined to a cavity.[1]

Polymeric micelles are another category of polymer-based NPs based on morphology. These consist of amphiphilic block copolymers that self-assemble into a micelle in the solution as their concentration increases above critical micelle concentration (CMC). Polymersomes are a class of artificial vesicles comprising a bilayer of amphiphilic block copolymers. Hydrophilic part of the polymer forms the inside and outside of the hydrated coronas with the hydrophobic part in the middle, separating the fluidic core from the outside medium as shown in Fig. 6.2. One classification of polymer-based NPs that has captured substantial attention recently is dendrimers, having a complex yet symmetric polymer network. Dendrimers possess unique characteristics that differentiate them from their polymeric counterparts; hence, the synthesis and properties of dendrimers are discussed in the section below.

Dendrimers

Dendrimers are unique three-dimensional, highly branched, nano-sized chemical polymers containing very high numbers of functional groups on their surfaces. These hyperbranched macromolecules can be easily functionalized whereby modifying their physicochemical properties. Despite their large molecular size, dendrimers are structurally well-defined with a low polydispersity when compared to traditional polymers. One of the important features of dendrimers is a high density of functionalities on the surface along with a small molecular volume. On a molecular level, the dendritic branching results in semi-globular to globular structures. A brief overview of the synthesis strategies employed is presented in this section, some of their interesting physicochemical properties pointing toward a myriad of applications.

As opposed to synthesis of traditional polymers, dendrimers are synthesized via a fully controlled step-by-step approach. This is one of the reasons why dendrimers have

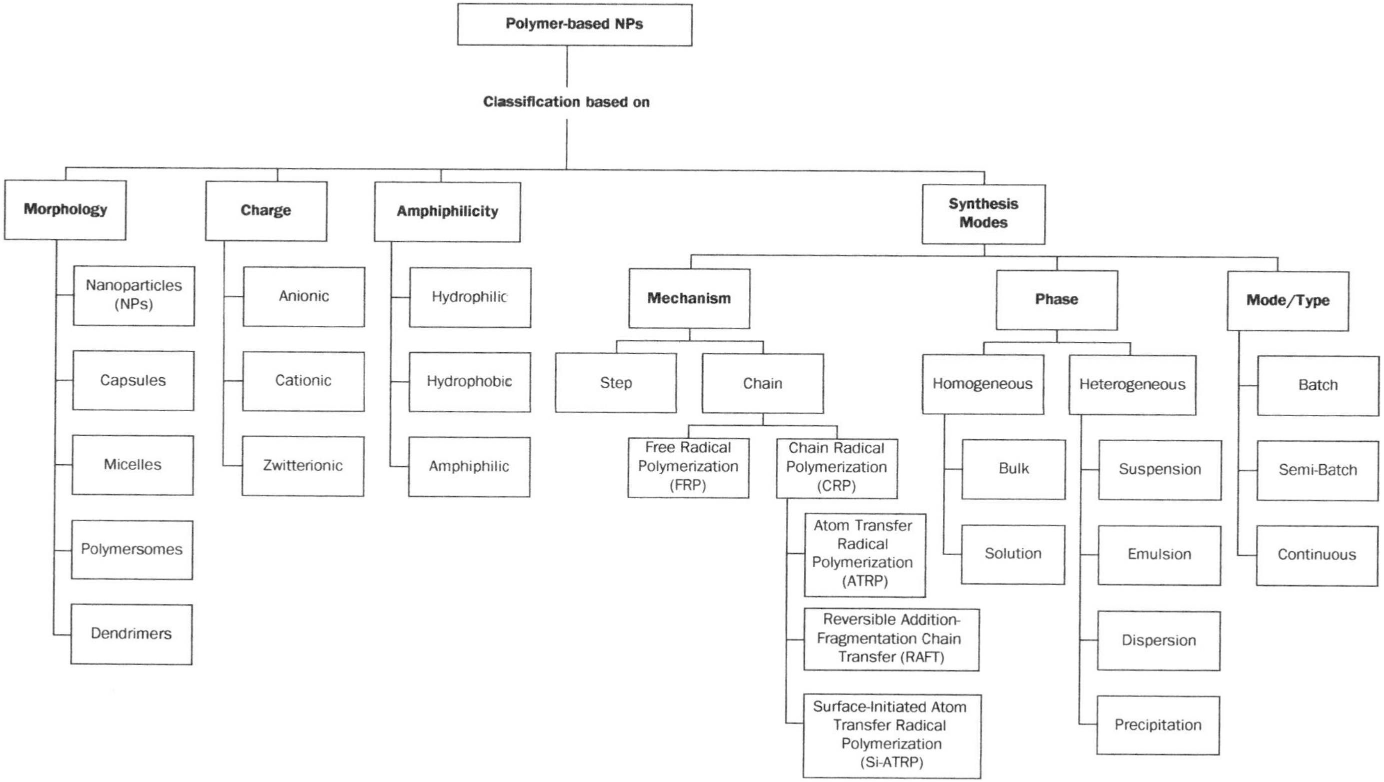

Figure 6.1 Flowchart showing different classifications of polymer-based NPs.

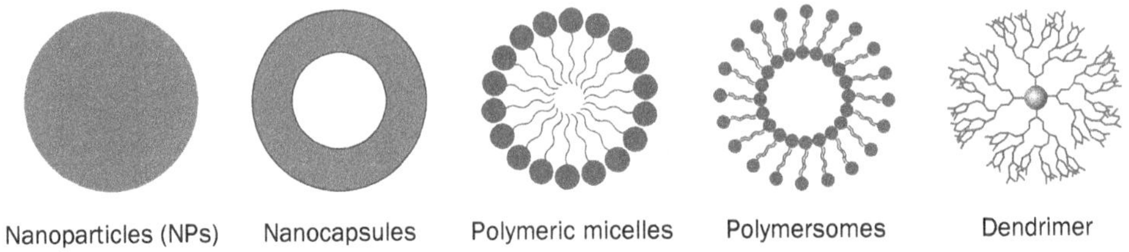

FIGURE 6.2 Different polymer-based NPs classified on the basis of morphology.

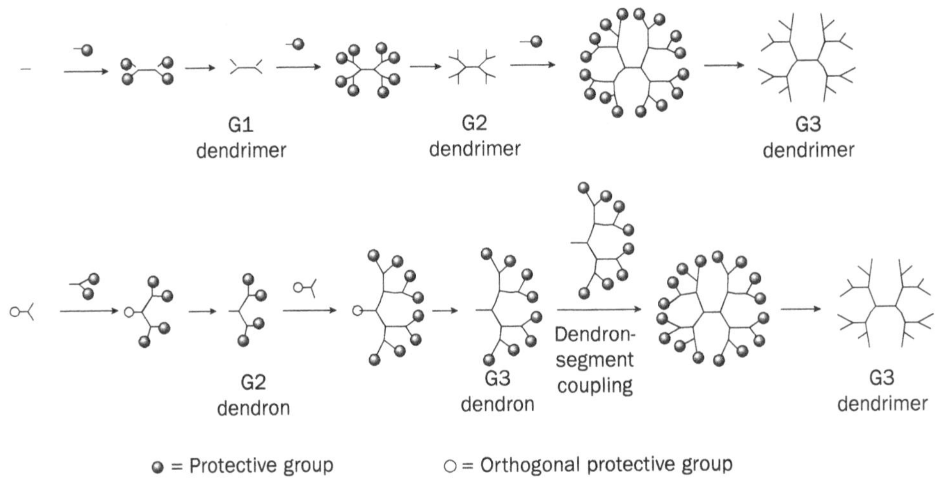

FIGURE 6.3 Divergent and convergent strategies of synthesis of dendrimers.[2]

well-defined molecular structures. The two commonly applied strategies include the divergent and convergent approaches.

In case of the divergent strategy, the dendrimer is built from the core outward, layer by layer. Each layer is often termed as a generation of growth that finally determines the generation of the dendrimer. Figure 6.3 shows the synthesis of a G3 dendrimer via divergent strategy. Although this method is conceptually straightforward, an exponentially increasing number of reactions need to be performed on a single molecule to achieve desired generation. This is because of the fact that as the dendrimer increases in size, the number of surface functionalities also increases. Even with very efficient reactions, the yield of perfect final product is low. Further, separation of the defect-free dendrimers from the other by-products is a difficult endeavor since their mass, size, and properties are very similar to the defect-free dendrimers.

However, in case of a convergent strategy, the growth starts from the surface and ends up at the core. This is schematically depicted in Fig. 6.3. At any given time, only two reactions are performed. In principle, two building blocks with their surface groups protected react with one unit whose focal group is protected. Upon reaction, the deprotection of the group at the focal point leads to the second-generation dendron. This second-generation dendron may get attached to a dendrimer core or be reacted to yield higher generation dendrons. Thus, in the convergent approach, only a small number of reactive sites are functionalized in each step, giving a small number of possible side-reactions and less flawed products. Each synthesized generation of dendrimer can therefore be purified, yielding dendrimers without defects. However, since the reactive group is buried at the focal point of the dendrons, steric hindrance can make the reactions slower and less efficient. This method also enables synthesis of asymmetric dendrimeric structures in a controlled fashion.

As pointed out in the discussion above, dendrimers are important polymeric NPs owing to their structure–properties relationship. The dendrimers are characterized by several compartments. The structure of dendrimers can be broadly classified into three parts:

1. The multivalent surface
2. The outer shell
3. The core

The main feature of the multivalent surface is the presence of a high number of potential reactive sites. This makes the surface to suit host–guest interactions that call for close proximity of a large number of species. The surface also determines the macroscopic properties of the dendrimer such as solubility, interaction, and so on. The outer shell, located just beneath the surface, comprises monomer layers. It possesses a well-defined microenvironment protected from the outside by the dendrimer surface. Encapsulation of molecular species is an important function of this part of the dendrimer. The core is buried within the dendritic branches and does not experience the surrounding bulk solvent. Instead, it is capable of sensing the unique encapsulating microenvironment created around it by the branching. Thus, the core provides a nanoenvironment for chemical species especially in higher generation dendrimers.

Dendrimers find wide applications ranging from dendritic sensors, drug carriers, and glycocarriers, to antiviral, antibacterial, or antitumor drugs, to protein denaturants, among others. These applications are beyond the scope of this chapter, but you may refer to some excellent reviews.[2-4]

However, before this discussion on dendrimers is closed, it is important to understand that most of the interesting applications of the dendrimers arise from one of the following advantages—a high number of surface functional groups, tailored sanctuary created within the branches, or special microenvironment at the dendrimer core. An example of manifestation of one of these features is the stimuli-dependent conformational changes of a class of widely studied dendrimers, polyamido amine (PAMAM) and polypropylene imine (PPI).

Amino-terminated PAMAM dendrimers display extended conformations upon lowering of pH. This happens because of electrostatic repulsion between the protonated tertiary amines in the interior as well as between the primary amines at the dendrimer surface leading to repulsion among the dendrimer branches. At pH > 9, back-folding occurs owing to hydrogen bonding between the interior protonated tertiary amines and the primary surface amines, resulting in a denser core. However, PPI dendrimers having surface carboxylic groups have the most extended conformations at pH 4 and 11. This is most likely due to electrostatic repulsion between the protonated cationic inner tertiary amines at low pH, and electrostatic repulsion between the negatively charged deprotonated carboxylates at the dendrimer surface at high pH. These conformational

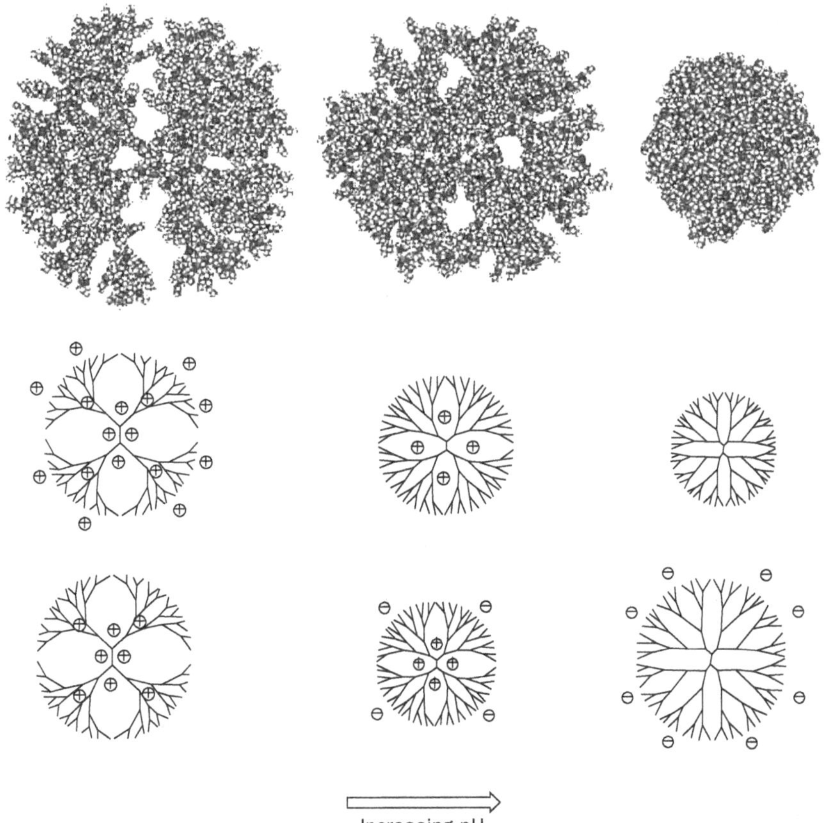

FIGURE 6.4 PPI dendrimers showing different conformational changes.[2]

changes are depicted schematically in Fig. 6.4. Thus, the pH-related conformational changes are dependent on the nature of the charged group on the dendrimer surface.

Similar to dendrimers, the structural conformation as well as surface properties of other polymer-based NPs also depend on the functional groups embedded in the polymer chains. Along with the functional groups, the solvent also plays a major role in influencing the properties of these NPs, which in turn depends on the hydrophilic and/or hydrophobic nature of the polymers. This forms another category of the classification of the polymer-based NPs based on their interaction with the solvent, as discussed in the section below.

6.1.2 Hydrophilicity

As polymer-based NPs are an integral part of this chapter, analyzing the interaction of polymers with the solvent molecules becomes important. Depending on the type of interaction, the polymer-based NPs can be categorized as hydrophilic (water loving) or hydrophobic (water repelling), the latter also referred to as *lyophilic* (oil loving). The interactive–repulsive nature of the polymers imparts different properties to the polymer-based NPs, which in turn dictates its end application, either in medicine, oil

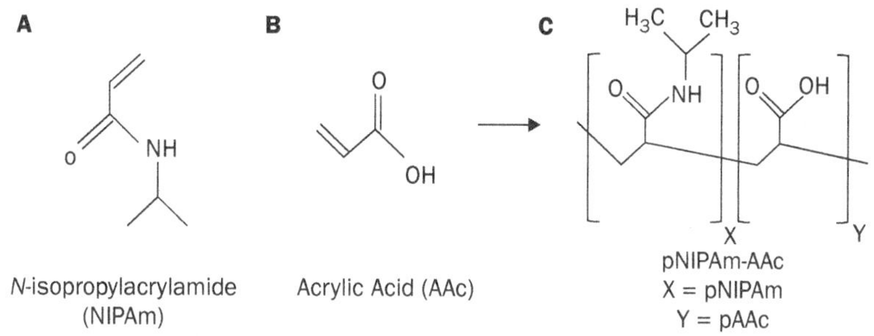

FIGURE 6.5 Chemical structures of (A) NIPAm monomer, (B) AAc monomer, and (C) pNIPAm-AAc comprising polymer-based NPs. X and Y refer to NIPAm and AAc contents in the final polymer-based NPs.

recovery, or water management, to name a few. Usually, surfactants are used during synthesis in order to stabilize the formed NPs by preventing agglomeration through steric or ionic repulsion. The hydrophilic-lipophilic balance (HLB) is a number ranging from 1 to 30, where lower and higher values refer to oil-soluble and water-soluble surfactants, respectively.[5] Therefore, by selecting the appropriate surfactant, the HLB of the NPs can be designed according to the desired application.

HLB is not only limited to just the surfactants but also can be achieved by altering the monomer units in the polymer chains that are forming the NPs. By increasing the amount of water-interacting groups in a polymer chain, one can achieve a higher hydrophilicity of the NP. For example, *N*-isopropylacrylamide (NIPAm), a widely used monomer for drug delivery applications due to its stimuli-response characteristic (discussed later),[6] comprises an acryl and an amide group, as shown in Fig. 6.5A, thereby making it quite hydrophilic at room temperature. Incorporation of acrylic acid (AAc) (Fig. 6.5B) as a co-monomer for synthesis of pNIPAm-AAc-based NPs (Fig. 6.5C) enhances the hydrophilicity of the system. Usually, hydrophobicity also increases with increasing molecular weight of the polymer chains, as the end group effects are diminished with growing chains.

6.1.3 Charge

As discussed above, HLB dictates the polymer-based NP–solvent interactions that further depend on the nature of surface charge carried by the NP. The NPs can be classified as non-ionic, anionic (negative), cationic (positive), or zwitterionic (negative/positive based on system pH), depending on the charge imparted by the surface functional groups. The surface charge on a particle is often measured in terms of zeta potential (ξ) and gives an account of the stability of the NP suspension in the solvent with values -30 mV $> \xi < +30$ mV being considered generally as a stable suspension. ξ can be measured using electrophoresis in which the charged particles are placed in an electric field and their velocity, being proportional to the degree of surface charge, is measured. Therefore, the higher the surface charge, the higher the electrophoretic mobility of the NPs.

The nature of the surface charge plays a vital role in understanding how the NP system will interact with the surrounding medium and whether it is suitable for that particular application. In the case of biomedical applications, the surface charge determines the efficacy of cellular interactions of the NPs as well as its circulation life (residence time) within the body (*in vivo* studies) before it is expelled out by the reticuloendothelial system (RES).[7] AAc has extensively been utilized in ocular (through eyes) drug delivery applications due to its very high adhesive bond strength in contact with tissues, enhancing the mucosal penetration of drugs. Since, AAc is a weak polyacid, its protonation highly depends on the system pH, which in turn would control the surface charge of the AAc-based NPs. Thus, AAc-based NPs behave in response to pH just like other stimuli-responsive systems. At higher pH, NPs show higher ξ due to higher degree of protonation of –COOH groups. This increases the size of the NPs as a result of the steric repulsion between the charged functional groups of the polymer chains. A reverse trend can be seen on decreasing the pH of the system. This pH-triggered stimuli-responsive behavior of AAc is used to achieve higher drug-release efficiency by lowering the pH of the medium.

Polymer-based NPs can be designed to have different morphologies, hydrophilicity, as well as surface charge according to the application, and this can be achieved by various synthesis techniques. These techniques also provide another way to classify the polymer NPs based on the mechanism of synthesis, the reaction phase during synthesis, or by the synthesis mode they utilize as already shown in Fig. 6.1. The preparation techniques employed in the synthesis of these NPs are discussed below.

6.1.4 Preparation Techniques

As discussed above, polymeric NPs can be synthesized using a wide array of techniques that allow for control of morphology, hydrophilicity, and charge, among other properties. For insights into classification of polymerization methods in traditional polymer chemistry, you can refer to standard text books in polymer science such as *Fundamentals of Polymer Science—An Introductory Text.* Although there are numerous ways to synthesize polymers, our discussion here is limited to only those methods that produce NPs. Therefore, even though living polymerization is an important category of polymer synthesis, it is beyond the scope of this chapter because of the aforementioned reason.

Mechanism Based

One of the preferred approaches of classifying polymer-based NPs is on the basis of polymerization mechanism, which can be step-growth or chain. The important feature of the step-growth mechanism involves systematic build-up of chains in a stepwise fashion. Here, high-molecular-weight polymers are produced at the end of polymerization, which takes place in a rather slow manner. Monomers combine with one another to give higher oligomers, as shown below.

$$\begin{aligned} M_1 + M_1 &\rightarrow M_2 \\ M_1 + M_2 &\rightarrow M_3 \\ M_2 + M_2 &\rightarrow M_4 \\ M_1 + M_3 &\rightarrow M_5 \\ M_3 + M_3 &\rightarrow M_4 \\ M_3 + &\ldots\ldots\ldots\ldots \end{aligned}$$

Here, M_1, M_2, and M_3 are monomer, dimer, and trimer, respectively.

Chain polymerization is characterized by the presence of an active site at the end of a growing chain, where monomers are added one after the other. During the polymerization process, only a few species are active that allow monomer addition to these sites and subsequent transfer of the active site to the chain end after monomer addition. In principle, chain polymerization involves the following steps—*initiation, propagation, chain transfer,* and *termination.*

Free radicals act as active species in free radical polymerization (FRP), and the chain growth propagates through growing macroradicals. For FRP, the rate of initiation is equal to the rate of termination at steady state, which are much (approximately 1000 times) slower than the rate of propagation. A radical initiator is quite often used to speed up the initiation rate; however, the unreacted initiator is left behind in the reaction mixture in most cases. The generation of free radicals results from the disproportionation of initiator radicals, as shown in Eq. (6.1):

$$I \Rightarrow 2R^* \tag{6.1}$$

These radicals go on to attach to the monomer molecule [M], thus initiating chain growth with additional monomers attaching to the propagating polymer chain and the free radical constantly moving to the chain end as depicted in Eqs. (6.2) and (6.3):

$$R^* + M \Rightarrow RM^* \tag{6.2}$$

$$RM^* + M \Rightarrow RM_2^* + M \Rightarrow RM_3^* \ldots\ldots \Rightarrow P_i^* \tag{6.3}$$

The termination of the chain can be achieved in the following ways:

$$P_i^* + P_i^* \Rightarrow P_i + I \qquad \text{Recombination} \tag{6.4}$$

$$P_i^* + P_i^* \Rightarrow P_i + P_i + 2R^* \qquad \text{Disproportionation} \tag{6.5}$$

$$P_i^* + R^* \Rightarrow P_iR \qquad \text{Termination by initiator radical} \tag{6.6}$$

Even though a majority of the commercial polymers are synthesized by FRP, obtaining polymers with desired architecture is not possible with this technique. Controlled radical polymerization (CRP) overcomes this limitation and enables you to synthesize block copolymers with precise control over polymer block lengths, molecular weights, polydispersity, and so on. Contrary to FRP, CRP is characterized by instantaneous initiation and absence of termination step. The primary principle underlying all CRP reactions is the establishment of a dynamic equilibrium between the propagating radicals and the dormant species, and can be categorized into Type I and Type II mechanisms, as depicted below.[8]

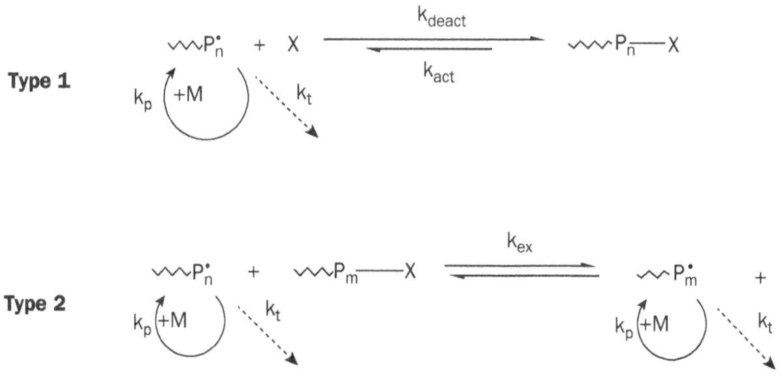

Figure 6.6 Schematic showing the two types of mechanism operational in case of chain radical polymerization (CRP).[8]

Atom Transfer Radical Polymerization (ATRP)

Discovered independently by Mitsuo Sawamoto and by Jin-Shan Wang and Krzysztof Matyjaszewski in 1995, this method involves the use of a transition metal complex (Mt^n/L) as a catalyst and an alkyl halide $(P_n - X)$ as an initiator, as shown in Fig. 6.7. Various metals, such as Cu, Fe, Ru, Ni, Os, etc., which are capable of expanding their coordination sphere, can be used as catalyst in ATRP. Free radical $(P_n^{\bullet})$ is generated upon homolytic cleavage of the alkyl halogen bond caused by the catalyst, wherein the halogen (X) attaches to the transition metal complex, thereby increasing its oxidation state in the formed halogen–metal complex $(X - Mt^n/L)$ (Fig. 6.7).[9] During the deactivation cycle, X breaks away from the halogen–metal complex and bonds again with the growing polymer (P_{n+1}), forming $(P_{n+1} - X)$. The polymer grows as the activation and deactivation cycles continue consuming the monomers simultaneously.

The commercial availability of various ATRP reagents (alkyl halides, ligands, and transition metals) makes this technique easy to use for the synthesis of polymer-based NPs. Additionally, the equilibrium between the dormant species and the propagating radicals can easily be tuned by selecting the appropriate catalyst.

ATRP proves to be highly successful in synthesizing polymers with defined architecture as well as highly targeted molecular weight ranges. However, a major drawback of ATRP is that it requires very high concentrations of the transition metal catalyst that remain in the product after completion of the reaction. In order to use the polymer commercially, you have to remove the catalyst, which proves to be a tedious and costly affair. To overcome this limitation, reversible addition-fragmentation chain transfer (RAFT) polymerization has been developed that uses a chain transfer agent in place of the catalyst.

Reversible Addition-Fragmentation Chain Transfer (RAFT)

First reported in 1998 by Chiefari et al.[10] this polymerization technique is one of the most widely used methods due to its applicability to an extensive array of monomers and solvents, including aqueous solutions. As mentioned previously, RAFT uses a chain transfer agent (CTA) as opposed to a catalyst in ATRP that stays a part of the synthesized polymer after the completion of the reaction. Typically, CTA is a thiocarbonylthio group (S=C–S) with *R* and *Z* substituents in either side (Fig. 6.8) that are responsible for determining the rate kinetics along with the degree of structural control. The *Z* group controls the reactivity of the C=S bond, where the other thiol group in the weekly bonded –S–C is attached to the hydrocarbon group (*R*).

RAFT typically follows a polymerization mechanism of Type II, where P_m becomes the free radical $P_m^{\bullet}$ upon cleavage with the $-S_1-C$ bond, as shown in Fig. 6.9A, forming $S_1=C$. Simultaneously, the other thiol group S_2 generates a free radical for the attachment of another monomer or a growing polymer unit $(P_m^{\bullet})$ to form $P_n-S_2-C=S_1$ transition complex (Fig. 6.9C). Thus, the free radicals ($P_m^{\bullet}$ and $P_n^{\bullet}$) generated in the forward and backward reactions continue to grow at a slow pace, giving a high control over the molecular weight and size distribution of the formed polymeric NPs, as shown in Fig. 6.9.

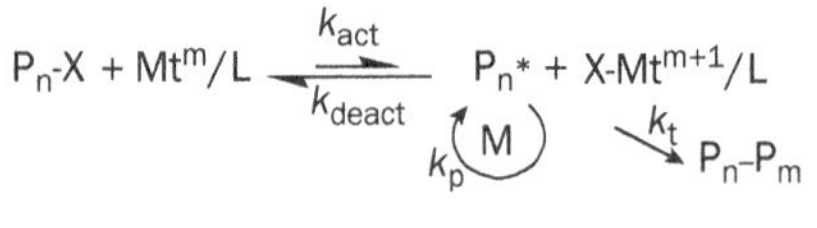

FIGURE 6.7 Schematic showing the mechanism of atom transfer radical polymerization (ATRP) process.[9]

So far, you have learned about the fundamentals of polymer formation and growth on the molecular level; however, no comments have been made about the effect of the solvent and the phase of the system. These are important aspects when it comes to understanding the mechanism on a macromolecular level and how they eventually affect the properties, namely molecular weight, of the system. Hence, polymer-based NPs can be classified based on the phase of polymerization.

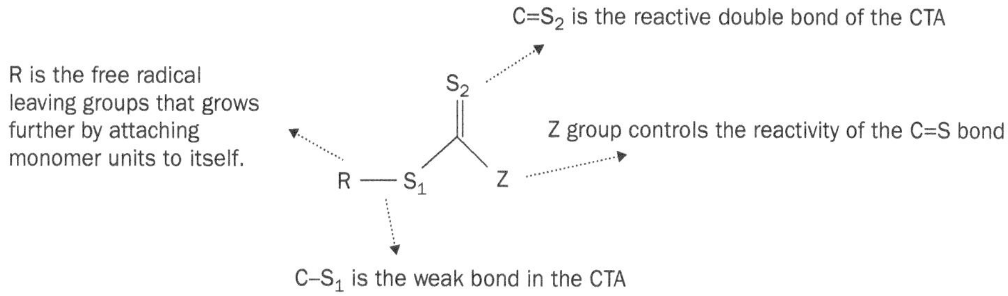

FIGURE 6.8 Schematic of a typical chain transfer agent (CTA).

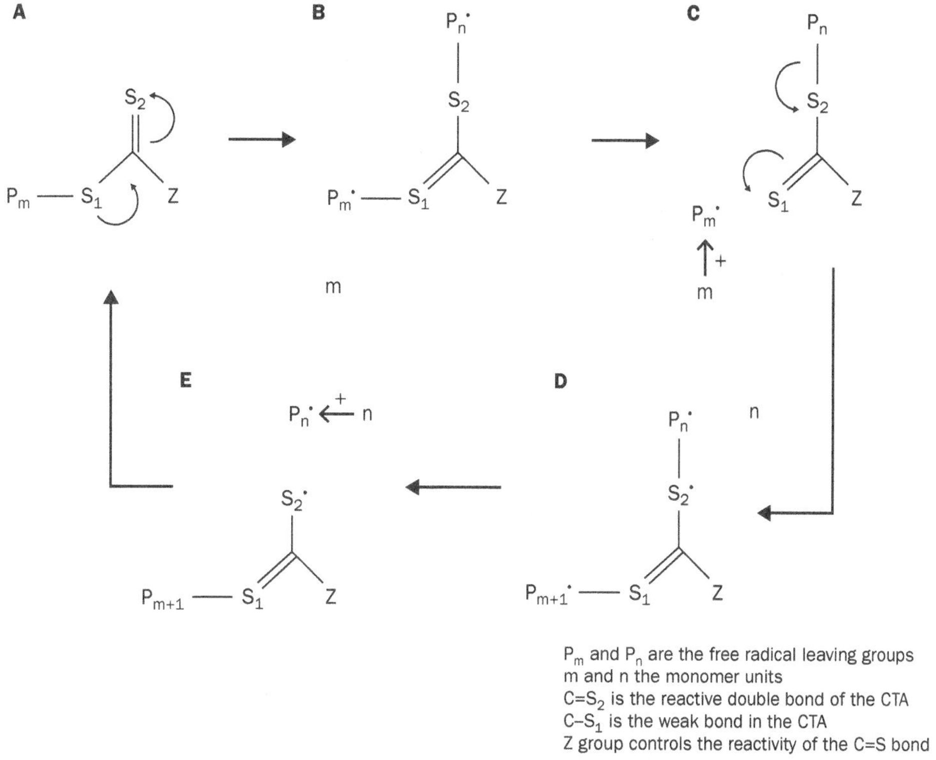

FIGURE 6.9 Schematic showing the mechanism of reversible addition-fragmentation chain transfer (RAFT) polymerization.

6.1.5 Phase-Based Classification

So far, you have seen how the reaction mechanism plays a dominant role in determining the end products after polymerization. Depending on the initial phase of the reaction medium in which the polymerization starts, polymerization can be classified as homogeneous or heterogeneous, commonly referred to as *single phase* and *multiphase* polymerizations, respectively. It might happen that the system starts as a homogeneous system with every component dissolved in the solvent. However as polymerization proceeds, the formed polymer might not be soluble in the solvent and phase separate out forming a heterogeneous mixture. Hence, the concept of homogeneity and heterogeneity refers only to the initial state of the system.

Bulk Polymerization

Bulk polymerization consists of only the monomer and an initiator. As there is no solvent present, polymerization occurs in the undiluted monomer resulting in the synthesized polymer that might be either soluble or insoluble in the monomer. The advantage of bulk polymerization is that it produces polymers of high molecular weights requiring no purification due to the lack of solvents or diluents. Poor heat control is a major limitation of this method due to which it can only be applied to processes with low heats of polymerization as such polymerizations that follow step-growth mechanism. As discussed earlier, in step-growth polymerization, high-molecular-weight polymer is produced only in the last stages of polymerization. This ensures that the viscosity of the medium stays low throughout most of the course of polymerization.

However in case of most chain polymerizations, viscosity and heat of polymerization are important parameters that need to be controlled to prevent disastrous consequences. These consequences might include formation of gels, hot spots, and eventual acceleration of reaction rates to run-away proportions. One simple way to control these problems is to perform the polymerization in solution.

Solution Polymerization

Solution polymerization uses a solvent that allows for efficient heat control, thereby making the process suitable for exothermic reactions. As polymerization proceeds, more solvent may be added to regulate the viscosity and reaction rate, thereby preventing a runaway reaction. The monomer, the initiator, as well as the formed polymer are all soluble in the solvent and thus solvent separation becomes an issue with this technique. However, if the application permits, the synthesized polymer in the solvent can directly be used as paints, adhesives, etc.

Although water is one of the preferred solvents to overcome viscosity and heat control–related problems during polymerization, most polymers or even monomers are insoluble in water. This results in multiphase processes where the monomers or the polymers are suspended or dispersed in water.

Suspension Polymerization

Suspension polymerization consists of a water-insoluble monomer along with a stabilizer, where a monomer-soluble initiator is added to initiate polymerization by chain growth. The monomer exists as droplets in water. In order to prevent the two immiscible phases from separating out, the process relies on extreme agitation as shown by Fig. 6.10A. These droplets act as reaction vessels that transform into rigid particles with the addition of more and more monomer units as polymerization continues. As mostly

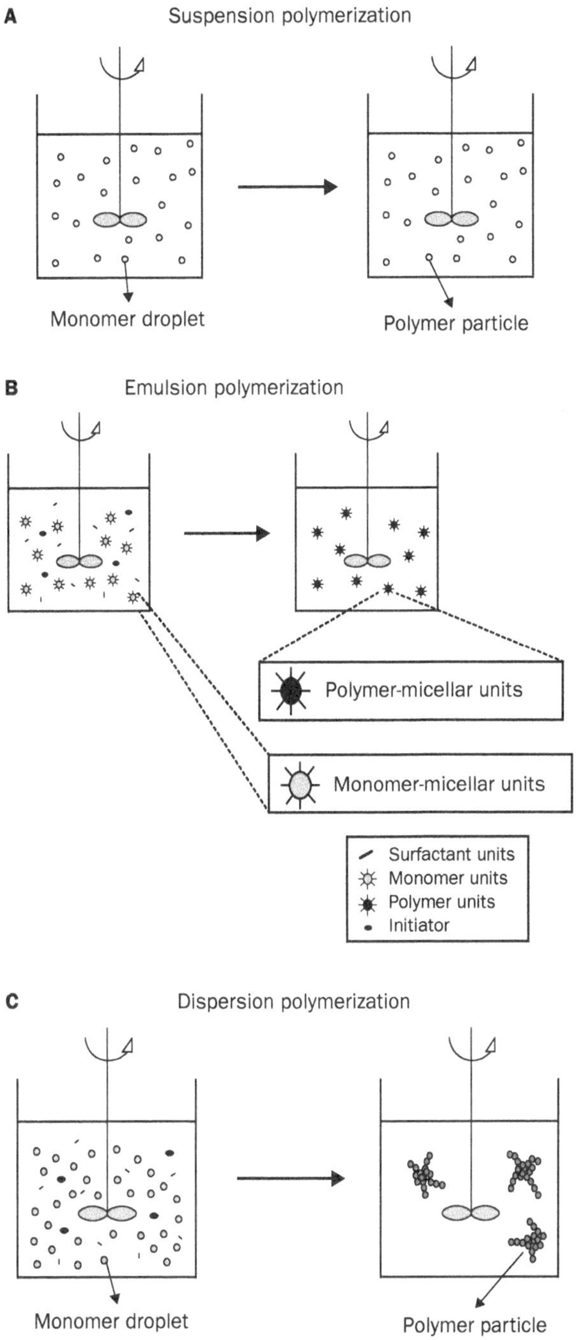

FIGURE 6.10 Schematic of (A) suspension polymerization, (B) emulsion polymerization, and (C) dispersion polymerization processes.

water is used as the continuous phase, it not only acts as an efficient heat transfer agent, but also makes controlling viscosity easy. Different morphologies can be obtained depending on the solubility of the polymer in the monomer. Smooth-translucent beads and porous-irregular beads can be obtained if the polymer is soluble and insoluble in the monomer, respectively.

Emulsion Polymerization

This process commences with an emulsion comprising the water-insoluble monomer, a stabilizer, and a water-soluble free radical initiator, all dispersed in a continuous water phase. In contrast to suspension polymerization, this method may not always require agitation. This is because the monomer, along with the stabilizer, forms micelles in water, mostly referred to as *latex* particles (aqueous dispersion of polymers), as shown in Fig. 6.10B. Polymerization starts once the initiator reaches the micelles in the form of a free radical. As the monomer inside the micelles is utilized in the polymer, more monomers continuously migrate from large monomer droplets into the micelles until all free monomers are consumed. These growing macroradicals are isolated from each other and act as emulsion reactors where typically one radical per reactor exists. Hence, high-molecular-weight polymers can be synthesized due to lower frequency of termination reactions besides achieving low polydispersity as compared to suspension polymerization.

Dispersion Polymerization

Dispersion polymerization belongs to one of the subcategories of heterogeneous polymerization wherein the monomer and the initiator are soluble, while the formed polymer is insoluble in the continuous phase. Therefore, the onset of polymerization occurs in the homogeneous phase and depending on the solubility of the macroradicals and macromolecules, phase separation occurs. This leads to nucleation and the formation of primary particles that are swollen by the polymerization medium, as shown in Fig. 6.10C.[11] Further polymerization proceeds within the individually formed particles resulting in spherical particles in the size range of 0.1–10 μm.

Precipitation Polymerization

The process commences similar to dispersion polymerization, that is, a homogeneous solution; however, the difference in the two processes lies in the fact that the formed particles do not swell in the medium, but literally precipitate out. Therefore, in dispersion polymerization, the locus of reaction lies within the surfactant stabilized particles formed that act as the reaction vessel for the growing polymer chain. While in the case of precipitation polymerization, the locus lies within the continuous phase as there is no formation of particles, due to the unavailability of surfactants or stabilizers, the process sometimes is referred to as surfactant-free emulsion polymerization. Thus, both initiation as well as polymerization occur in the homogeneous medium resulting in continuous nucleation followed by polymer growth. The polymer collapses upon itself on reaching a critical length generating precursor particles that grow by aggregation with other precursor particles, by being captured by existing particles, by capturing growing oligoradicals, and by monomer addition. On reaching a critical size, the particles are stabilized by the charge imparted by the initiator. This method gives a high control over particle sizes that can be achieved by using an additional ionic surfactant that provides ionic stabilization to even smaller precursor particles resulting in smaller polymer-based NPs.

In order to have control and optimization of the reaction, one needs to have a good understanding of the mechanism as well as the rate kinetics of precipitation polymerization.

The mechanism of precipitation polymerization is more complicated than for solution polymerization as the system undergoes a shift from a homogeneous to a heterogeneous regime as polymerization proceeds. The mechanism of precipitation polymerization with an intent to derive the rate of polymerization for such a process is discussed below.

Rate Model

As mentioned previously, precipitation polymerization consists of two distinct phases: the continuous phase (solvent) and the dispersed phase (growing/formed polymer), as shown in Fig. 6.11.[12] The main assumption lying behind the model development is that the monomer consumption in the continuous phase has been neglected. This is because it only accounts for a minute fraction (<1%) of the total monomer conversion. Another assumption is the locus of polymerization, which is considered to be in the heterogeneous polymer phase.

The following model development has been adapted from the excellent work of Avela et al.[13] Starting with the rate equation of homogeneous polymerization, the polymerization rate in the polymer phase can be defined as:

$$r_{P,s} = k_P \times c_{R^\cdot,s} \times c_{M,s} \tag{6.7}$$

where k_P is the propagation rate constant, $c_{R^\cdot,s}$ is the radical concentration in the polymer phase, and $c_{M,s}$ is the monomer concentration in the polymer phase, which is considered same as the monomer concentration in the homogeneous phase $c_M (c_{M,s} = c_M)$.

The difference between the radical initiation ($\dot{n}_i$) and termination ($\dot{n}_t$) rates gives the rate of change of number of radicals $\left(\frac{dn_{R^\cdot,s}}{dt}\right)$, defined as:

$$\frac{dn_{R^\cdot,s}}{dt} = \dot{n}_i - \dot{n}_t \tag{6.8}$$

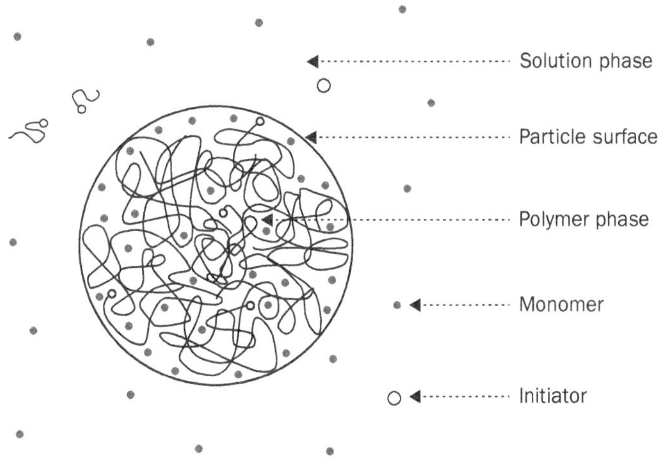

FIGURE 6.11 Schematic showing different phases during precipitation polymerization.[12]

As the initiator is dissolved in the homogeneous phase at the start, the radical initiation rate is dependent on the volume of the continuous phase, which is represented by:

$$V_R \times (1 - \Phi) \tag{6.9}$$

where V_R is the volume of the reaction mixture, while Φ is the volume fraction of solid polymer.

Therefore, the radical initiator rate can be defined as:

$$\dot{n}_i = 2 \times f \times k_d \times c_I \times V_R \times (1 - \Phi) \tag{6.10}$$

where f is the initiator efficiency, k_d the initiator decomposition rate constant, and c_I the initiator concentration in homogenous phase.

Equation (6.10) is similar to the radical generation rate in case of chain polymerization (which you may have come across in polymer-related courses) except for the volume term:

$$\frac{d[R^\bullet]}{dt} = 2k_d C_I \tag{6.11}$$

That can be transformed in terms of radical initiation rate as:

$$R_i^\bullet = 2fk_d C_I \tag{6.12}$$

Similarly, as the polymerization occurs in the dispersed phase, radical termination is dependent on the volume of polymer phase, which is represented by:

$$V_R \times \Phi \tag{6.13}$$

Therefore, the radical termination rate can be written as:

$$\dot{n}_t = k_t \times c^2_{R^\bullet,s} \times V_R \times \Phi \tag{6.14}$$

where k_t is the termination rate constant.

When applying steady-state approximation in Eq. (6.8), we assume that there is no net change in the number of radicals, i.e.:

$$\frac{dn_{R^\bullet,s}}{dt} = 0 \tag{6.15}$$

Therefore, on combining Eqs. (6.10) and (6.14), we arrive at:

$$2 \times f \times k_d \times c_I \times V_R \times (1 - \Phi) = k_t \times c^2_{R^\bullet,s} \times V_R \times \Phi \tag{6.16}$$

Equation (6.16) can be rewritten in terms of the radical concentration in the polymer phase:

$$c^2_{R^\bullet,s} = \Phi^{-0.5} \times (1 - \Phi)^{0.5} \times \left(2 \times f \times k_d \times \frac{c_I}{k_t}\right)^{0.5} \tag{6.17}$$

With the monomer concentration in the polymer phase being defined as:

$$c_{M,s} = c_M = c_{M,0} \times \left(\frac{1 - X_M}{1 - \Phi} \right) \tag{6.18}$$

we can write the rate of polymerization in the same phase as:

$$r_{P,s} = k_p \times \Phi^{-0.5} \times (1 - \Phi)^{0.5} \times \left(2 \times f \times k_d \times \frac{c_I}{k_t} \right)^{0.5} \times c_{M,0} \times \frac{(1 - X_M)}{(1 - \Phi)} \tag{6.19}$$

Also, the molar volume of the polymer V'_p, in terms of monomer molar mass (M_{AA}) and polymer density (ρ_P), can be written as:

$$V_P' = \frac{M_{AA}}{\rho_P} \tag{6.20}$$

Therefore, the volume fraction of the solid polymer results as:

$$\Phi = c_{M,0} \times X_M \times V_P' \tag{6.21}$$

The rate of polymerization for the entire system can be defined as:

$$r_P = r_{P,s} \times \Phi \tag{6.22}$$

Hence, by combining Eqs. (6.19), (6.20), (6.21), and (6.22), the rate of polymerization for precipitation polymerization can be defined as:

$$r_P = \frac{k_P}{k_t^{0.5}} \frac{c_{M,0}^{1.5} \times X_M^{0.5} \times (1 - X_M) \times V_P'^{0.5}}{(1 - c_{M,0} \times X_M \times V_P')^{0.5}} \times \frac{(2 \times f \times k_d \times c_{I,0} \times \exp(-k_d \times t))^{0.5}}{(1 - c_{M,0} \times X_M \times V_P')^{0.5}} \tag{6.23}$$

The only adjustable parameter in the rate equation above is $\left(\frac{k_P}{k_t^{0.5}} \right)$, which can be determined from the rate versus time data curves. In terms of chain polymerization, the rate of polymerization can be described as the rate of monomer disappearance as follows:

$$-\frac{d[M]}{dt} = R_P = k_P[M][M^\bullet] = \left(\frac{k_P}{k_t^{0.5}} \right) [M](f \times k_d \times [I])^{0.5} \tag{6.24}$$

The negative sign indicates the monomer is being consumed as polymerization proceeds, where $[M]$ and $[M^\bullet]$ represent the monomer and monomer radical concentrations, respectively. Therefore, both Eqs. (6.23) and (6.24) have the $\left(\frac{k_P}{k_t^{0.5}} \right)$ dependence in the rate expressions, indicating that precipitation polymerization proceeds via chain polymerization.

As discussed, depending on the solubility of the monomer, initiator, and polymer, polymerization can be carried out via emulsion, suspension, or dispersion. However, as a chemical engineer you might be interested in production volume of polymer-based

NPs. In short, distinction can be made based on reactor modes in which polymerization is carried out.

6.1.6 Mode/Type-Based Classification

Polymer-based NPs can also be classified based on the mode of synthesis used, namely batch, semi-batch, and continuous. Using batch polymerization is the simplest form that involves introducing all the reactants at once at the start and polymerization proceeds over time. However, this method allows for poor heat control and might require installation of cooling jackets to maintain the reaction temperature, as in the case of bulk polymerization.

Semi-batch polymerization overcomes this limitation as it allows for the addition of solvent or diluent as polymerization proceeds. This gives a better temperature control due to heat loss by convection in addition to maintaining the viscosity of the system, as in the case of solution polymerization. These two polymerization modes are mostly used in areas such as research laboratories, small-scale industries, and commercial polymer products. Yet, the ever-growing industrial demand globally makes it desirable to switch to continuous polymerization techniques for shorter turnaround times.

As you might have seen, even though polymer-based NPs can be classified on the basis of their structure, surface functionality, and synthesis techniques as discussed above, there is a correlation among all the different categories. For instance, precipitation polymerization may be carried out as a batch process and can yield hydrophobic polymer-based NPs. In addition, it might be possible to synthesize a particular NP system using different synthesis techniques, resulting in the alike or improved structural properties and surface functionalities. Therefore, it is impractical to limit the classification of polymer-based NPs to only a limited set of categories, and you might find different sets of classification in other references.

Taking this aspect into consideration, observe some polymeric systems that can be synthesized using different techniques, although resulting in similar properties. The stimuli-sensitive pNIPAm-based hydrogels, micelles, and nanocapsules are discussed next.

6.1.7 Examples of Polymer-Based Nanoparticles: pNIPAm-Based Systems

pNIPAm-based stimuli-responsive hydrogels form one of the categories of polymer-based NPs that utilize HLB and surface charge to alter their physicochemical properties. Hydrogels comprise a 3D network of cross-linked polymers, capable of absorbing large amounts of water making them extremely hydrophilic in nature. Just as with the previously discussed dendrimers, hydrogels have caught the attention of researchers globally, especially in the medical industry owing to traits such as high water content, softness, flexibility, and biocompatibility. Additionally, they undergo stimuli-triggered response (swell or collapse) on exposure to changes in their surroundings such as temperature, pH, light, and electricity, to name a few.

Our group has been successfully synthesizing pNIPAm-based hydrogels over several years, employing precipitation polymerization using AAc as a co-monomer, *N,N'*-methylenebis(acrylamide) (BIS) as a cross-linker, sodium dodecyl sulfate (SDS) as a surfactant and an initiator, and potassium persulfate (KPS) at 70°C. This elevated temperature is above the lower critical solution temperature (LCST) of the polymer;

thus, the formed hydrogel becomes insoluble in water and hence precipitates out from the solvent.[6]

If you recall previous discussions, hydrophilicity of the polymeric chains can be altered by selecting appropriate functional groups or playing with the molecular weight of the polymer chains. Xia et al. managed to synthesize well-controlled molecular weight pNIPAm with low polydispersity, using ATRP technique at room temperature. Instead of using another co-monomer, as discussed in Sec. 6.1.2, the hydrophilicity of pNIPAm was altered using four different initiators having different functional groups, as shown in Fig. 6.12, with the hydrophilicity decreasing from left to right.[14]

RAFT polymerization can also be used for the synthesis of pNIPAm-based hydrogels using 4-cyanopentanoic acid dithiobenzoate as the CTA along with BIS as the cross-linker in order to achieve increased stimuli response as compared to the conventional hydrogels obtained from FRP techniques (Fig. 6.13).[15] The gelation time of the system is much higher in RAFT polymerization than in radical polymerization methods due to slow-growing chains in the former case. This in turn permits greater intermolecular and reduced intramolecular cross-linking in the RAFT gels, thus resulting in a hyperbranched structure as shown in Fig. 6.14. While in the case of conventional polymerization technique, the chain propagation is so fast that it leaves no time for the polymer chains to relax, resulting in cross-linking within the same molecule. As polymerization proceeds, intermolecular cross-linking occurs combining the chains into larger molecules forming microgels, as shown in the figure.[15]

Moving on to a different morphology for pNIPAm-based NPs, polymeric micelles can be synthesized using ATRP technique having a hydrophobic core and a hydrophilic shell. This example uses hydroxyl-terminated polybutadiene (HTPB) as well as

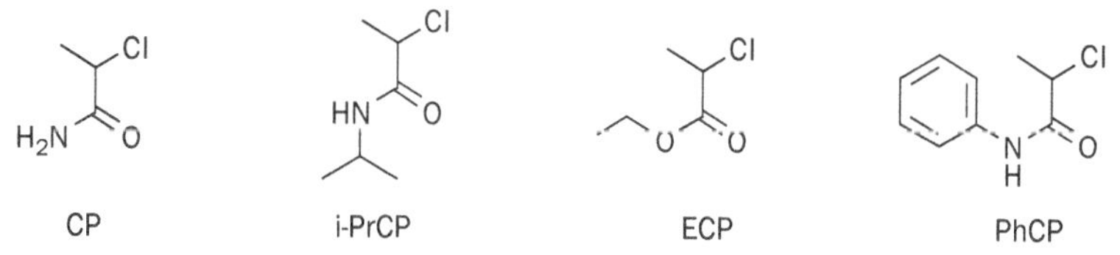

FIGURE 6.12 Chemical structures of different initiators used in the synthesis of pNIPAm-based NPs via atom transfer radical polymerization (ATRP).

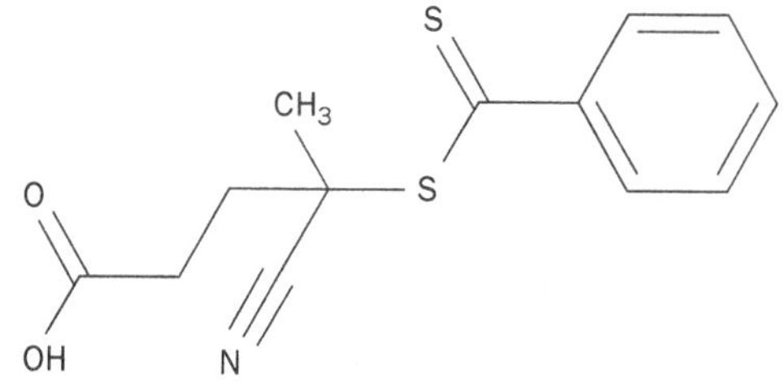

FIGURE 6.13 Chemical structure of 4-cyanopentanoic acid dithiobenzoate used as the chain transfer agent (CTA) in the synthesis of pNIPAm-based NPs via reversible addition-fragmentation chain transfer (RAFT) polymerization.

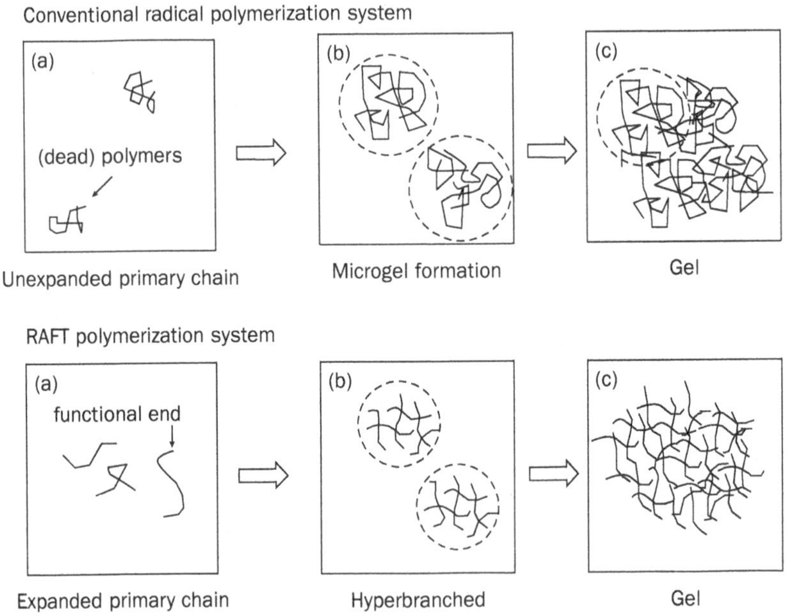

FIGURE 6.14 Schematic showing gel formation through conventional radical polymerization and RAFT polymerization.[15]

pNIPAm in the synthesis of polymeric core-shell micelles, in which the hydrophobic center is made up of water-insoluble HTPB, while pNIPAm forms the hydrophilic core of the copolymeric micelle. The advantage of having a hydrophobic core is that you can load hydrophobic drugs onto the NPs, which would otherwise prove to be a challenge in the case of hydrogels.

Camptothecin shows effective anti-tumor activity; however, it is water insoluble, thereby resulting in lower stability and higher cytotoxicity for *in vivo* purposes. All these limitations can be dealt with if camptothecin is encapsulated inside the above-mentioned co-polymeric micelle, having stimuli-response capabilities for thermally triggered targeted drug release. The authors were able to attain different micelle morphology of four-armed multiblock copolymers and linear triblock copolymers using ATRP, by varying pNIPAm's molecular weight.

Another way to encapsulate hydrophobic moieties inside NPs can be achieved by using nanocapsules having a hollow hydrophobic core protected by a polymeric shell. Emulsion polymerization technique has been successfully used in order to synthesize these nanocapsules, which utilizes mini or micro emulsion droplets acting as reactors for polymerization at the water–oil interface. In another case, hollow core-shell polymeric nanocapsules were prepared by polymerizing the monomer on a surface of a preformed template that was later removed by dissolution or other means, leaving behind the polymeric shell. The template can either be of an organic[16] or an inorganic[17] origin, as shown in Figs. 6.15 and 6.16, respectively.

No matter what synthesis procedure you might choose for your NPs or hydrogels, keep in mind the final application and what properties should be incorporated that

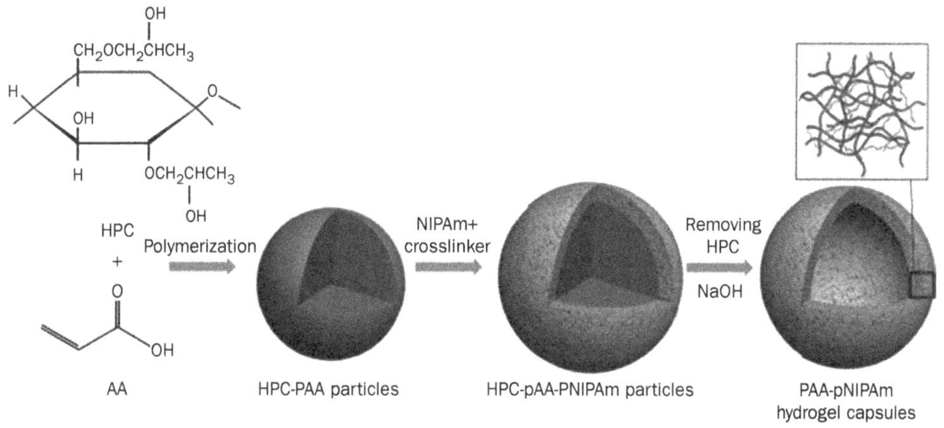

Figure 6.15 Template-assisted NP growth of pNIPAm-AAc hollow core-shell nanocapsules using organic template.[16]

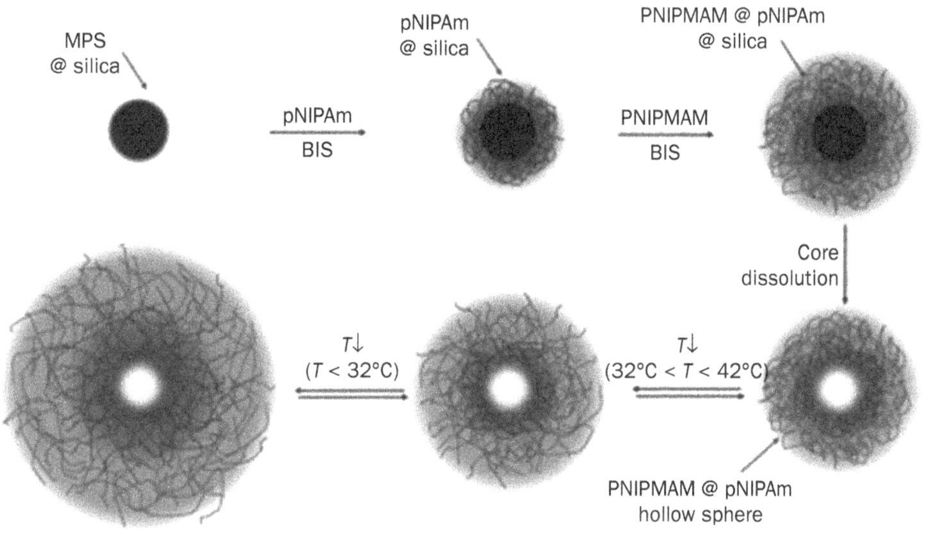

Figure 6.16 Template-assisted NP growth of pNIPAm-AAc hollow core-shell nanocapsules using inorganic template.[17]

might be useful for the system. Once that is finalized, you can move on to the designing and selection of the reactant species such as CTA, catalysts, and various functional groups that will optimize the system physicochemical properties. Since hydrogels have such a great importance in the biomedical industry, the nature of these hydrogels and its effect on the physicochemical properties are discussed in detail in the section below.

6.2 Stimuli-Sensitive Hydrogels

The term lower critical solution temperature (LCST) is predominantly used for polymers in solutions that describes the miscibility of the polymer in a particular solution, where the polymer is miscible at temperatures below its LCST and immiscible above it. As discussed earlier, hydrogels are polymeric units cross-linked together into a 3D network. Above or below the LCST, the polymer chains either swell or collapse depending on the nature of solvent. However, when the effect of all the polymer chains swelling or collapsing is combined in a hydrogel unit, it results in a change of volumetric proportion. This analogous term used for hydrogels is known as volume phase transition temperature (VPTT).

pNIPAm, short for poly(*N*-isopropylacrylamide), is one of the most widely studied temperature-responsive polymers that takes advantage of HLB of its polymeric chains. First synthesized in the 1950s from *N*-isopropylacrylamide by FRP, pNIPAm shows a characteristic coil-globule phase transition (Fig. 6.17) above its volume phase transition temperature (VPTT). This behavior is analogous to the lower critical solution temperature (LCST) of a polymer unit.

At temperatures lower than the VPTT, pNIPAm is hydrophilic and shows high polymer–solvent interactions, owing to the enthalpy contribution (ΔH) from hydrogen bonding between the amide groups and water molecules. Upon increasing the temperature above VPTT, there is an entropically driven release of bonded water, increase in the entropic term (ΔS), thereby causing the polymer chains to collapse into a globule as the change in the Gibb's free energy change tends toward negative, as Eq. (6.25) depicts.

$$\Delta G = \Delta H - T\Delta S \tag{6.25}$$

VPTT is a function of the cross-linking density, nature of solvent, and nature of hydrogel. By varying the hydrophilicity of the polymer, one can alter the VPTT of the polymeric system. Usually, adding more hydrophobic groups causes a decrease in the

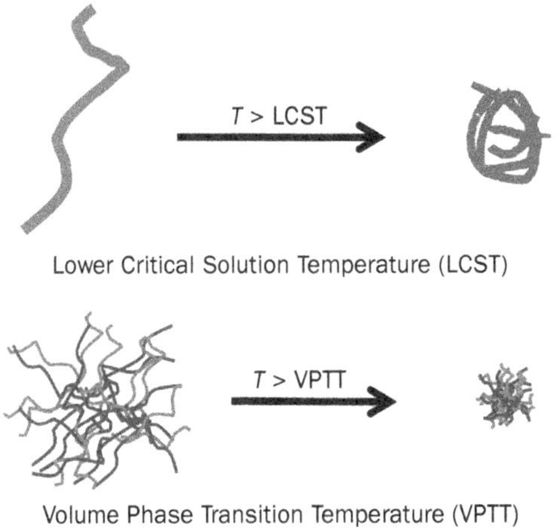

Figure 6.17 Schematic showing coil to globule transition of a polymer (LCST) and polymeric network (VPTT).

VPTT, whereas ionic functional groups and moieties capable of hydrogen bonding increase solubility and thus can increase the VPTT. This swelling-collapse behavior of pNIPAm-based NPs has been used for applications of targeted drug delivery, which will be discussed in detail in Chap. 10.

Similar to the temperature-based stimuli response depicted by pNIPAm, volumetric phase transition can be triggered by altering other physical conditions such as pH, light, concentration, electricity, etc. For simplicity, let us continue with the example of pNIPAm as our base polymeric moiety.

pNIPAm can be copolymerized with monomers containing an acid functional group such as acrylic acid (AAc) that can provide pH-based sensitivity to the hydrogel. By addition of acrylic acid monomers to NIPAm, both temperature as well as pH-based stimuli response can be achieved, as has been shown by studies from our group.[6] pNIPAm and polyacrylic acid (pAAc)-based hydrogels show swelling in basic conditions (using sodium hydroxide—NaOH), whereas a collapse can be seen in acidic conditions when using a stronger acid for lowering the pH (in our case hydrochloric acid—HCl). However, as mentioned earlier, by incorporating another hydrophilic group (AAc), an increase in VPTT of the system can also be observed.

The next chapter will deal with the physicochemical properties of hydrogels, most of which depend on the fundamentals discussed above. Here, we have briefly visited the various synthesis methods for polymer-based NPs that will form the basis of functionalization in order to ascribe advanced properties. Our focus henceforth will be on hydrogels, since these exhibit interesting stimuli-sensitive properties that make them interesting candidates for applications such as drug delivery, biosensing, water management, and others.

6.3 Study Questions

6.3.1 Concept-Based Questions

6.1 What are the major differences between step-growth and chain polymerization?

6.2 Describe the two main pathways to synthesize dendrimers and list some of the pros and cons for each method.

6.3 Mention the different compartments in a dendrimer molecule that may be beneficial for drug delivery.

6.4 Compare and contrast suspension and emulsion polymerization processes for production of polymer-based NPs. Also comment on how sizes of the final products may be controlled.

6.5 Discuss the underlying principles for controlled radical polymerization and atom transfer radical polymerization.

6.6 What is a chain transfer agent? In regards to RAFT, discuss the roles of functional moieties in a typical chain transfer agent.

6.7 Schematically show the mechanism underlying RAFT polymerization.

6.8 The temperature at which a polymeric chain undergoes coil to globule transitions in response to stimuli is known as LCST. Discuss from an energy standpoint what happens in case of cross-linked systems comprising similar polymers in response to temperature.

6.3.2 Research-Based Questions

6.9 Different samples of pNIPAm-AAc nanogels (NGs) were synthesized using precipitation polymerization with varying concentrations of a surfactant, sodium dodecyl sulfate (SDS), and a cross-linker, *N,N'*-methylenebis(acrylamide) (BIS).[6] The NGs were observed to have a stimuli response with respect to temperature, swelling, and collapsing at temperatures above and below the VPTT. The effect of SDS, BIS, and reactant concentrations on the size of the NGs is shown in the figure below.

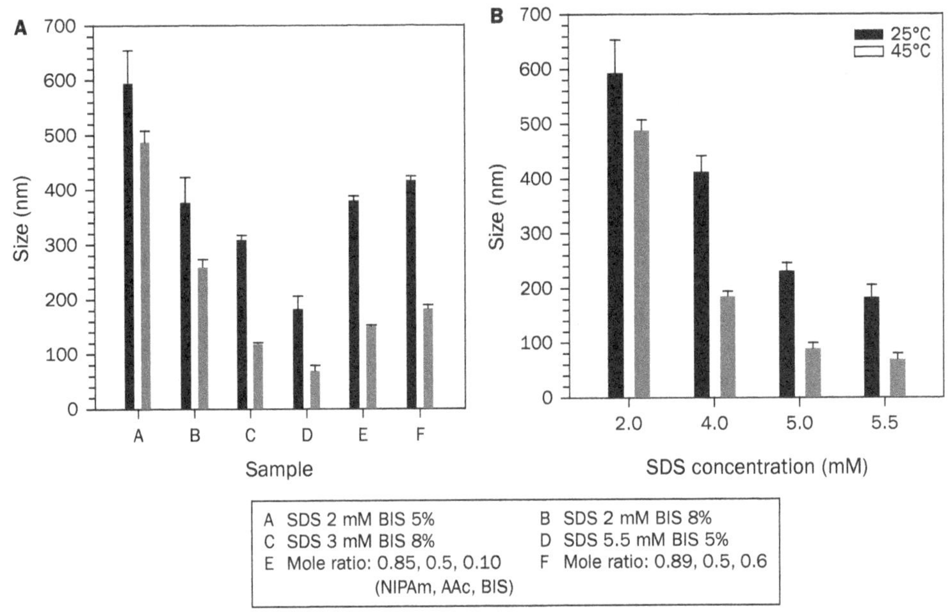

a. Explain the variation in NG size with respect to the concentration of the surfactant (SDS) and the cross-linker (BIS).

b. The NGs have been shown to have stimuli response to temperature. Can any other stimulus be used to trigger a response from these materials in your opinion?

c. Is it possible to vary the VPTT of these NGs? If yes, how?

d. Could you name other techniques that may be used for the synthesis of pNIPAm-based hydrogels?

6.10 pNIPAm-based NPs were synthesized using a two-stage seeded polymerization method as shown in the following schematic, where NIPAm and poly(ethylene glycol) methacrylate (PEGMA) are the monomer and co-monomer moieties, respectively, along with poly(ethylene glycol) dimethacrylate (PEGDMA) as the cross-linker. X is formed as the main product after the first stage, while main product Y is generated in the second stage.

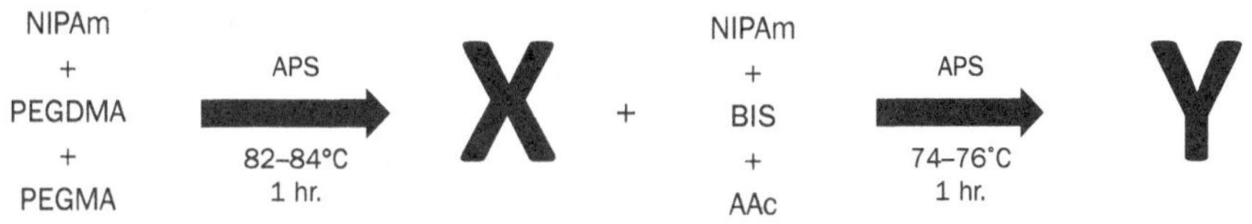

a. Predict the main products, X and Y, that will be formed as per the figure above. Can you comment on the morphology of the products from a polymer perspective?

Different average molecular weights of PEGDMA (M_n) were used along with varying mole ratios of the monomer, co-monomer, and cross-linker and the NPs formed after stage 2 (product Y) were characterized using dynamic light scattering (DLS). The variation of size with M_n and the mole ratios of monomer, co-monomer, and cross-linker, and the polydispersity index (PDI) with respect to the cross-linker are both plotted in the figure below.

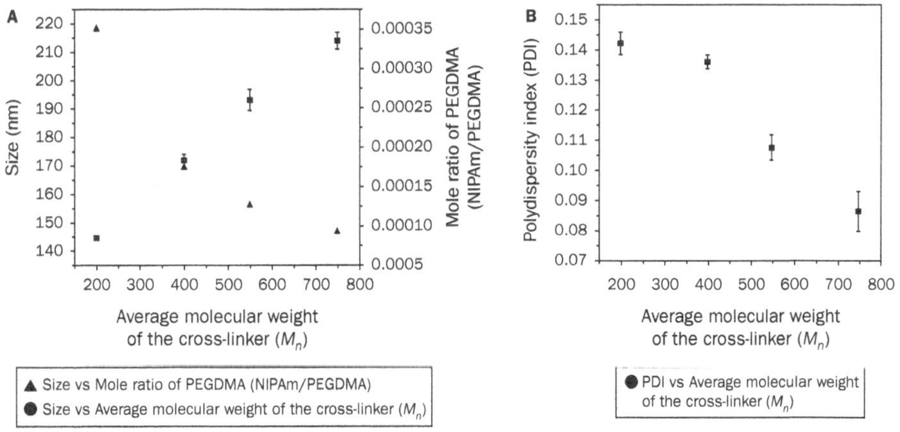

b. The sizes of the hydrogels increase with M_n and the mole ratios of the reactants. Give possible reasons for this trend highlighting back on the synthesis method and structure–property relationships.

c. Can you explain the decrease of PDI with M_n in relation to the reasons mentioned in part b?

6.4 References

1. Guterres, S. S.; Alves, M. P.; Pohlmann, A. R., Polymeric nanoparticles, nanospheres and nanocapsules, for cutaneous applications. *Drug Target Insights.* **2007**, *2*, 147–157.

2. Boas, U.; Heegaard, P. M., Dendrimers in drug research. *Chemical Society Reviews.* **2004**, *33*(1), 43–63.

3. Dykes, G. M., Dendrimers: a review of their appeal and applications. *Journal of Chemical Technology and Biotechnology.* **2001**, *76*(9), 903–918.
4. Abbasi, E.; Aval, S. F.; Akbarzadeh, A.; Milani, M.; Nasrabadi, H. T.; Joo, S. W.; Hanifehpour, Y.; et al., Dendrimers: synthesis, applications, and properties. *Nanoscale Research Letters.* **2014**, *9*(1), 247.
5. Sengel Türk, C.; Sezgin Bayindir, Z.; Badilli, U., Preparation of polymeric nanoparticles using different stabilizing agents. *Journal of Faculty of Pharmacy of Ankara University.* **2009**, *38*(4), 257–268.
6. Bandyopadhyay, S.; Andersen, M. K.; Alvi, M. A. A.; Sharma, A.; Raju, R.; McDonagh, B. H.; Glomm, W. R., Incorporation of Fe@Au nanoparticles into multiresponsive pNIPAM-AAc colloidal gels modulates drug uptake and release. *Colloid and Polymer Science.* **2016**, *294*(12), 1929–1942.
7. Sadat, S. M.; Jahan, S. T.; Haddadi, A., Effects of size and surface charge of polymeric nanoparticles on in vitro and in vivo applications. *Journal of Biomaterials and Nanobiotechnology.* **2016**, *7*(02), 91.
8. Braunecker, W. A.; Matyjaszewski, K., Controlled/living radical polymerization: features, developments, and perspectives. *Progress in Polymer Science.* **2007**, *32*(1), 93–146.
9. Matyjaszewski, K., Atom transfer radical polymerization (ATRP): current status and future perspectives. *Macromolecules.* **2012**, *45*(10), 4015–4039.
10. Chiefari, J.; Chong, Y.; Ercole, F.; Krstina, J.; Jeffery, J.; Le, T. P.; . . . Moad, G. J. M., Living free-radical polymerization by reversible addition–fragmentation chain transfer: the RAFT process. **1998**, *31*(16), 5559–5562.
11. Arshady, R., Suspension, emulsion, and dispersion polymerization: a methodological survey. *Colloid and Polymer Science.* **1992**, *270*(8), 717–732.
12. Liu, T.; DeSimone, J. M.; Roberts, G. W., Kinetics of the precipitation polymerization of acrylic acid in supercritical carbon dioxide: the locus of polymerization. *Chemical Engineering Science.* **2006**, *61*(10), 3129–3139.
13. Avela, A.; Poersch, H. G.; Reichert, K. H., Modelling the kinetics of the precipitation polymerization of acrylic acid. *Die Angewandte Makromolekulare Chemie* [*Applied Macromolecular Chemistry and Physics*]. **1990**, *175*(1), 107–116.
14. Xia, Y.; Burke, N. A.; Stöver, H. D., End group effect on the thermal response of narrow-disperse poly (*N*-isopropylacrylamide) prepared by atom transfer radical polymerization. *Macromolecules.* **2006**, *39*(6), 2275–2283.
15. Liu, Q.; Zhang, P.; Qing, A.; Lan, Y.; Lu, M., Poly (*N*-isopropylacrylamide) hydrogels with improved shrinking kinetics by RAFT polymerization. *Polymer.* **2006**, *47*(7), 2330–2336.
16. Nan, J.; Chen, Y.; Li, R.; Wang, J.; Liu, M.; Wang, C.; Chu, F., Polymeric hydrogel nanocapsules: a thermo and pH dual-responsive carrier for sustained drug release. *Nano-Micro Letters.* **2014**, *6*(3), 200–208.
17. Ramli, R. A., Hollow polymer particles: a review. *RSC Advances.* **2017**, *7*(83), 52632–52650.

CHAPTER 7

Functionalization and Properties of Hydrogels

Although a number of synthesis strategies of polymeric NPs have been discussed in Chap. 6, this chapter discusses their properties. We will discuss the importance of synthesis parameters in modulating the physicochemical properties of a class of polymer-based NPs, namely hydrogels. Through this discussion you will learn how various synthesis parameters such as monomer ratios, cross-linkers, stabilizers, and reaction conditions influence the size, swelling-collapse, surface charge, and other properties of the resultant hydrogels.

Interest in hydrogels has seen a dramatic increase in the last 50 years owing to their flexibility in retaining water without dissolving in it. As already introduced in the last chapter, hydrogels are water-swollen, cross-linked polymeric networks produced by the simple reaction of one or more monomers. These are capable of swelling and changing their dimensions depending on the solvent and environmental parameters such as temperature, pH, ionic strength, and other stimuli. This makes them interesting candidates for stimuli-controlled release of a cargo molecule.

One of the important aspects of functionalization of hydrogels is to ascribe multiple properties to a single construct besides providing stability. For instance, imagine how useful a multifunctional Swiss army knife may be on a trip. Tasks are often easier when you use the right tool. A Swiss army knife serves as a toolbox in your pocket. You can use it for various functions such as opening a bottle, tightening screws, opening cans, self-administering an emergency haircut, or the like. In a similar fashion, hydrogels can be functionalized with different entities such as contrast agents, biomarkers, and targeting ligands to make them multifunctional. In addition to the stimuli-directed responses, these may be tracked, imaged, or actively targeted to a desired location. Figure 7.1 shows such a hypothetical nanomaterial that may be formed by combining the knowledge gained about gold NPs in Chap. 5 and polymer-based NPs in Chap. 6.

In the above situation, plasmonic properties from the Au NPs may be used for optical tracking, while the stimuli-sensitive hydrogels may be used to release a bound drug in response to temperature and/or pH. The targeting ligand on the surface of the nanomaterial may be used to guide the construct to the site of action, where it is expected to deliver its function. Unlike the Swiss army knife example, our nanomaterials can perform one or several of the functions simultaneously. This means it will be possible to track the nanomaterials while it is delivering the drug at the target site via the collapse

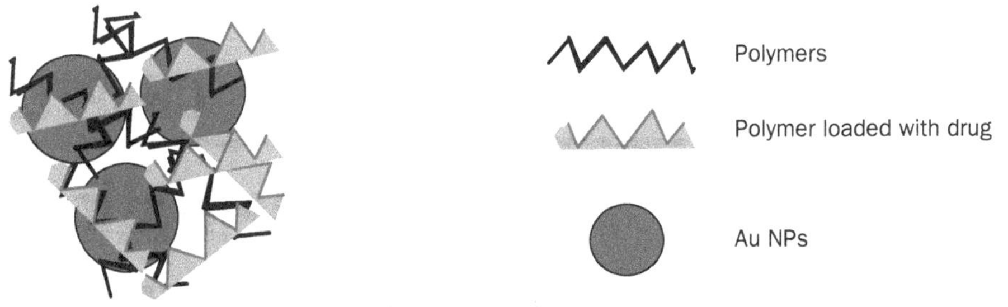

Figure 7.1 Schematic representation of multifunctional NPs.

of the hydrogels. Now, you will learn different ways to modify the properties of hydrogels and the resultant nanomaterials.

7.1 Functionalization

Functionalization (as already introduced in Chap. 5) is the process of altering, adding, or replacing functional groups—in the present case, on hydrogels. Like a Swiss army knife whose application depends on the tools it has, functionalization determines the properties and applications of hydrogels. The hydrogels can be functionalized either when they are synthesized or after their synthesis. Consider the previous example of multifunctional nanomaterials: if the inorganic NPs are added to the reaction mixture in which hydrogels are forming, it is called *in situ* functionalization. However, if the inorganic NPs are added to hydrogels synthesized beforehand, it is called post-synthesis (or *post situ*) functionalization.

7.1.1 *In Situ* Functionalization

To better understand how *in situ* functionalization can help achieve resultant physicochemical properties of the nanomaterials, consider the synthesis of a class of multifunctional nanomaterials comprising stimuli-responsive hydrogels and inorganic NPs. During the synthesis of the hydrogels, inorganic NPs are added along with monomers and surfactants. In a typical reaction, as shown in Fig. 7.2, temperature sensitive *N*-isopropylacrylamide (NIPAm) monomer is added along with the cross-linker *N,N′*-methylenebis(acrylamide) (BIS) and the surfactant sodium dodecyl sulfate (SDS) at a temperature above the lower critical solution temperature (LCST) of NIPAm.[1] The NPs are added into the system along with SDS. The reaction mixture is allowed to stir in order for the monomers to collapse and facilitate precipitation polymerization. To further add pH sensitivity to the resultant nanomaterials, acrylic acid (AAc) monomers are added and polymerization is initiated by the addition of potassium persulfate (KPS). Nanogel obtained from such a reaction has a morphology in which inorganic NPs are in a mesh/matrix of polymeric NPs.

Similar *in situ* techniques can be used for naturally occurring hydrogels as well. Alginate, a naturally occurring polyanion, chelates with divalent cations such as Ca^{2+} to form hydrogels. Due to their biological source, alginates are highly desirable for tissue engineering. One of their disadvantages is their inertness; as a result, they cannot be absorbed by many biological macromolecules such as proteins. Such problems can be

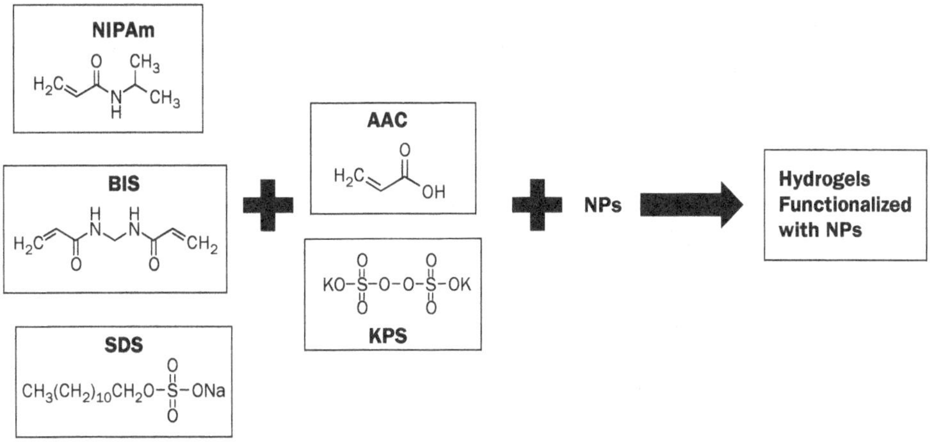

FIGURE 7.2 Synthesis routine for *in situ* functionalization of hydrogels with NPs.

overcome by functionalizing these alginates with cell-adhesive peptides, which helps the macromolecules to survive in the alginate matrix. Alginates possess monomers with carboxylic acid groups, which utilize carbodiimide chemistry to form peptides. As you may know, a peptide bond is formed when a carboxylic acid reacts with an amine group while water is released. The carboxyl group from the mannuronic acid present in alginate is modified by *in situ* reaction with peptides; this is then introduced to Ca^{2+} ions that form hydrogels. The functionalized hydrogels do not affect the morphology of encapsulated cells and its viability increases.

7.1.2 Post-Synthesis Functionalization

In the post-synthesis functionalization method, hydrogels are functionalized after their synthesis. The multifunctional nanomaterials can be synthesized by coating Au NPs with hydrogels. In a typical routine, the stimuli-sensitive hydrogels are synthesized by the method discussed in Sec. 7.1.1 without introducing the inorganic NPs. These hydrogels are thereafter mixed with inorganic NPs and stirred at high speed to allow coating. The as-formed nanomaterials are then centrifuged to obtain hydrogels functionalized with inorganic NPs.[2]

Hydrogels that respond to ionic composition or light can be used for detecting target molecules such as proteins. Such hydrogels are functionalized to imbibe detection capability. The ability to sense in such biosensors depends on the molecule that is used for functionalizing. MacConaghy et al. developed such functionalized hydrogels that can detect proteins in cells.[3] In short, acrylamide-based hydrogels are synthesized by photocatalysis in the presence of a colloidal suspension of polystyrene spheres. Purified hydrogels are functionalized by reacting with a type of modified peptide substrate in the presence of cupric sulfate, *tris*(3-hydroxypropyltriazolylmethyl) amine and sodium ascorbate for 24 hours. These peptides can recognize certain proteins and trigger a morphological change in hydrogels that can be observed through a change in color. This is yet another example of post-synthesis functionalization.

One of the key aspects of the NIPAm-based hydrogels is its ability to release a targeted molecule such as a drug. Depending on the nature of hydrogels, different techniques can be used to load the hydrogels with the desired drug. The drug can be either on the surface of the hydrogel or it can be inside its porous matrix. Chemical or physical methods can be used to bind the drug. Consider the case in which the drug is chemically bound to the hydrogel. Here, the drug may be covalently bound to the hydrogel structure and the release may be triggered by breaking the bond between the hydrogel and drug. An alternative method for drug loading is through physical methods. The physical method to load a drug is usually through the diffusion of the drug into hydrogels causing them to swell, as shown in Fig. 7.3. As can be imagined, the release of the drug in such a case will be primarily caused by solvent–hydrogel interactions.

This process of loading the drug into the hydrogel matrix is called *breathing-in mechanism*. Lyophilized hydrogel solutions are imbibed with a concentrated drug solution, allowing for swelling of the hydrogel matrix with the drug. The release of the drug is triggered by the collapse of the hydrogel that happens in response to external stimuli such as temperature or pH. As the size of hydrogels decreases, due to its interaction with external stimuli, the drug is squeezed out. The application of hydrogels, such as a drug delivery agent or biosensor, depends on its properties, and functionalization can change these properties. Some of the important properties of hydrogels are their size, response to stimuli, water content, and mechanical strength, among others. Some of these properties are discussed in the subsequent sections.

7.2 Properties of Hydrogels

You know how to synthesize different types of polymeric NPs, including hydrogels, and how to functionalize them. Now we will discuss about the properties of these hydrogels, as these properties make them ideal candidates for several applications. The physicochemical properties of hydrogels are governed by one or more of the following factors: type of monomers, co-monomers and cross-linkers used, synthesis method employed, and functionalization, among others. We will also discuss about the effect of interplay among some of these factors on physicochemical properties of hydrogels.

7.2.1 Size

Consider an example of an NP-based drug delivery system where the interaction with cells is the key function. In such situations, although the synthesis of NPs needs to focus on imparting biological and physicochemical properties to have specific responses, the ability to access the cell primarily depends on the final size of the NPs. Also, the size

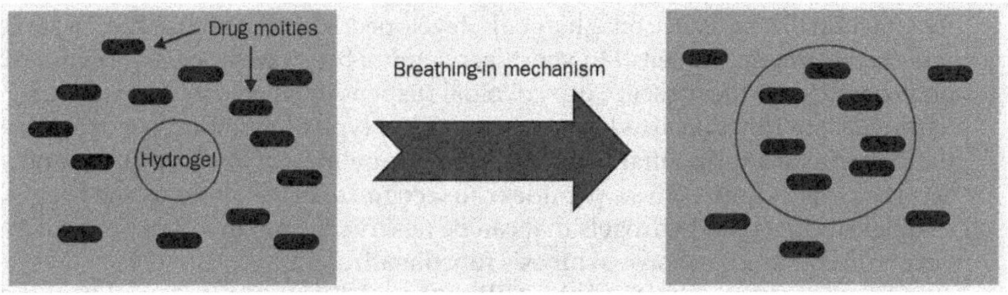

FIGURE 7.3 Breathing-in mechanism to load hydrogels with drug.[4]

of NPs should be monitored so that they can be cleared by kidneys as you don't want a build-up of NPs in your body. Therefore, size control of the NPs or hydrogels, in this case, is a critical step.

One of the advantages of polymeric NPs is the ease of control of size. The size of hydrogels can be influenced by varying one or several parameters that include concentrations of co-monomers, cross-linkers, and surfactants. The concentration of monomers affects the size by influencing the number of monomeric units that are incorporated in the hydrogels. Usually, an increase in the concentration of monomers, when concentrations of other reactants are kept constant, increases the size of the hydrogels. Similar observations can be seen when there is an increase in the concentration of co-monomers. However, such increase in size occurs only up to a certain concentration due to depletion of other chemicals such as initiator or surfactant.

If you can recollect some of the polymerization methods discussed in Chap. 6, you may assert that surfactant plays a major role in precipitation polymerization by providing charge stabilization to the growing hydrogels. Therefore, an increase in the concentration of surfactant will stabilize precursor particles and will result in smaller hydrogels. A similar trend can also be observed when the initiator concentration is increased. Chain polymerization follows the conventional steps of initiation, propagation, chain transfer, and termination (already discussed in detail in Chap. 6). The increase in initiator will lead to higher/faster initiation, thereby influencing the availability of monomers for propagation. This will result in smaller size and the same concept can be extended to hydrogel formation.

Consider the pNIPAm/AAc hydrogels—an interesting class of stimuli-sensitive polymeric NPs, where the effect of these parameters can be seen.[1] A higher amount of NIPAm or AAc added in the reaction mixture is known to increase the size of the hydrogels. When the concentration of SDS is increased but maintained under the CMC, there is a steady decrease in the size of the hydrogels as shown in Fig. 7.4. This could be due to charge stabilization provided to the nuclei onto which the chains collapse.

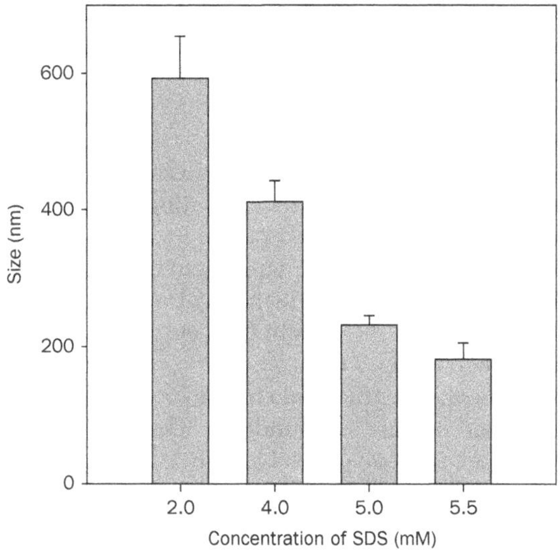

FIGURE 7.4 Effect of surfactant concentration on the size of hydrogels.[1]

Cross-linking is another parameter that can influence the size of hydrogels. Both physical and chemical cross-linking can be used: processes such as UV radiation, hydrophobic interaction, hydrogen bonding, and strong covalent bonding or small molecule cross-linking can be used. In NIPAm hydrogels, BIS is used as a cross-linker and an increase in BIS causes a decrease in the size. A similar observation can be made in PEG-acrylate-based hydrogels where acryloyl carbonate is used for cross-linking. This effect is common in all types of cross-linking and is caused due to an increase in the cross-linking densities between the monomers. Although the size can be controlled by changing the concentration of cross-linkers or surfactants, its effect can be seen only when there are sufficient amounts of other reactants. The absence of these reactants can lead to formation of other products; for example, if SDS concentration is extremely high as compared to the concentration of NIPAm, all the monomers can be stabilized by SDS and might not lead to precipitation polymerization, thereby yielding no resultant hydrogels.

The size of hydrogels can also be altered by functionalization. Consider a hydrogel that responds to temperature alone. Such synthesized hydrogels can be further functionalized by using it as a seed for further polymerization. pNIPAm/AAc hydrogels, when used as seeds in the presence of NIPAm monomers along with a cross-linker and surfactant, will give a bigger hydrogel with a core-shell structure. The core will have NIPAm/AAc moieties while the shell will mainly consist of NIPAm units. This is an example of controlling size by post-synthesis functionalization. In case of *in situ* functionalization of hydrogels with NPs, the NPs will interact with the growing hydrogels and will affect the final size. For example, the NPs can provide a stable surface for the NIPAm to collapse or the NPs can be on the surface and act as cross-linkers. Although the size can be tuned with relative ease, its effects can be seen in other properties of the hydrogels which you will study later.

For the time being, consider two hydrogels A and B of similar size, around 300 nm at 25°C. However, when the temperature is increased to 50°C, A collapses to 150 nm and B collapses to 100 nm. Thus, although the size of the hydrogels is an important parameter, it is not a stand-alone property of the hydrogels that can describe their behavior completely. Therefore, swelling collapse of the hydrogel is the next important parameter that has to be assessed before using the hydrogels.

7.2.2 Swelling Collapse

On studying the application of hydrogels, you can arrive at the conclusion that its ability to swell-collapse is perhaps the most exciting property. Hydrogels, as you already know by now, are highly cross-linked polymers and exhibit properties similar to polymers. Stimuli-responsive polymers in a solvent remain in a random coil-like structure. A change in the solvent property, which may be triggered by stimuli such as temperature (T), pH, the electric field (E), light ($h\nu$), etc., results in a collapsed globule state of the polymer, as shown in Fig. 7.5.

This conformational or structural change observed is caused due to the polymer's interaction with itself and its surroundings. The interactions between polymer–polymer, solvent–polymer, and solvent–solvent dictate the swell-collapse of such polymers. In the following section, you will learn about swell-collapse of hydrogels followed by swell-collapse of functionalized hydrogels, and some techniques to determine the degree of responsiveness of these hydrogels.

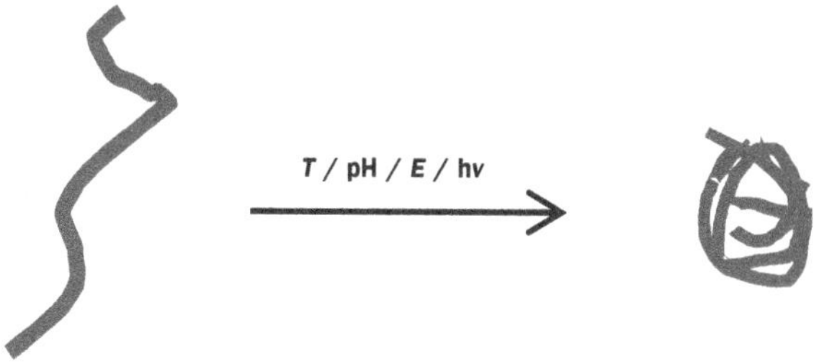

Figure 7.5 Behavior of stimuli-sensitive polymer in response to external stimuli.

Swell-Collapse of Hydrogels

Effect of Temperature Consider the widely used temperature-responsive pNIPAm that in water forms hydrogen bonds with water, while the carbon backbone offers hydrophobicity. Water molecules align based on these interactions and result in an ordered arrangement. In pNIPAm, hydrogen bonds exist between the amide groups in the polymer and water molecules; however, the isopropyl groups tend to create hydrophobic structuring of water. You already know that conformation of the polymer depends on different interactions between the polymer and the solvent. When the polymer–solvent interaction is the dominant one, the pNIPAm will be in a random coil state. At higher temperature, the hydrogen bonds break and the polymer–polymer interactions increase, which leads to an entropically favored release of the bound water, resulting in a globule structure of the polymer. The temperature at which this occurs in a polymer is known as the *lower critical solution temperature* (LCST), already introduced in Chap. 6. The LCST of pNIPAm is around 32°C.

When the temperature of the pNIPAm is above LCST, the solution turns turbid with two distinct phases due to the release of bound water. In the case of cross-linked hydrogels, the critical temperature above which similar coil-to-globule transition happens is called *volume phase transition temperature* (VPTT). The volumetric collapse efficiency (VCE) is a parameter that can be used to analyze these hydrogels and quantify their stimuli responsiveness. The VCE is defined as the relative change in the volume of the hydrogels between the swollen and collapsed states.

$$VCE = \frac{(V_{SH} - V_{CH})}{V_{SH}} \times 100 \tag{7.1}$$

where V_{SH} and V_{CH} are the volumes of swollen and collapsed hydrogels, respectively. Similar terms can be used for other stimuli if the volumes of the hydrogel in swollen and collapsed states are known.

Although, as discussed previously, the LCST and VPTT are the terms used for polymers and cross-linked polymeric networks, respectively, researchers use these terms interchangeably while defining either of these systems.

One physical property that is directly affected during phase transition that can be measured is the size of the hydrogel. In aqueous solutions, an increase in the system's hydrophilicity will cause the system to swell and consequently increase its size, whereas an increase in the system's hydrophobicity will lead to the collapse of the system, thereby decreasing the system's size. Therefore, accurate measurement of the system's size can provide us with the VPTT of the system and help us optimize the system according to our needs and intended applications.

Generally, cloud point (CP) measurement is used to determine LCST of a linear thermo-responsive polymer, where the onset of CP is taken as the point of phase transition. The LCST of a polymer depends on factors such as the chain length, hydrophilicity or hydrophobicity of functional groups, and solvent, to name a few. If we assume a highly monodisperse polymeric system, then ideally all the polymer chains should undergo phase transition (either swelling or collapse) simultaneously generating a single point of LCST. However, since a hydrogel is a cross-linked network consisting of multiple polymer chains with different chain lengths, they respond differently to stimuli resulting in a transition temperature range instead of a single transition temperature. Thus, it becomes more challenging while commenting on the VPTT of a system rather than LCST of a polymer.

Characterization techniques such as dynamic light scattering (DLS) and UV-Vis spectroscopy can be used to determine the VPTT of pNIPAm-based hydrogels. Typically, the respective spectroscopic parameters are plotted against temperature, and the inflection point on the curve marks the VPTT of the system.

For instance, poly(acrylic acid-co-acrylamide) hydrogels have been studied extensively by Yang et al.[5] The effect of adding hydrophobic content has been investigated using UV-Vis spectroscopy, where the transmittance has been measured against temperature at 600 nm. The study observed that addition of hydrophobic content (urea in this case) decreased the phase transition temperature, as shown in Fig. 7.6.

Since the size of the hydrogels changes with respect to the applied stimuli, the size can be tracked using DLS. Our group has worked extensively on pNIPAm-AAc-based hydrogels with an attempt to understand their swell-collapse properties. Figure 7.7 shows a plot of swelling ratio α of pNIPAm-AAc hydrogels as a function of temperature. α is defined by the following equation:

$$\alpha = \left(\frac{D}{D_0}\right)^3 \tag{7.2}$$

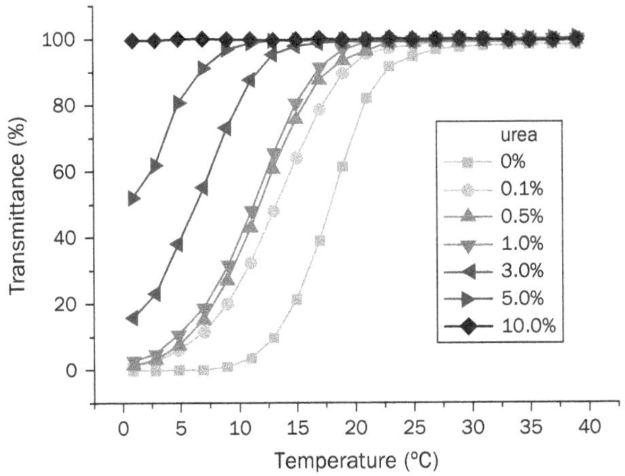

Figure 7.6 The effect of increasing hydrophobic content (urea) on the VPTT of poly(acrylic acid-co-acrylamide) hydrogels.[5]

where D represents the size of the hydrogel at any temperature and D_0 represents the size of the hydrogel at the base temperature.

The VPTT from the DLS measurements shows a close resemblance to the inflection point on the differential scanning calorimetry (DSC, another technique often employed to determine phase transition temperatures) curve, which is shown in Fig. 7.7.[1] Thus, there are several ways of measuring the VPTT of the system using different characterization techniques that show a variation in any physical property of the hydrogel system with the instrument parameter.

Even though researchers have used the inflection point of the curves as an effective way to determine VPTT, it lacks preciseness when deciding where the inflection point is. Therefore, our group developed a conceptual model that determines the VPTT of stimuli-responsive systems with greater preciseness. This model is schematically shown in Fig. 7.8. In a typical example, the variation of hydrogel sizes can be measured using DLS. The hydrogels collapse as the temperature increases and it equilibrates after a particular temperature. The curve follows a sigmoidal pattern; therefore a five-parameter Gaussian curve fit can be used (Fig. 7.8B). The fit and the area under the curve is analyzed using a MATLAB® code. The phase transition curve is divided into two regions, such that the region below the VPTT has a higher ratio of swelled hydrogel states as compared to collapsed hydrogel states, whereas the region above the VPTT has a higher ratio of collapsed hydrogel states when compared with swelled hydrogel states. The areas under the curve for the two regions are calculated and equated to each other (Fig. 7.8C). The point or the temperature value where the two areas become equal, that is, the ratio of swelled and collapsed hydrogel states is equal, is given as the VPTT of the system (Fig. 7.8D).

The pNIPAm-AAc hydrogels synthesized by precipitation polymerization have a VPTT around 39°C. The method is suitable for characterizing and extracting information of any hydrogel system that is undergoing swelling or collapse under the influence of a stimulus.

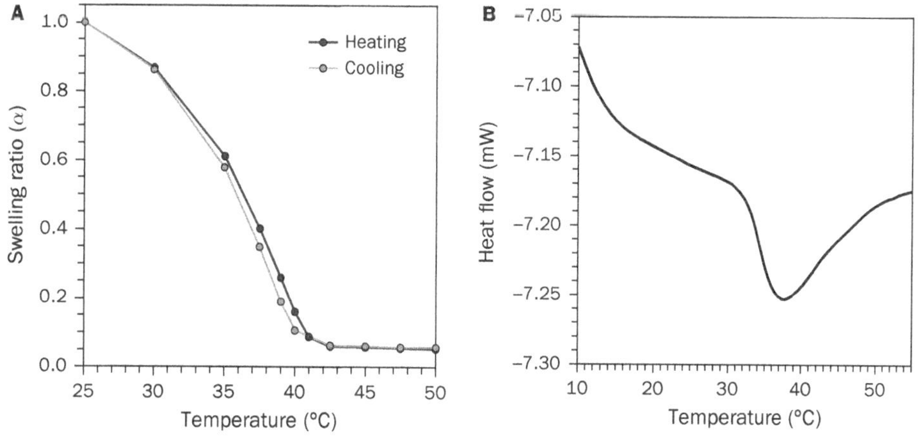

FIGURE 7.7 (A) The collapse of pNIPAm-AAc-based hydrogels above the VPTT as shown by DLS measurements.[1] (B) DSC curve of pNIPAm-AAc-based hydrogels showing the influx point close to the VPTT as obtained by DLS measurements. (Reproduced from Bandyopadhyay, S.; Andersen, M. K.; Alvi, M. A. A.; Sharma, A.; Raju, R.; McDonagh, B. H.; Glomm, W. R., Incorporation of Fe@ Au nanoparticles into multiresponsive pNIPAM-AAc colloidal gels modulates drug uptake and release. *Colloid and Polymer Science.* **2016**, *294*(12), 1929–1942.)

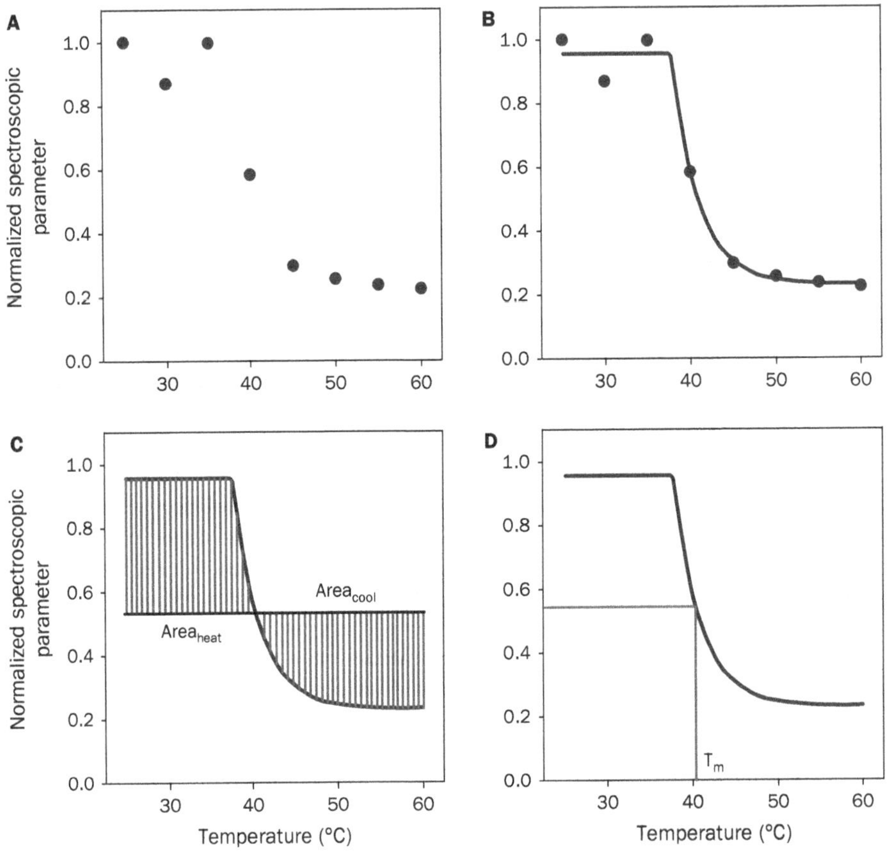

FIGURE 7.8 A schematic showing the different steps involved in calculating the VPTT of stimuli-responsive polymeric NP systems.[6]

Effect of pH So far, we have discussed about systems that respond to temperature alone. Now, we will discuss how hydrogels respond to pH. Hydrogels prepared by cross-linking polyelectrolytes undergo a conformational change when the pH of the solution is altered. These polymers have an abundance of functional groups such as carboxyl, amine, sulfonic, phosphate, and acryl groups that can change charges as a function of pH. The swell-collapse of the pH-sensitive hydrogels can be understood using the thermodynamic model proposed by Siegel to find the total free energy ΔG.[7] The major contributors to this are the hydrogel-solvent system, hydrogel-solvent mixing (ΔG_{hsm}), deformation of the polymer (ΔG_{pd}), and osmotic pressure (ΔG_{op}) of ions in the hydrogel. The hydrogel–solvent mixing depends on the hydrophobic–hydrophilic interactions, and the deformation of polymer and osmotic pressure depends on the mobility of ions.[8]

$$\Delta G = \Delta G_{hsm} + \Delta G_{pd} + \Delta G_{op} \tag{7.3}$$

Take the example of poly(acrylic acid) (pAAc) hydrogels at high pH, in which case the carboxylic groups lose protons by dissociation. This will cause Coulombic repulsion between carboxylate ions on the surface of the hydrogels and provide space for water to enter. Dissociated AAc is more hydrophilic than the non-dissociated ones and therefore decreases the free energy of mixing. However, an increase in hydrophilicity will cause swelling of the hydrogel. The interaction between associated and dissociated pAAc parts acts as a physical cross-linking. The extent of this depends on the concentration of AAc monomers or the degree of dissociation of the pAAc, which contributes to the deformation of the polymer. At this pH, the concentration of OH^- and Na^+ ions (from NaOH added to increase the pH) is more in bulk than within the hydrogel. The osmotic pressure inside the hydrogel prevents the flow of ions into pAAc. When the pH is high enough to overcome the osmotic pressure, the concentration gradient forces the counterions to diffuse into the hydrogels and cause a counter ion shielding effect that leads to collapse. When the osmotic pressure barrier is crossed, the free energy of hydrogel-solvent mixing increases along with de-cross-linking of associated and dissociated pAAc parts leading to further swelling. Therefore, when pH increases, the overall free energy of the system increases leading to swelling of pAAc.

Response to Ionic Strength Ionic strength (I) of a solution is defined in terms of the sum of products of concentration of ions and square of the charges of the respective ions. It is helpful to understand the interaction between different ions in an electrolyte solution. The formula of ionic strength is:

$$I = \frac{1}{2}\sum C_i Z_i^{\,2} \tag{7.4}$$

where C_i and Z_i are the concentration and charge of the ith ion, respectively.

Hydrogels that respond to pH usually respond to the ionic strength of the solution. However, for instance, hydrogels that respond to different stimuli can be made sensitive to ions by using an ionic co-monomer during synthesis.

In general, the swelling of hydrogels in water has swelling pressure contributions from polymer–solvent mixing, and elastic contributions from the deformed polymer and from the bound and mobile ions. When ionic strength is increased, these hydrogels continue to collapse. At lower ionic strength, the concentration of ions in the hydrogels usually known as *bound charges/ions* that arise from its functional group is more than the concentration of ions in the solutions. The high pressure inside the hydrogel caused by these bound ions causes the hydrogel to swell. When the ionic strength is increased by the addition of salt, the hydrogel continues to collapse until the concentration of ions within the hydrogel equals the concentration of ions in the solution. This collapse phenomenon can also be explained by Coulombic interactions. At lower ionic strength, the functional groups repel each other causing the hydrogel to swell to minimize the Coulombic repulsions. At higher ionic strength, mobile counter ions provide a shielding effect, allowing the functional groups to come closer to one another causing the hydrogels to collapse. Most of the study in this field has been with monovalent ions (using NaCl, LiCl, etc.). However, Horkay et al.[9] studied the effect of monovalent and divalent ions on swelling of pAAC hydrogels. Both monovalent and divalent ions showed a collapse at higher ionic strength. However, the collapse was faster when divalent ions were used, as one divalent cation replaces two monovalent ions.

Since the hydrogels that respond to pH also respond to ionic strength, the response to ionic strength depends on pH and vice versa. Consider a hydrogel with a polymer containing carboxylic group such as pAAc, at neutral pH. Ions are introduced into the system by addition of NaCl. H^+ from the acid will then be replaced by Na^+ ions. This increase in dissociation will increase the osmotic pressure inside the hydrogel. When all the acrylic acid groups are dissociated, an additional increase in salt concentration will lead to the collapse of the hydrogel. However, at basic pH, pAAc hydrogel continues to collapse because the acid groups are already dissociated and an increase in salt concentration will directly contribute to the collapse. When the ionic strength is varied for hydrogels that respond to pH, the maximum swelling decreases. For example, hydrogel containing (diethylamino)ethyl methacrylate showed lesser response to pH change at higher ionic strength; however, at lower ionic strength the hydrogel collapsed easily.[10] Such effects are common when one hydrogel can respond to two different stimuli, and when such hydrogels are used with other stimuli-responsive hydrogels, a combined effect can be observed. This sets the stage to discuss combined effect of stimuli on the behavior of hydrogels.

Combined Effect of Temperature and pH The preparation of multifunctional hydrogels with the ability to respond to temperature or pH has been discussed before. Although pNIPAm-based hydrogels are one of the widely studied polymeric systems, they are not biodegradable and have a slow swell collapse, and this forms a hydrophobic skin surrounding it. These are highly non-desirable for applications such as drug delivery and can be overcome by incorporating hydrophilic blocks into the hydrogel matrix. Using AAc imbibes pH responsiveness along with the easy flow of water thereby forcing an efficient collapse and reducing the possible formation of hydrophobic skin. The presence of pNIPAm is advantageous for pH response, as the hydrophobic–hydrophilic interaction (one of the contributing factors to total free energy) is profound due to the availability of hydrophobic isopropyl groups. Such pNIPAm/AAc cross-linked hydrogels undergo conformational changes when the temperature is above the VPTT and/or when its pH is below the pK_a of AAc. Figure 7.9 shows the response of such hydrogels when either pH or temperature is changed or when both the stimuli are used.

Response to Light Polymers that respond to light by undergoing chemical or physical changes have been widely used in several fields. Such polymers, when used to synthesize hydrogels, can be advantageous over other stimuli-sensitive polymers due to their

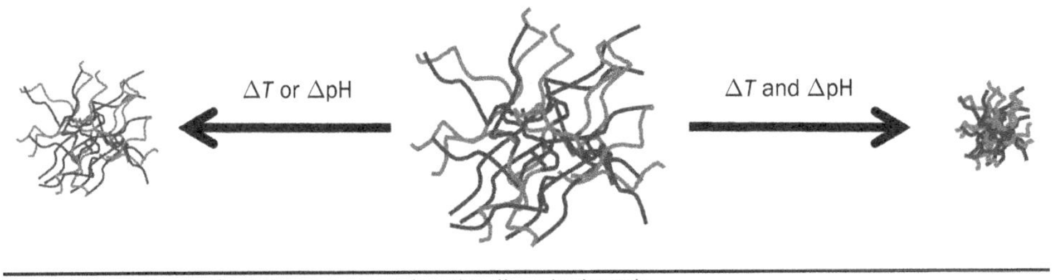

Figure 7.9 Effect of single and multiple stimuli on hydrogels.

ability to respond over a wide wavelength spectrum. The response to photo stimulus varies with the type of photoreactive moiety that is used. These photoreactive moieties capture the optical signal and convert the photo signal into a chemical signal. Such photo reactions may proceed via:

1. Isomerization
2. Cleavage
3. Dimerization

In photo-isomerization, photo-reactive groups will undergo changes in their structures upon irradiation. There can be structural changes between *cis* and *trans* or between an open and closed structure. The wavelength used to induce such changes depends on the type of the functional group that is used. Some of the commonly used photoreactive groups that cause isomerization are azobenzene, spiropyran and spironaphthoxazine groups. For example, NIPAm-containing spirobenzopyran moieties cause swell-collapse of the hydrogel by undergoing an isomerization when irradiated with light,[11] as shown in Fig. 7.10.

Some of the functional groups that induce cleavage upon irradiation are *o*-nitrobenzyl ether, 4,5-dimethoxy-2-nitrobenzyl, and azosulfonate, among other moieties. NIPAm copolymerized with phenylmethane derivative gives a dark green color when irradiated with UV light. This is due to the generation of triphenyl methyl cations, which will affect the osmotic pressure of such hydrogels and affect the swell-collapse properties. A measurable change in the response to temperature in the presence of UV light was also reported.[12]

Coumarin is one of the commonly used photoreactive groups that helps in controlling the size of the hydrogel via dimerization. It acts as a reversible photo cross-linker in the hydrogel. The cross-linking density, which determines the size, is controlled by adjusting the UV radiation used.[13] A few more examples of chemicals that can undergo photodimerization are cinnamylidene acetate and anthracene.

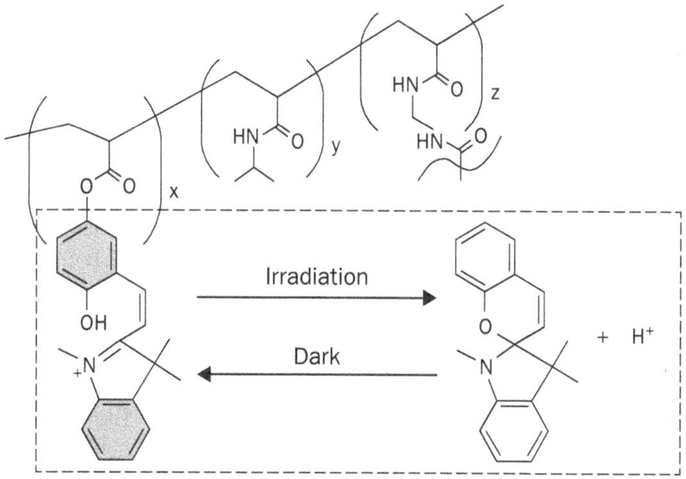

FIGURE 7.10 Response of pNIPAm-spirobenzopyran to light.[11]

As mentioned in the preceding discussion, responses to more than one stimuli can result in multifunctional hydrogels possessing a wide array of interesting properties. Alternatively, the combination of hydrogels with other NPs or functional molecules through different modes of functionalization can also impart multifunctionality.

Functionalized Hydrogels

In this section, you will learn how functionalization will add new stimuli response such as response to a magnetic field or alter the response to a stimulus such as a temperature change.

Response to Magnetic Field Hydrogels that respond to a magnetic field are swollen with a fine, colloidal network of magnetic nanoparticles (NPs). When a magnetic field is applied, the hydrogel changes its shape and size due to the deformation of the polymeric matrix as the NPs move toward the field. Superparamagnetic NPs such as Fe_2O_3 are magnetized in the presence of a magnetic field. These are ideally chosen to synthesize such hydrogels. When a non-homogenous magnetic field is applied, the NPs move toward the stronger field and force the hydrogel to change its conformation because of the cross-linking between the polymers. The shape and response can be fine-tuned by adjusting the strength or position of the magnetic field. The advantages of these hydrogels are their quick response time, reversibility, and high mechanical strength.

The swelling of these hydrogels depends on the amount of NPs used as they assist in physical cross-linking, as discussed in Sec. 7.2.1. An increase in the concentration of NPs results in a decrease in the swelling of the hydrogels.[14] In the presence of a magnetic field, the NPs move toward the field and force the polymer to move along with it, thereby forcing a collapse of the hydrogel. The movement of NPs can also force a decrease in pore size of the hydrogel. Such properties are highly desirable in applications such as drug delivery.

NIPAm-based hydrogels functionalized with magnetic NPs show swell-collapse in the presence of a high-frequency alternating magnetic field. Owing to magnetic hysteresis, an alternating field in such cases causes local heating and a subsequent increase in the local temperature in the vicinity of the NPs. This is then transferred to the hydrogel matrix, which results in collapse of the hydrogel. Such properties open up applications for hyperthermia in cancer treatment, where the resultant heat may lead to cell apoptosis (death).

Response to Temperature You have learned the synthesis routine to functionalize hydrogels with NPs such as Au NPs. As discussed earlier, these will interact with the hydrogels and these interactions can cause a change in the properties of the hydrogels. When pNIPAm-AAc hydrogels are functionalized with NPs, its VCE and VPTT is affected and this depends on the type of functionalization as well.

For example, pNIPAm-AAc hydrogels coated using *in situ* method with NPs shows an increase in its VCE. Further, VPTT increased from 37°C to 38.3°C ± 0.5°C. However, when the same hydrogels are functionalized using post-synthesis technique, the hydrogels swell when the temperature is increased. These functionalized hydrogels behave like NPs at high temperature, which shows an increase in size due to aggregation. The corresponding VPTT of these is 38.9°C ± 0.8°C. The change in the behavior of hydrogels is because of the NPs which can act as cross-linker and the location of these NPs. It has been hypothesized that in the *in situ* method, the NPs are present in the hydrogel matrix, whereas in the coating technique the NPs surround the hydrogels. These lead to changes in their behavior with respect to increase in temperature.[1]

Response to Biological Stimuli Hydrogels can be functionalized to react with certain biological stimuli and can induce a change in its structure. Depending on the functionalized group and the polymer used, the hydrogel can respond by swelling, collapsing, changing cross-linking density, or self-assembling. Some of the biological stimuli are glucose, different types of enzymes, proteins, etc.

Glucose-sensitive hydrogels contain groups such as glucose oxidase or galactose-binding protein; these can detect the presence of glucose and undergo chemical reactions. For example, glucose oxidase forms gluconic acid in the presence of glucose. This underlying reaction is used to synthesize methacrylic acid-based hydrogels that respond to glucose.[14] These hydrogels, in the presence of a high quantity of glucose, will form gluconic acid. This results in increase in pH and triggers a collapse of the hydrogel. Similar techniques are used in developing hydrogels that respond to enzymes, where albumin is squeezed out of the polyethylene glycol (PEG)-based hydrogels when in contact with a specific enzyme (thermolysin).[15]

A few other biological stimuli are peptides and DNA, in which polyacrylamide (pAAm) hydrogels can be used. pAAm can be functionalized with single-stranded DNA (ssDNA) and has been shown to undergo a change in size when in contact with complementary ssDNA. This happens due to alteration of cross-linking of such hydrogels leading to swelling or collapse of the hydrogel.[16]

The example discussed in Sec. 7.1 is also an example of the enzyme-responsive hydrogel, in which acrylamide-based hydrogels are functionalized with peptides. When in contact with the specific protein, they swell up and change color. The change in color of hydrogels, in general, is an easier way to detect response as it is visible to the naked eye. For systems that cannot be seen, for example, hydrogels in the blood system, or to detect the hydrogels without inducing a collapse, other spectroscopic methods can be used that sense the optical properties of the hydrogels.

7.2.3 Optical Properties

A visible change in color as a response can be introduced by functionalization. However, hydrogels that are not functionalized can show some changes in their color as a response to a stimulus. When NIPAm hydrogel collapses at high temperature, a visual observation can be made as the solution turns more turbid. This can be observed even when NIPAm is used along with other co-monomers, such as in the case of pNIPAm/AAc hydrogels. However, hydrogels functionalized with Au NPs can be tracked using a UV-Vis spectrometer. The Au NPs, when exposed to certain wavelengths, can lead to very high absorbance values, which can be tracked using UV-Vis. Hold on to your questions about how Au NPs can lead to such a high values; it will be explained in Chap. 8. Such properties are used in applications for targeted drug delivery, where the hydrogels can deliver the drug and Au NPs can be used to track their intravenous movements.

Water serves as an excellent transport medium for dissolved species to diffuse in and out of the hydrogel, thereby making the hydrogels ideal candidates for drug delivery. In addition, it acts as a surface energy bridge between the hydrogel and the external environment, thus measuring and controlling the water content in hydrogels are important for its applications.

7.2.4 Water Content

High water content is undoubtedly the most sought-after characteristic of hydrogels, which makes them appropriate for biomedical applications. The amount of water incorporated by the hydrogel is described using equilibrium water content (*EWC*), which is defined by Eq. (7.5).

$$EWC = \left(\frac{W_{wh}}{W_w}\right) \times 100\% \tag{7.5}$$

where W_{wh} is the weight of water in the hydrogel and W_h is the total weight of the hydrogel.

Water can either be bonded to the hydrogel via hydrogen bonding or be free, having higher mobility. These are, at times, referred to as freezing and non-freezing water, respectively, and the binding can be studied using techniques such as differential scanning calorimetry (DSC) and nuclear magnetic resonance (NMR). The properties of the hydrogels are dependent on the ratio of freezing and non-freezing water.

Hydrogels are hard and brittle in a dry state; however, they become soft and rubber-like in the swollen state after absorbing water. Having mostly water in the matrix also means that hydrogels retain refractive index closer to that of water, thus rendering them quite transparent. The polymeric chains of the hydrogel favor solubility in the solvent, whereas the cross-linker prevents these chains to be lost to the solvent thereby keeping the hydrogel matrix intact. Therefore, there exists a balance between the degree of swelling of the hydrogel and the elastic forces due to the cross-linker. As hydrogels primarily comprise water, they show high permeability to dissolved species diffusing in and out of the hydrogel matrix. Dissolved gases, for instance, oxygen, can permeate through the hydrogel moieties. All these attributes taken together make them ideal candidates for applications such as contact lenses, where the lens needs to be flexible, transparent, permeable to oxygen, and have high water content and good refractive index. The properties or behavior of hydrogels are highly dependent on the environment in which they are being used. Therefore, studying the mechanical properties of the hydrogels becomes crucial and can help improve the performance of the hydrogels.

7.2.5 Mechanical Properties

Hydrogels, unlike polymers, are water-swollen and are difficult to be tested for mechanical properties. In general, hydrogels possess weak mechanical strength. Several techniques have been developed to study specific properties such as elastic modulus, Young's modulus, creep behavior, adhesion, etc. In this section, the focus will be on altering and controlling the mechanical properties of the hydrogels. Some of the parameters that can affect the mechanical properties are monomer concentration, cross-linking density, measurement conditions, and degree of swelling.

In the case of hydrogels, the effect of co-monomer depends on the type of co-monomer being used. Mechanical strength can be increased by increasing the concentration of the monomer that is physically stronger. For example, if acrylate is replaced by methacrylate, the overall strength can be increased.[17] However, if *N*-vinyl-2-pyrrolidone is introduced as co-monomer, the Young's modulus of the hydrogels decreases.[18]

The mechanical strength of hydrogels is mostly due to its cross-linking; hence, the overall mechanical strength of hydrogels can be altered by changing the cross-linking density. In NIPAm hydrogels, Young's modulus increases when the amount of cross-linker used is increased. A similar trend can also be seen in AAc-based hydrogels. For more details you can refer to the excellent work by Matzelle et al.[19]

When performing these measurements, the environment and the state of the hydrogels must be consistent as they will affect the measurement. Hydrogels in the swollen state will have better elastic properties as they contain more water; however, collapsed hydrogels will show a higher mechanical strength due to the increased polymer–polymer interactions. Young's moduli of NIPAm increases with increase in temperature as it responds to the change in temperature by collapsing. However, AAc-based hydrogels, which do not respond to changes in temperature like NIPAm, show a slight decrease in Young's moduli. These results iterate the importance of test conditions when mechanical strength is to be determined. Another method to increase the mechanical strength of the hydrogels is by functionalizing them with inorganic NPs. Different NPs such as clay, Au, and Fe have been used to increase the mechanical strength. These hydrogels are soft, flexible, but possess very high tensile strength. A systematic study of functionalization of NIPAm hydrogels with Au NPs showed an increase in the viscoelastic property when compared with NIPAm hydrogels.[20] An in-depth understanding of the mechanical properties of the hydrogels will help in real life applications where the mechanical properties are important, like wound healing. On the other hand, it can also help assessing the behavior of hydrogels in a scenario where it is subjected to mechanical stress.

You may have already noticed by now that there exist in principle a multitude of methods that may be employed to functionalize hydrogels or more generally speaking polymer-based NPs. This chapter has given you an overview of some of the most relevant techniques. Applications of hydrogels can range from drug delivery, to imaging, to therapy, to sensors, to other environmental applications, and so on. Hence, a detailed understanding of their properties will help to create a library of tailor-made hydrogels. However, the question is how will you know that you have achieved proper functionalization of your hydrogels? This and many other related questions about characterization of NPs will be answered in the next chapter.

7.3 Study Questions

7.3.1 Concept-Based Questions

7.1 Explain the importance of polymer functionalization.

7.2 Distinguish between *in situ* and post-synthesis functionalization strategies using suitable examples.

7.3 What do you mean by LCST for homopolymers? What is the difference between LCST and VPTT?

7.4 What are the advantages of stimuli-responsive hydrogels over conventional hydrogels?

7.5 Why is it important to know the water content of hydrogels and how is the water content calculated?

7.6 Based on Siegel's theory, explain the factors that are responsible for collapse of pH-sensitive hydrogels at acidic pH values.

7.7 A batch of pNIPAm-AAc hydrogel was synthesized at Ugelstad Lab and loaded with a model drug using breathing-in technique. The pH of the aqueous solution of the hydrogel loaded with drug was measured to be 5. The release of the drug from the hydrogel matrix was performed at pH 3 at room temperature. The drug release after 36 hours was found to be 25%. Can you comment on the reason for slower release of the drug and what could be done to enhance the release?

7.3.2 Research-Based Question

7.8 Hybrid nanomaterials comprising Au NPs coated with pNIPAm nanogels were synthesized using two different methods, namely, coating and *in-situ* methods. In the former method, the nanogels and NPs were synthesized separately and mixed. In the latter method, pre-synthesized NPs were added during the formation of nanogels. The gold concentration used for the synthesis was kept constant at 0.15 mg/mL in both cases. Temperature response of the nanomaterials was analyzed using DLS by heating the sample from 25°C to 55°C and cooling it back from 55°C to 25°C at intervals of 5°C. The size of the particles was measured for every change in temperature and is reported in the table below.

Heating			**Cooling**		
	Coating	***In Situ***		**Coating**	***In Situ***
Temperature (°C)	**Size (nm)**	**Size (nm)**	**Temperature (°C)**	**Size (nm)**	**Size (nm)**
25	89	305	55	93	141
30	84	296	50	98	132
35	87	230	45	101	129
40	94	129	40	100	125
45	107	131	35	78	205
50	103	136	30	80	285
55	93	141	25	78	295

The data were fitted using sigmoidal five-parameter (depicted by the following equation) curves, and the parameter values for both the nanomaterials are as follows.

$$y = y_0 + \frac{a}{\left[1 + e^{-\left(\frac{x - x_0}{b}\right)}\right]^c}$$

	Coating		***In Situ***	
Parameters	**Heating**	**Cooling**	**Heating**	**Cooling**
a	0.5216	0.9635	0.9233	0.9218
b	0.2897	0.0006	−2.2221	−2.1493
c	0.1888	1.9967	157.5914	51.4028
x_0	41.1209	38.2433	46.2713	42.8213
y_0	0.9211	1.0447	0.0866	0.0900
R^2 **value**	0.6743	0.7272	0.9997	0.9981

a. As per the data presented above, the size of the hydrogel measured at the same temperature points during heating and cooling cycles is observed to be different. Discuss the reversibility of both the systems using concepts discussed in the research work by Bandyopadhyay et al.[1]

b. Comment on the validity of the data curve-fits used for the nanomaterials. Would you recommend or suggest a better curve-fit model for the data?

7.4 References

1. Bandyopadhyay, S.; Andersen, M. K.; Alvi, M. A. A.; Sharma, A.; Raju, R.; McDonagh, B. H.; Glomm, W. R., Incorporation of Fe@Au nanoparticles into multiresponsive pNIPAM-AAc colloidal gels modulates drug uptake and release. *Colloid and Polymer Science.* **2016,** *294*(12), 1929–1942.
2. Bandyopadhyay, S., *Smart and Multifunctional Core-Shell Nanoparticles (NPs) for Drug Delivery*. Doctoral thesis. **2016**.
3. MacConaghy, K. I.; Chadly, D. M.; Stoykovich, M. P.; Kaar, J. L., Optically diffracting hydrogels for screening kinase activity in vitro and in cell lysate: impact of material and solution properties. *Analytical Chemistry.* **2015,** *87*(6), 3467–3475.
4. Sharma, A. Different shaped hybrid nanocarriers for theranostic applications. NTNU, 2016.
5. Yang, M.; Liu, C.; Li, Z.; Gao, G.; Liu, F., Temperature-responsive properties of poly (acrylic acid-co-acrylamide) hydrophobic association hydrogels with high mechanical strength. *Macromolecules.* **2010,** *43*(24), 10645–10651.
6. Bandyopadhyay, S.; Sharma, A.; Alvi, M. A. A.; Raju, R.; Glomm, W. R., A robust method to calculate the volume phase transition temperature (VPTT) for hydrogels and hybrids. *RSC Advances.* **2017,** *7*(84), 53192–53202.
7. Siegel, R., *Pulsed and Self-Regulated Drug Delivery ed J Kost*. Boca Raton, FL: CRC Press, 1990.
8. Ni, H.; Kawaguchi, H.; Endo, T., Preparation of pH-sensitive hydrogel microspheres of poly (acrylamide-co-methacrylic acid) with sharp pH–volume transition. *Colloid and Polymer Science.* **2007,** *285*(7), 819–826.
9. Horkay, F.; Tasaki, I.; Basser, P. J., Effect of monovalent–divalent cation exchange on the swelling of polyacrylate hydrogels in physiological salt solutions. *Biomacromolecules.* **2001,** *2*(1), 195–199.
10. Feil, H.; Bae, Y.; Feijen, J.; Kim, S. W., Mutual influence of pH and temperature on the swelling of ionizable and thermosensitive hydrogels. *Macromolecules* **1992,** *25*(20), 5528–5530.
11. Sumaru, K.; Ohi, K.; Takagi, T.; Kanamori, T.; Shinbo, T., Photoresponsive properties of poly(*N*-isopropylacrylamide) hydrogel partly modified with spirobenzopyran. *Langmuir.* **2006,** *22*(9), 4353–4356.
12. Mamada, A.; Tanaka, T.; Kungwatchakun, D.; Irie, M., Photoinduced phase transition of gels. *Macromolecules.* **1990,** *23*(5), 1517–1519.
13. He, J.; Tong, X.; Zhao, Y., Photoresponsive nanogels based on photocontrollable cross-links. *Macromolecules.* **2009,** *42*(13), 4845–4852.
14. Wnek, G. E.; Bowlin, G. L., *Encyclopedia of Biomaterials and Biomedical Engineering*. New York, NY: Informa Healthcare USA, 2008.

15. Thornton, P. D.; Mart, R. J.; Webb, S. J.; Ulijn, R. V., Enzyme-responsive hydrogel particles for the controlled release of proteins: designing peptide actuators to match payload. *Soft Matter.* **2008,** *4*(4), 821–827.
16. Murakami, Y.; Maeda, M., DNA-responsive hydrogels that can shrink or swell. *Biomacromolecules.* **2005,** *6*(6), 2927–2929.
17. Anseth, K. S.; Bowman, C. N.; Brannon-Peppas, L., Mechanical properties of hydrogels and their experimental determination. *Biomaterials.* **1996,** *17*(17), 1647–1657.
18. Davis, T. P.; Huglin, M. B., Studies on copolymeric hydrogels of *N*-vinyl-2-pyrrolidone with 2-hydroxyethyl methacrylate. *Macromolecules.* **1989,** *22*(6), 2824–2829.
19. Matzelle, T.; Geuskens, G.; Kruse, N., Elastic properties of poly (*N*-isopropylacrylamide) and poly (acrylamide) hydrogels studied by scanning force microscopy. *Macromolecules.* **2003,** *36*(8), 2926–2931.
20. Marcelo, G.; López-González, M.; Mendicuti, F.; Tarazona, M. P.; Valiente, M., Poly (*N*-isopropylacrylamide)/gold hybrid hydrogels prepared by catechol redox chemistry. Characterization and smart tunable catalytic activity. *Macromolecules.* **2014,** *47*(17), 6028–6036.

CHAPTER 8

Characterization of Metal Nanoparticles

You might have experienced that there exists in reality a large number of methods to synthesize and functionalize nanoparticles (NPs)—both metal and polymer based. The rapid development of new nanomaterials for a myriad of applications has also advanced the field of characterization techniques. Let us imagine that you have synthesized some colorful Au NPs after working hard for several days in the lab. Although it is known that Au NPs may find wide applications ranging from catalysis to biomedicine, your NPs will not make it that far unless you know for instance how big they are, are they of the same size, what is their surface charge, do they fluoresce, how do they behave in different solvents, how do they interact with biomolecules, and so on. Characterization techniques are means of answering some of these questions. They serve the purpose of assigning characteristic attributes to the nanomaterials in the same way as barcodes help classify and identify various products and give them their uniqueness. Another way to understand the importance of characterization techniques is by imagining how we use our five senses to see, hear, touch, smell, or taste things around us and evaluate their usefulness; these attributes of the nanomaterials (revealed using various characterization techniques) help them to stand out from one another and prove their usefulness.

With this plot in mind, the present chapter covers some of the characterization techniques that are widely used for metal NPs. While some of them require suspensions of the metal NPs in a solvent of choice, others can be done using powder samples. The distinction between the solution-based or dry-state characterization is generally made based on the final applications of the resultant nanomaterials. For the above example of Au NPs, if the final aim is to use them as novel materials for bioimaging, their solution properties will be more important to measure. However, if they are designed for use in catalysis, some of the dry-state characterization techniques will be deemed necessary. Therefore, the choice of characterization techniques not only depends on the synthesis method applied but also varies with the intended application of the nanomaterial. Thereby, these form a link between the fabrication and application of nanomaterials.

This chapter first discusses the optical properties of noble metal NPs, which makes their characterization using various spectroscopic methods a preferred alternative. Thereafter, different characterization techniques, both in solution as well as dry states, are discussed—the principle of operation, pros and cons, and specific examples of nanomaterials that can be studied using these techniques. The discussion of the characterization techniques is tailored to metal NPs; however, some of the same characterization techniques can be easily used for polymer-based NPs discussed in Chaps. 6 and 7.

8.1 Optical Properties

Although the concept of nanoscience and nanotechnology surfaced in the mid-1990s, metal NPs in the form of colloidal gold was in use as early as the 4th century. The manifestation of the same can be seen in the fresco art and in colored glass; for example, colloidal gold was used to impart red color while colloidal silver was used for yellow. The earliest scientific observation of the optical properties of NPs was made by Michael Faraday; his interest in the interaction of matter with light made him produce thin sheets of gold. These thin sheets were produced by washing thicker golden sheet with chemicals. He obtained a ruby colored liquid after washing the gold sheets and suggested that the color was produced by tiny gold particles that were difficult to observe with the tools available at that time but were of enough size to scatter light.

8.1.1 Scattering and Absorption

A number of scientists, along with and long after Faraday, were fascinated with the interaction of light or electromagnetic waves with matter. Depending on the type of material and its interaction with light, a number of phenomena (Fig. 8.1) can occur.

Of these interactions, scattering, absorption, and transmission are of interest to us as they help in characterizing the NPs. These phenomena form the basis in characterization tools such as UV-visible (UV-Vis) spectrometer, dynamic light scattering, electron microscope, etc.

The interaction of light with a roughened surface can cause it to deflect in all directions with a loss in coherence and phase. This phenomenon is called scattering and

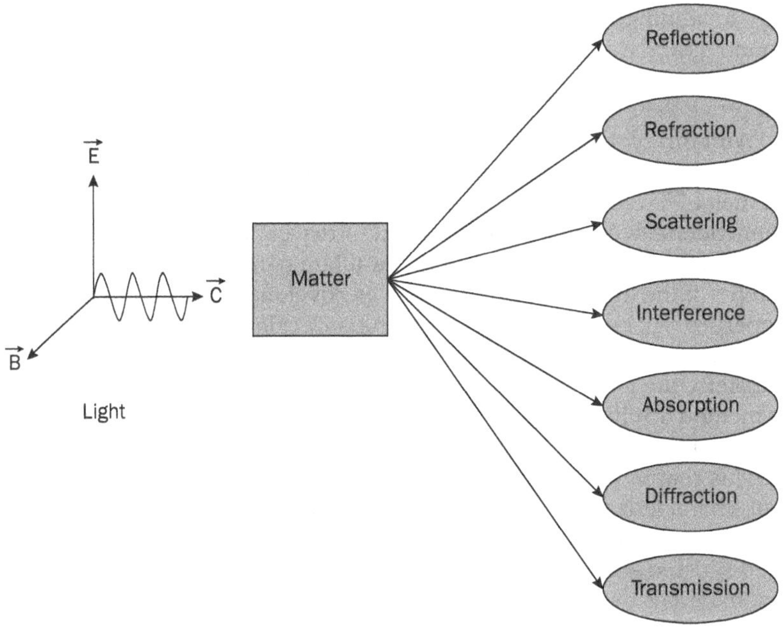

FIGURE 8.1 Interaction of light (an electromagnetic wave) with matter.

it depends on the size and the surface of the material interacting with light and the wavelength of light. Depending on the conservation of kinetic energy, scattering can be classified as elastic scattering (where the kinetic energy is conserved) and inelastic scattering. In elastic scattering, the energy is conserved, but the direction of the incident wave is altered. Based on the incident wavelength and size of the particle interacting with the wave, elastic scattering can further be classified into Rayleigh scattering and Mie scattering. When the characteristic length or diameter of the particle is around 0.1 times the incident wavelength (i.e., 0.1λ), Rayleigh scattering method is applicable, and for bigger particles, Mie scattering method is used. A common example of Rayleigh scattering is the blue sky that we see. Rayleigh scattering intensity scales inversely with the fourth power of the wavelength. Since the wavelength of blue light is half that of red light, it scatters 16 times more than red light. A typical example of Mie scattering is the white color of mist or fog produced by scattering of light by larger droplets of water.

In inelastic scattering, the energy of the photon incident on a molecule is lost after interacting with it, a typical example of which is Raman scattering. When an atom or a molecule is irradiated with photons, some of the photons are absorbed and re-emitted at a lower frequency or energy—the phenomenon called *Raman scattering.* The lost energy is used by the atom or molecule to reach higher energy levels. When emitted photons have a lower energy than an absorbed photon, it is called Stokes–Raman scattering. However, if the energy of the emitted photon is higher, it is called Anti-Stokes–Raman scattering.

The process of absorbing the energy of light by NPs is called *absorption.* Absorption is dependent on the NPs and the interacting electromagnetic wave. The absorbed energy helps the atoms or molecules of the NPs to reach higher energy states. The definite set of energy states in which the atoms can exist is called absorption band. This transition between energy states can be electronic transitions, vibrational transitions, or rotational transitions. Techniques such as UV-Vis or luminescence spectroscopy are based on electronic transitions occurring in the NPs while interacting with an electromagnetic field. However, infrared and Raman spectroscopy are based on the vibrational and rotational transitions occurring in the NPs. The ability of an electromagnetic wave to pass through a material is called *transmission.* When an electromagnetic wave interacts with metal NPs, one or several of the following phenomena can happen simultaneously—absorption, scattering, or transmission.

This chapter deals with characterization of metal NPs; therefore, the concepts succeeding the example below are discussed for metal NPs in water unless stated otherwise. The surface of such NPs has a number of atoms with electrons on the surface that interact with one another. When the surface is irradiated with an electromagnetic wave, these electrons are excited to the conduction band. This displacement of electrons from nuclei of atoms creates surface charge separations and leaves behind uncompensated positive charges (positive kernel) on the surface. The electrons in turn experience a restoring force due to Coulombic attraction by the positive kernels on the surface. The collective plasma oscillation of the electron cloud between surface and dielectric medium (water in this case) is called *surface plasmon.* The plasmon is defined as one quantum of plasmon oscillation. These oscillations are in phase with the wave and when their frequency matches, it leads to a resonance. This is called *surface plasmon resonance* (SPR).

In case of a conductor or metal strip, the surface electrons are excited leading to oscillations at the interface between the metal and the surrounding dielectric medium. These oscillations lead to the formation of an electric field that decays into the surrounding medium but can propagate perpendicularly along the surface of the metal. These oscillating electric dipoles that propagate along the metal surface in a waveguide-like fashion until released at some distance from their origin are called *surface plasmon polaritons* (SPP). However, a non-propagating excitation of conduction electrons within metal NPs is called *localized surface plasmon* (LSP). For a detailed understanding of these concepts, you are urged to refer to the chapter *Nano Optics* in the book *Introduction to Nanoscience and Nanotechnology* by Hornyak et al.[1]

8.1.2 Localized Surface Plasmon Resonance

Consider an example where noble metal NPs such as spherical Au or Ag NPs interact with light. When the radius of these NPs is of the order of less than 0.1 times the wavelength of the incident light, the NPs experience Rayleigh scattering. These NPs experience LSP, i.e., the electron clouds on the surface experience a collective excitation upon interaction with light. The restoring force by the positive nuclei results in an oscillation of the electrons. When the frequency of this oscillation matches with the frequency of the incident light, a resonance is observed. This phenomenon is called *localized surface plasmon resonance* (LSPR), which is depicted in Fig. 8.2. The absorbance of the NPs is maximum at this frequency and can be tracked by UV-Vis spectrometer as an absorbance maximum (or a maximum peak) at a particular wavelength.[2]

In LSPR, the oscillating electric dipoles are within the NPs. They do not propagate and decay exponentially close to the surface of the NPs. However, in SPP, the oscillating electric dipoles propagate along the surface of the metal. They also decay exponentially but over a space larger than in LSPR. The resonance of Au and Ag NPs occur in the visible part of the electromagnetic spectrum and the LSPR gives these NPs their bright colors.

The ability to control the size of metal NPs is detailed in Chap. 5, along with the ability to absorb or scatter light; metal NPs are interesting candidates for a number of applications that utilize their optical properties. Some of the applications are in the

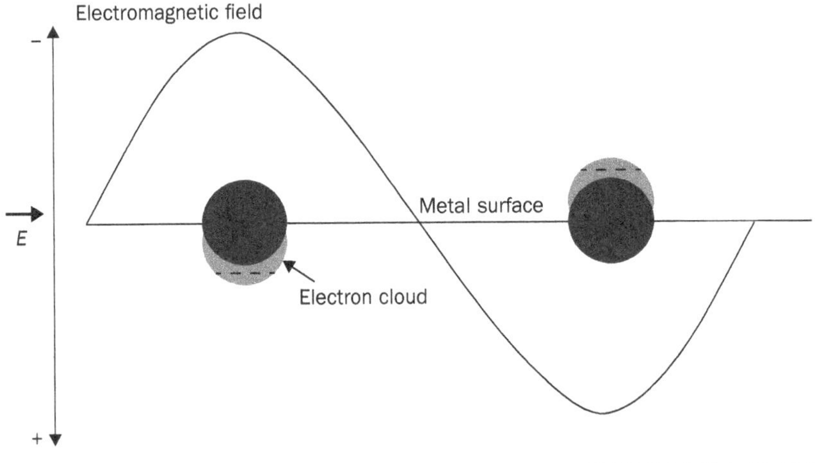

FIGURE 8.2 Localized surface plasmon resonance on the metal surface.[3]

fields such as optics, biosensors, chemical sensors, nano-optical devices, solar cells, and optical amplifiers.

LSPR depends on the following factors:

1. Type of metal NPs
2. Size and shape of NPs
3. Surface functionalities of NPs
4. Chemical environment

In general, the size and shape of NPs along with local dielectric properties determine the location of the LSPR. Mie–Drude theory describes the position of LSPR quantitatively. The following derivation to find the extinction coefficient and change in LSPR wavelength is derived from the research by Willets and Van Duyne,[4] Mayer and Hafner,[5] Link and El-Sayed,[6] Mulvaney,[7] and Okamoto and Imura.[8]

The behavior of charged particles in magnetic or electric field can be described using Maxwell's equations. You know that the surface plasmon resonance is caused by unbalanced charges and the phenomenon of LSPR occurs at a specific wavelength of electromagnetic wave. Gustav Mie solved Maxwell's equation with boundary conditions of spherical NPs using quasi-state approximation (no time derivatives in the equation).

Consider a spherical NP of radius a which is irradiated by polarized light (z) of wavelength λ, where a/λ is less than 0.1 (within Rayleigh scattering). The electric field caused by unbalanced charge appears static around the surface of the NP. When Maxwell's equation is solved, the following solution is obtained for the electromagnetic field present around the NPs.

$$\overrightarrow{E_{out}}(x,y,z) = E_0\hat{z} - \left[\frac{\varepsilon_{in} - \varepsilon_{out}}{\varepsilon_{in} + \varepsilon_{out}}\right] a^3 E_0 \left[\frac{\hat{z}}{r^3} - \frac{3z}{r^5}\left(x\hat{x} + y\hat{y} + z\hat{z}\right)\right] \tag{8.1}$$

where ε_{in} and ε_{out} are dielectric constants of the NP and the medium, respectively. The cross-sectional area in which energy loss contributed by absorption and scattering takes place is called extinction cross section and is defined as sum of scattering and absorption cross sections. In the above-mentioned case the extinction cross section can be obtained as follows:

$$\sigma_{ext} = \frac{18\pi\varepsilon_{out}^{\frac{3}{2}}V}{\lambda}\frac{\varepsilon_2(\lambda)}{\left[\varepsilon_1(\lambda) + 2\varepsilon_{out}\right]^2 + \varepsilon_2(\lambda)^2} \tag{8.2}$$

This equation holds true when dipole oscillations contribute significantly to the extinction cross section. ε_1 and ε_2 are real and imaginary parts of the complex metal dielectric function.

$$\varepsilon_{in} = \varepsilon_1 + i\varepsilon_2 \tag{8.3}$$

LSPR happens when ε_{in} is equal to $-2\varepsilon_{out}$, and this results in enhanced electromagnetic field with respect to the incident field. The wavelength at which LSPR occurs (λ_{max}) has a dependence on the dielectric function of the medium. Drude model describes the damping ω_p^2 effect experienced by the free electrons and this is used to define ε_1:

$$\varepsilon_1 = 1 - \frac{\omega_p^2}{\omega^2 + \gamma^2} \tag{8.4}$$

where ω_p is the plasma frequency, γ is the damping parameter of the bulk metal, and $\gamma << \omega_p$ at visible and infrared frequencies. At resonance, Eq. (8.4) simplifies to give $\omega_{\max}$ which is the LSPR frequency.

$$\omega_{\max} = \frac{\omega_p}{\sqrt{2\varepsilon_{out} + 1}} \tag{8.5}$$

It is related to $\lambda_{\max}$ as follows:

$$\lambda_{\max} = \frac{2\pi c}{\omega_{\max}} \tag{8.6}$$

where c is the velocity of light. When ε_{out} is substituted with n_{out}^2 in the equation, the following expression is obtained:

$$\lambda_{\max} = \lambda_p \sqrt{2n_{out}^2 + 1} \tag{8.7}$$

where λ_p is the wavelength corresponding to the plasma frequency of the bulk metal and n_{out} is the refractive index of the medium.

Introduction of a thin surface layer will change the extinction cross section by

$$\sigma_{ext} = 4\pi R^2 k^* \operatorname{Im}\left[\frac{(\varepsilon_s - \varepsilon_{out})(\varepsilon_c - 2\varepsilon_s) + (1-g)(\varepsilon_c - \varepsilon_s)(\varepsilon_{out} - 2\varepsilon_s)}{(\varepsilon_s - 2\varepsilon_{out})(\varepsilon_c - 2\varepsilon_s) + (1-g)(2\varepsilon_s - 2\varepsilon_{out})(\varepsilon_c - \varepsilon_s)}\right] \tag{8.8}$$

where ε_c and ε_s are complex dielectric functions of the core and the shell, respectively, g is the volume fraction of the shell, and R is the radius of the coated particle.

When a chemical species is adsorbed on the NPs, the local environment is changed and causes a shift in $\lambda_{\max}$ and this change is depicted by $\Delta\lambda_{\max}$:

$$\Delta\lambda_{\max} = m\Delta n\left[1 - \exp\left(\frac{-2d}{l_d}\right)\right] \tag{8.9}$$

where m is the bulk refractive index and Δn is the change in refractive index due to the adsorbate, d is the thickness of the adsorbate layer, and l_d is the electromagnetic decay length.

The extinction cross section changes for anisotropic materials. For elongated ellipsoids (nanorods), the extinction can be found using dipole approximation:

$$\sigma_{ext} = \frac{2\pi\varepsilon_{out}^{\frac{3}{2}} V}{3\lambda} \sum_j \frac{(1/P_j^2)\varepsilon^2}{\left\{\varepsilon_1 + \left[(1-P_j)/P_j\right]\varepsilon_{out}\right\}^2 + \varepsilon_2^2} \tag{8.10}$$

where P_j is the depolarization factor along the three axes of the nanorods (A, B, C, where $A > B = C$), depicted as:

$$P_A = \frac{1-e^2}{e^2}\left[\frac{1}{2e}\ln\left(\frac{1+e}{1-e}\right)-1\right] \tag{8.11}$$

$$P_B = P_C = \frac{1-P_A}{2} \tag{8.12}$$

$$e = \left[1-\left(\frac{B}{A}\right)^2\right]^{\frac{1}{2}} = \left(1-\frac{1}{(AR)^2}\right)^{\frac{1}{2}} \tag{8.13}$$

The length-to-diameter ratio is the aspect ratio of the NPs (AR). In such materials, the extinction spectrum shows two peaks, one for transverse and one for longitudinal axis, respectively. A red shift (bathochromic shift) in the LSPR can be observed when the AR is increased. The change in peak (red or blue shift) can also be caused by functionalizing the surface with other chemical species. For example, coating NPs with polyethylene glycol (PEG) is very common as it is inert, biocompatible, and nontoxic. These stealth and cloaking properties are advantageous when NPs are used in harsh chemical conditions and need to be protected from the environment.

Consider an example where you synthesize two spherical Au NPs, Au-1 and Au-2, of diameters 46 nm and 115 nm, respectively. You also synthesize Au nanorods (Au NR). Later you coat Au-1 with PEG (Au-PEG) for studying certain *in vivo* interactions with cells. You obtain UV-Vis spectra (Fig. 8.3) and arrive at the following

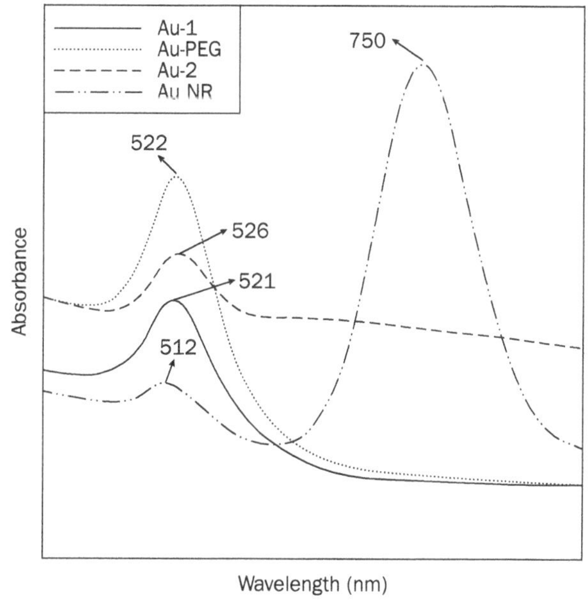

FIGURE 8.3 LSPR peaks of different Au NPs. Data from experiments performed in our laboratories at Department of Chemical Engineering, NTNU.

inferences. The shift in LSPR between Au-1 and Au-2 from 521 to 526 nm could be due to the increase in size. Using Eq. (8.9), you can explain the shift in LSPR from 521 to 522 nm when coated with PEG. The two peaks observed could be due to the split in plasmon band as a result of anisotropy. These inferences are based on measurement obtained by UV-Vis spectroscopy, which helps in locating the LSPR peaks of the metal NPs.

8.1.3 UV-Vis Spectroscopy

UV-Vis spectroscopy is the study of interaction between matter and ultraviolet-visible region of the electromagnetic waves. This technique measures the absorption of the wave with respect to the intensity of the incident wave, which is depicted by Beer–Lambert's law. Apart from finding LSPR peaks of noble metal NPs, it can also be used to determine the concentration of NPs or for that matter other solutes as well. The technique tracks the electronic transitions when irradiated with UV-Vis spectra by measuring the change in intensity of the spectra before and after irradiating the materials.

Beer–Lambert's Law

The absorbance (A) of a solution is directly proportional to the concentration of the emitters (C) and path length of the sample (L) and is given by the following equation:

$$A = -\log_{10}\left(\frac{I}{I_0}\right) = \alpha CL \tag{8.14}$$

I and I_0 are the intensities of incident and transmitted light, respectively, and α is the extinction coefficient or molar absorptivity. Although Beer–Lambert's law is extensively used in spectroscopic studies, it can fail under certain conditions. Hence, for the right use of the law, a set of conditions must be valid.

Assumptions

1. Absorbers must act independently to each other.
2. The absorbing medium is homogenous.
3. The absorbing medium should not be turbid.
4. The incident light should not influence the atoms or molecules under study.

Consider a scenario, where you prepared a batch of Au NPs using a protocol given by a fellow researcher. However, you are not sure about the shape of Au NPs. In this scenario, UV-Vis spectroscopy is the technique that can help you to reveal some important information regarding the same. Before you use the UV-Vis spectrometer, you must make sure all the assumptions are satisfied. The Au NPs are in a colloidal state and are in a uniform solution. Turbid solutions can cause the incident light to scatter. Au NPs, being a colloidal solution, cannot agglomerate and scatter light under these conditions. With these facts in mind start measuring the absorbance of Au NPs. The UV-Vis spectrometer shows only one peak at 520 nm and you conclude that the particles are spherical in shape and your colleague confirms it. To confidently suggest that these are spherical Au NPs, other characterization techniques are needed to visualize the shape. Such techniques are discussed later in this chapter.

For the time being, let us focus on another scattering-based characterization technique that can reveal information about the size of NPs in solution.

8.2 Dynamic Light Scattering

As discussed in the previous section, spectroscopy can be an interesting way to understand some of the fundamental physicochemical properties of NPs; however, not all kinds of NPs can be characterized by the UV-Vis spectroscopy method. NPs need to possess optical properties that are distinguishable. However, a more general characterization technique to have a first impression of NPs in suspension is dynamic light scattering (DLS). Although the physics behind the technique is not grossly intuitive, the final information that may be achieved using this characterization method might encourage you to know the physics as well. So, what are the parameters you can infer after measuring your sample? DLS allows you to know the particle size, particle size distribution, such as polydispersity among others, stability of the dispersion, phenomenon of aggregation, emulsion properties, and a lot of interesting parameters about proteins and other macromolecules.

Next, we discuss the underlying principle of operation of DLS.

8.2.1 Theory

The fundamental principle under which DLS works is based on measuring the radius of a hypothetical hard sphere that diffuses with the same velocity as the particle under examination. The reported hydrodynamic size is indicative of the apparent size of the hydrated particle. This means that the size of the NP measured using this method will be very different from the dry size. This is not only because hydrodynamic size accounts for the associated water with the NP, but also because some NPs can aggregate when taken out of solution or behave differently. Thus, whether to report the hydrodynamic size or dry size of NPs is a question that is primarily governed by the end use of the NPs. If some of the NPs will be used as dispersions, their hydrodynamic sizes would be more important to report than if the same NPs would be used on chips (for instance, in biosensing applications).

NPs when suspended in a solution undergo Brownian motion resulting from collisions. Further, NPs of particle size $<0.1\lambda$, where λ is the wavelength of incident light, scatter light primarily as Rayleigh scatterers. Owing to the random motion, the intensity of the light scattered by them undergoes time-based fluctuations as the distance between the scatterers continuously changes. These time-based fluctuations can be mapped using auto-correlation functions that contain information about the diffusion behavior of the NPs. From these, the average diffusion coefficients can be determined and hydrodynamic sizes can thereafter be calculated using the Stokes–Einstein equation. Thus, DLS enables you to calculate particle sizes and distributions from diffusion behaviors of particles suspended in solution by mapping time-based scattered intensity fluctuations.

8.2.2 Autocorrelation Function

The first step to translate time-based fluctuations to particle sizes is by employing an auto-correlation function. An auto-correlation function compares a signal with itself at increasing time intervals. In case of DLS, the intensity of scattered light when compared with itself decays as a function of time since the suspended NPs are in continuous motion. These result in the auto-correlation curve shown as a function of the delay time (τ) in Fig. 8.4. In case of short time delays, the correlation is high as the particles are less probable to move appreciably from the initial states that they were in. At high delay times, the curve tails off asymptotically, meaning that after a long time, there is

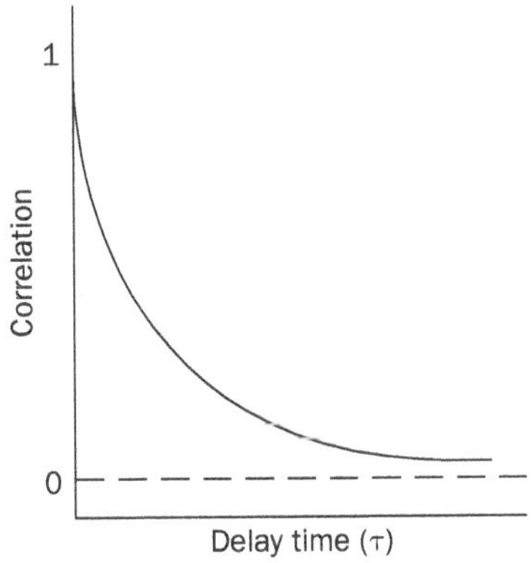

FIGURE 8.4 Correlation curve as a function of the delay time shows that the curve decays exponentially.

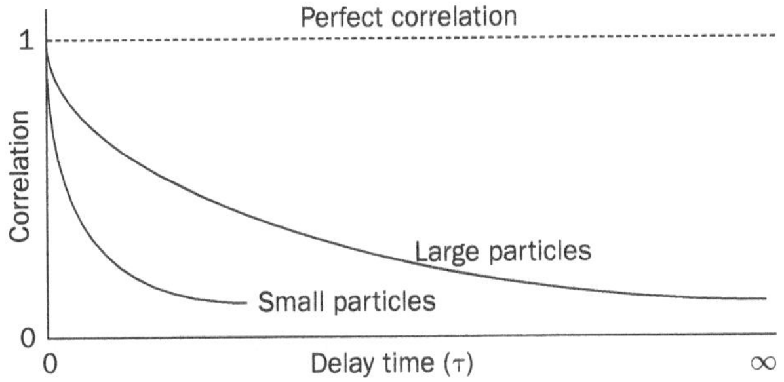

FIGURE 8.5 Relative decay rates of the auto-correlation curves for small and large NPs.

no correlation between the scattered intensity of the initial and final states. These auto-correlation curves contain information about the diffusion of the NPs in the dispersion and are used to estimate the average diffusion coefficients (D).

To have a first visual impression, it might be interesting to see that the curve decays much faster for smaller NPs than for larger NPs (Fig. 8.5). This is because, smaller particles move faster (higher average velocity) than the larger particles and hence the auto-correlation curve decays faster.

The time when decay starts indicates mean size, whereas the gradient of decay reflects the polydispersity of the sample—both of which will be discussed in detail in this chapter after the equation is developed. The auto-correlation function $C(s,t_d)$, therefore, models the decay between the highest intensity value $[i(s,0)]^2$ at $t_d = 0$ and its lowest asymptotic value $[\bar{i}(s)]^2$ as a function of delay time (t_d)\, where $\bar{i}(s)$ denotes the average intensity over time. The ratio of the auto-correlation function to its asymptotic value $[\bar{i}(s)]^2$ can be written in the form of Eq. (8.15), known as the *Siegert relation*.[9]

$$\frac{C(s,t_d)}{[\bar{i}(s)]^2} = g_2(s,t_d) = B + \xi |g_1(s,t_d)|^2 \tag{8.15}$$

In principle, the Siegert relation relates the second-order auto-correlation function $g_2(s,t_d)$ with the first-order auto-correlation function $g_1(s,t_d)$. B is the baseline (~1) and ξ is the coherence factor that depends on detector area, optical alignment, and scattering properties of particles.

$$g_2(s,t_d) = 1 + \xi|g_1(s,t_d)|^2 \tag{8.16}$$

The above relation is not valid if there is a very small number of scatterers or when their motion is limited. Before discussing further translation of the auto-correlation data into diffusion coefficients and hence hydrodynamic sizes, let us discuss how exactly modern DLS instruments work.

The fundamental system components of a DLS include a monochromatic laser source, an attenuator, the cuvette containing sample, the photon detector, the digital auto-correlator, and a computer where the software analyzes the data to yield hydrodynamic sizes. The laser shines a monochromatic wavelength on the sample contained in a cell. The light scattered by the particles in suspension can thereafter be measured by the detector placed at different angles (to measure backward scattering or forward scattering dependent on the particle size). In order to prevent detector saturation, the attenuator reduces the intensity of the laser source and thus the intensity of scattering. The detector then passes the scattering intensity signal to the correlator, which calculates the intensity auto-correlation function automatically and obtains the results in terms of $g_1(s,t_d)$. Thus, you should be mainly concerned about the interpretation of $g_1(s,t_d)$. Let us discuss how the function looks for two situations—monosized spherical particles and a polydisperse suspension.

Monodisperse NPs in Suspension

The assumption underlying such a treatment requires interparticle spacing to be so large that there are no interparticle interactions. Simply speaking, the treatment is valid for dilute dispersions. The field correlation function $g_1(s,t_d)$ decays exponentially with a decay rate of Γ. Therefore,

$$g_1(s,t_d) = \exp(-\Gamma t_d) \tag{8.17}$$

where

$$\Gamma = Ds^2 \tag{8.18}$$

D is the diffusion coefficient of the particles, and S is magnitude of the scattering wave vector, defined as the difference between the incident and the scattered wave vectors.

$$s = \frac{4\pi n}{\lambda_0}\sin\left(\frac{\theta}{2}\right) \tag{8.19}$$

where n is the refractive index of the solvent, λ_0 is the wavelength of the laser in vacuum, and θ is the scattering angle.

After combining Eqs. (8.15) and (8.16), Eq. (8.17) can be rewritten as:

$$g_2(s,t_d) = 1 + \xi\exp(-2Ds^2t_d) \tag{8.20}$$

Polydisperse NPs in Suspension

This is a more realistic situation where the suspension contains particles of different sizes. In such a case, $g_1(s,t_d)$ cannot be represented as a single exponential, since the overall decay is determined as a sum or an integral over a distribution of decay rates.[10]

$$g_1(s,t_d) = \int_0^\infty G(\Gamma)\exp(-\Gamma t_d)d\Gamma \tag{8.21}$$

$G(\Gamma)$ is normalized such that

$$\int_0^\infty G(\Gamma)d\Gamma = 1 \tag{8.22}$$

It is problematic to find the precise functional form for the distribution of decay rates $G(\Gamma)$ because the correlation function is measured discretely only over an incomplete range of t_d and there is always noise associated with the data. One of the simplest approaches to do this is to employ the method of cumulants,[11] which is a series expansion of $\ln g_1(s,t_d)$, depicted by:

$$\ln g_1(s,t_d) = -\bar{D}s^2 t_d + \sigma^2 s^4 \frac{t_d^2}{2!} + \ldots \tag{8.23}$$

Here, the first-order cumulant is related to the average diffusion coefficients of particles of all sizes ($\bar{D}$) and the second-order cumulant to the standard deviation σ of the distribution of diffusion coefficients.

Therefore, Eq. (8.16) becomes:

$$\ln[g_2(s,t_d) - 1] = \ln\xi + 2(-\bar{D}s^2 t_d + \sigma^2 s^4 \frac{t_d^2}{2!} + \ldots) \tag{8.24}$$

Neglecting higher order terms, the above equation becomes:

$$\ln[g_2(s,t_d) - 1] = a_0 - a_1 t_d + a_2 t_d^2 \tag{8.25}$$

Most often, the DLS sizes reported are the Z-average particle sizes D_z calculated as shown below:

$$D_z = \frac{1}{a_1}\frac{k_B T}{3\pi\eta}\left[\frac{4\pi\eta}{\lambda_0}\sin\left(\frac{\theta}{2}\right)\right]^2 \tag{8.26}$$

Wondering, what is special about Z-average particle sizes? Z-average particle sizes can be expressed as the intensity-based harmonic mean, as depicted in the following equation:[12]

$$D_z = \frac{\sum S_i}{\sum S_i/D_i} \approx \frac{\sum D_i^6}{\sum D_i^5} \tag{8.27}$$

Here, S_i is the scattered intensity from particle i and D_i is the diameter of the particle. The mean is in principle calculated from the intensity-weighted distribution, leading to the fact that the Z-average size is the intensity-weighted harmonic mean size. As the particles behave as Rayleigh scatterers (small sizes), $S_i \sim D_i^6$. The reason why Z-average size is the most commonly reported hydrodynamic sizc is because it is relatively insensitive to experimental noise. This is so, since the technique of cumulants described above relies on numerically stable least-squares fitting. Other distributions also available for analysis include volume-weighted and number-weighted distributions, which are not discussed in detail here. However, the following example would give you a clear distinction between the three kinds of distributions most often obtained from a standard DLS software.

Example

Let us consider a suspension containing an equal number of 5 and 50 nm particles. Figure 8.6 shows how the number-, volume-, and intensity-weighted distributions would look like for the same suspension. In the number-weighted distribution, the two peaks representative of the two different sizes would represent equal areas. However, in case of volume-weighted distribution, the bigger particles (50 nm) will be represented by an area 1000 times bigger than that for the smaller particles (5 nm). This is because the volume scales to the third power of the diameter. In case of the intensity-weighted distribution, the same difference will be six orders of magnitude higher. This is based on the assumption that the intensity of scattered light scales to the sixth power of the diameter as long as the particles are small compared to the wavelength of light (Rayleigh scattering). Thus, it is important to analyze the hydrodynamic sizes obtained from a DLS software, since intensity-weighted distributions magnify the effect of the particles having higher scattering cross sections.

Coming back to the discussion on hydrodynamic sizes obtained from a DLS measurement, another important parameter reported along with the size is the polydispersity index (PDI), which gives a measure of the size distribution:

$$PDI = \frac{2a_2}{a_1^2} \tag{8.28}$$

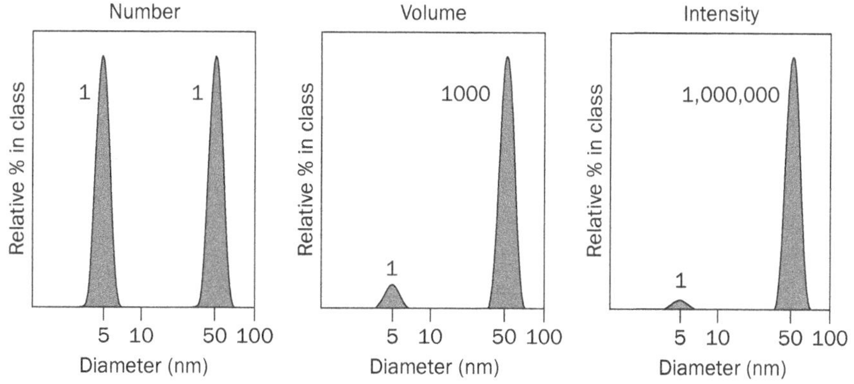

Figure 8.6 Number-, volume-, and intensity-based size distributions of a suspension containing equal number of 5 and 50 nm particles, respectively. (Courtesy DLS technical notes, Malvern Instruments.)

Up till now, we have arrived at expressions that enable us to compute the average diffusion coefficient of a particle size distribution either using Eq. (8.20) (for monodisperse suspension) or Eq. (8.24) (for polydisperse suspension). The next and final step is to relate this to the average hydrodynamic diameter of the particles. This is carried out using the famous Stokes–Einstein equation, which is reviewed in Sec. 8.2.3.

8.2.3 Stokes-Einstein Equation

If you have a background in chemical engineering or physical chemistry, you would have encountered this equation in some previous course. But we will discuss here the basics of the equation so that when employing it for determination of hydrodynamic diameter based on diffusion coefficients obtained using light scattering, we are fully aware of the boundaries where this equation can be used.

To start with, we will use the definition of chemical potential of species $i\,(\mu_i)$, the fundamental driving force for mass transfer. It is defined as partial molal Gibb's free energy change with respect to component i at constant temperature and pressure (T, P), represented by the following equation:

$$\mu_i \triangleq \left(\frac{\partial G}{\partial n_i}\right)_{T,P,n_{j\neq i}} \tag{8.29}$$

Generally speaking, any kind of force can be defined as negative gradient of potential. For instance, the electric field force on a charged particle can be defined as the negative gradient of the electric potential. Similarly, if we define diffusive force $\left(F_{diff}\right)$ for a particle, the concerned potential should be chemical potential. Therefore,

$$F_{diff} = -\frac{1}{N_A}\left(\frac{\partial \mu_i}{\partial x}\right) \tag{8.30}$$

where N_A is the Avogadro's number that arises from how chemical potential is defined above. The expression above represents the thermodynamic force for diffusion. The chemical potential of species $i\,(\mu_i)$ can be related to measurable properties of a system by the following expression:

$$\mu_i = \mu_i^0 + RT\ln a_i = \mu_i^0 + RT\ln \gamma_i C_i \tag{8.31}$$

where μ_i^0 is the standard chemical potential of species i, R is the universal gas constant, a_i is the activity, γ_i is the activity coefficient, and C_i is the concentration of species i.

For dilute solutions, $\gamma_i = 1$ and hence Eq. (8.31) becomes:

$$\mu_i = \mu_i^0 + RT\ln C_i \tag{8.32}$$

Partially differentiating Eq. (8.32) with respect to x, we obtain:

$$\frac{\partial \mu_i}{\partial x} = \frac{RT}{C_i}\frac{\partial C_i}{\partial x} \tag{8.33}$$

Substituting Eq. (8.33) in Eq. (8.30), we arrive at:

$$F_{diff} = -\frac{RT}{N_A C_i}\frac{\partial C_i}{\partial x} = -\frac{k_B T}{C_i}\frac{\partial C_i}{\partial x} \tag{8.34}$$

where $k_B = \frac{R}{N_A}$ is the Boltzmann constant. Dropping the subscript i, Eq. (8.34) can be re-written as:

$$v = -\frac{k_B T}{fC}\frac{\partial C}{\partial x} \tag{8.35}$$

So, we have arrived at an expression for the diffusive force containing the concentration gradient, a parameter difficult to measure. The force that counteracts diffusion is the viscous force (F_{visc}). Hence, at equilibrium,

$$F_{diff} = F_{visc} \tag{8.36}$$

The viscous force (F_{visc}) is defined as the product of friction factor (f) and velocity (v)

$$F_{visc} = fv \tag{8.37}$$

Combining Eqs. (8.37) and (8.35), Eq. (8.36) becomes:

$$-\frac{k_B T}{C}\frac{\partial C}{\partial x} = fv \tag{8.38}$$

Therefore,

$$v = -\frac{k_B T}{fC}\frac{\partial C}{\partial x} \tag{8.39}$$

From definition and knowledge of mass transfer, diffusive flux (J) can be written as the product of concentration and velocity.

$$J = Cv \tag{8.40}$$

Fick's first law of diffusion relates diffusive flux to the concentration gradient via diffusion coefficient.

$$J = -D\frac{\partial C}{\partial x} \tag{8.41}$$

Combining the above two equations,

$$v = -\frac{D}{C}\frac{\partial C}{\partial x} \tag{8.42}$$

Comparing Eqs. (8.39) and (8.42), we obtain a relationship between the diffusion coefficient and the friction factor.

$$D = \frac{k_B T}{f} \tag{8.43}$$

So far nothing has been assumed in regards to the shape of the particle. The friction factor f needs to be measured for the particles in question. However, for spherical particles, an empirical correlation exists:

$$f = 6\pi\eta R_h \tag{8.44}$$

Here, η is the viscosity of the solvent and R_h is the hydrodynamic radius of the particle. Inserting Eq. (8.44) in Eq. (8.43), we obtain what is famously known as the *Stokes–Einstein equation.*

$$D = \frac{k_B T}{6\pi\eta R_h} \tag{8.45}$$

Note that the hydrodynamic radius calculated from Eq. (8.45) is different from the dry radius of the particles, since the former accounts for the hydration layer around the particles. In some cases, it can also reflect the passivating ligand covering the particle surface partly or fully. Hence, we should be careful about which radius to report, hydrodynamic or dry, depending on intended applications.

8.3 Differential Scanning Calorimetry

Another way to characterize NPs to understand structural alterations is by performing differential scanning calorimetric measurements since such alterations are almost always accompanied by heat exchanges. First developed by E. S. Watson and M. J. O'Neill in 1962, differential scanning calorimetry (DSC) is a thermoanalytical characterization technique that relates the system's temperature to the specific physical property of the substance. In simpler terms, it is a technique that measures the heat flow necessary to establish a nearly zero temperature difference between a substance and an inert reference material, as the two specimens are subjected to identical temperature regimes in an environment, i.e., either heated or cooled at a controlled rate. Therefore, the instrument measures and plots the temperature and heat inflow/outflow associated with material transitions as a function of time and temperature. During a change in temperature, DSC measures a heat quantity, which is radiated or absorbed excessively by the sample on the basis of a temperature difference between the sample and the reference material. This technique is particularly used to understand biochemical reactions and is useful in determining the following:

- Glass transition temperatures of polymers
- Degree of crystallinity, purity, and structural stability of materials
- Phase transitions by measuring changes in specific heat capacities (C_p)

Besides being fast, highly sensitive, and convenient in usage, DSC results in the direct determination of the system's enthalpy associated with the point of interest by measuring the temperature difference between the sample and a reference.

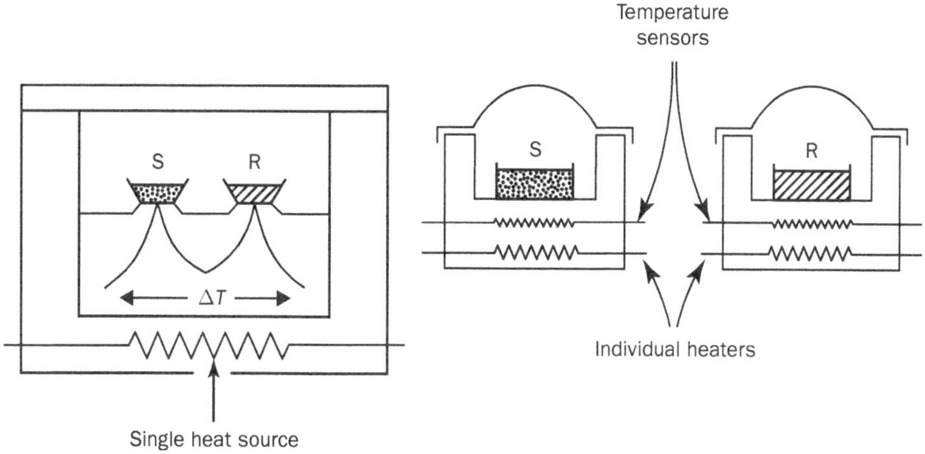

FIGURE 8.7 Schematic showing the heat-flux DSC (left) and the power-compensated DSC (right).[13]

DSC can broadly be categorized as heat-flux DSC (HF-DSC) and power-compensated DSC (PC-DSC) based on the mode of operation (Fig. 8.7). The former uses a single power source to heat up both the sample and the reference and the difference in the temperatures of the two are recorded. The difference in temperatures is created owing to different specific heat capacities (C_p) of the sample and the reference pans, which result in different heat flux. In the case of PC-DSC, the sample and reference pans are heated by separate heaters and therefore their temperatures can be controlled individually. The sample and reference pans are maintained at the same temperature by varying the energy input to the two heaters. Therefore, the energy required to achieve this is a measure of the enthalpy or heat capacity changes in the sample relative to the reference. In this case, the difference in thermal power required to maintain them at the same temperature is measured and plotted as a function of temperature or time.

8.3.1 Mathematical Background

In order to better understand the governing equations involved in DSC measurements refer to Fig. 8.8A, B, representing the sample and reference pan areas.

The governing law behind all heat flows within the system discussed above is depicted by Newton's law of cooling, stated as:[14]

$$\left(\frac{dq}{dt}\right) = k\Delta T \tag{8.46}$$

where $\frac{dq}{dt}$ represents the heat flow between two points that are separated by a barrier having a thermal conductivity k and temperature difference ΔT.

Considering the sample side first, the heating element will heat both the sample monitoring station as well as the sample itself. Assuming no heat loss, we can arrive at:

$$\frac{dq_s}{dt} = C_{sm}\frac{dT_{sm}}{dt} + C_s\frac{dT_s}{dt} \tag{8.47}$$

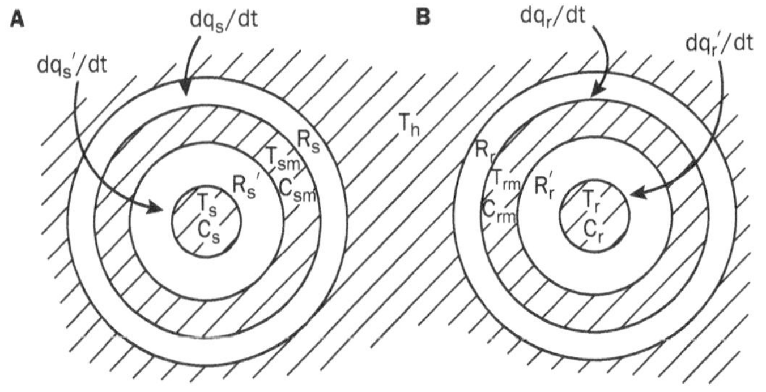

FIGURE 8.8 Schematic representing a conventional differential thermal instrument: (A) sample and (B) reference pan.[14]

where $\frac{dq_s}{dt}$ is the total heat flow toward the sample side. C_{sm} and C_s are the heat capacities of the sample-temperature monitoring system and the sample material itself, respectively. T_{sm} and T_s are the temperature at the monitoring point on the sample side and actual temperature of the sample material, respectively.

Also, since

$$\frac{dq_s'}{dt} = C_s \frac{dT_s}{dt} \tag{8.48}$$

where $\frac{dq_s'}{dt}$ is the heat flow between T_s and T_{sm}.

Then,

$$\frac{dq_s}{dt} = C_{sm} \frac{dT_{sm}}{dt} + \frac{dq_s'}{dt} \tag{8.49}$$

where $\frac{dq_s}{dt}$ is the heat flow between T_h and T_{sm}. T_h is the temperature of the heat source.

On applying Newton's law to Eqs. (8.47) and (8.48), the equations transform into:

$$\frac{1}{R}(T_h - T_{sm}) = C_{sm} \frac{dT_{sm}}{dt} + C_s \frac{dT_s}{dt} \tag{8.50}$$

$$\frac{1}{R'}(T_{sm} - T_{sm}) = C_s \frac{dT_s}{dt} \tag{8.51}$$

Here, we assume that the sample side and the reference pan side are constructed identically; therefore: $R_s = R_r \equiv R$ and $R_s' = R_r' \equiv R'$. R_s is the heat resistance between T_h and T_{sm}. R_s' is the heat resistance between T_s and T_{sm}. R_r and R_r' are analogous to the above terms in the reference side.

Likewise, we can obtain a similar set of equations for the reference side as well:

$$\frac{dq_r}{dt} = C_{rm}\frac{dT_{rm}}{dt} + C_r\frac{dT_r}{dt} \tag{8.52}$$

$$\frac{dq_r'}{dt} = C_r\frac{dT_r}{dt} \tag{8.53}$$

$$\frac{1}{R}(T_h - T_{rm}) = C_{rm}\frac{dT_{rm}}{dt} + C_r\frac{dT_r}{dt} \tag{8.54}$$

$$\frac{1}{R'}(T_{rm} - T_r) = C_r\frac{dT_r}{dt} \tag{8.55}$$

The HF-DSC measures the temperature difference between the reference and the sample that can be obtained by subtracting Eq. (8.55) from Eq. (8.50):

$$T_{rm} - T_{sm} = R\frac{dT}{dt}(C_s - C_r) \tag{8.56}$$

where R is a function of the instrument parameters only and does not depend on the properties of either the reference pan or sample. On rearranging Eq. (8.56), we obtain the fundamental equation of heat flux DSC where the output signal has already been adjusted to the units of power by the instrument:

$$\text{Signal} = \frac{1}{R}(T_{rm} - T_{sm}) = \frac{dT}{dt}(C_s - C_r) \tag{8.57}$$

In the case of PC-DSC, the reference and the sample are maintained at the same temperature; therefore, $T_h = T_{sm} = T_{rm}$. The system constantly measures the differential power in order to maintain the two specimens at the same temperature, which can be carried out by subtracting Eq. (8.55) from Eq. (8.51).

$$\frac{dq_s'}{dt} - \frac{dq_r'}{dt} = C_s\frac{dT_s}{dt} - C_r\frac{dT_r}{dt} \tag{8.58}$$

The fundamental PC-DSC can be given as:

$$\text{Signal} = \frac{dq_s'}{dt} - \frac{dq_r'}{dt} = \frac{dT}{dt}(C_s - C_r) \tag{8.59}$$

The signal is proportional to the difference in the heat capacities in the case of both the HF-DSC and PC-DSC; however, the latter does not have a thermal resistance dependence term in the expression.

Even though the DSC has proved to be a powerful analytical tool in recent years, it poses limitations when it comes to analyzing commercial materials that are generally blends of various substances. In these cases, the transition temperatures of different materials might overlap resulting in erroneous analysis. In addition, conventional DSC generates a single signal (*dq/dt*) that is the sum of all thermal events occurring at that specific temperature or time during the experiment resulting in difficulty to detect smaller transitions during data acquisition and interpretation. Modulated DSC (MDSC) improves upon this as it measures the total heat flow along with the heat capacity components in one simultaneous run, thereby resolving complex transitions into specific components and improving data interpretation.

8.3.2 Modulated Differential Scanning Calorimetry

A conventional DSC uses a specified linear heating rate (°C/min), whereas a MDSC utilizes a linear and a sinusoidal heating rate (Fig. 8.9). The linear temperature changes allow the heat flow measurement (conventional DSC) and the modulated change allows the heat capacity calculation.

The heat capacity from a DSC experiment can be calculated as:

$$C_p = K_{CP} \times \frac{\Delta q_f}{\Delta q_r} \tag{8.60}$$

where K_{CP} is the calibration constant. q_f and q_r are the heat flow and heating rate, respectively.

The MDSC allows the user to calculate the heat capacity from one single experimental run. In the case of the DSC, the user would have to run two different experiments at different heating rates, thus spending more time and energy. The total heat flow signal in the MDSC can be split into two primary components: (1) reversing heat flow or

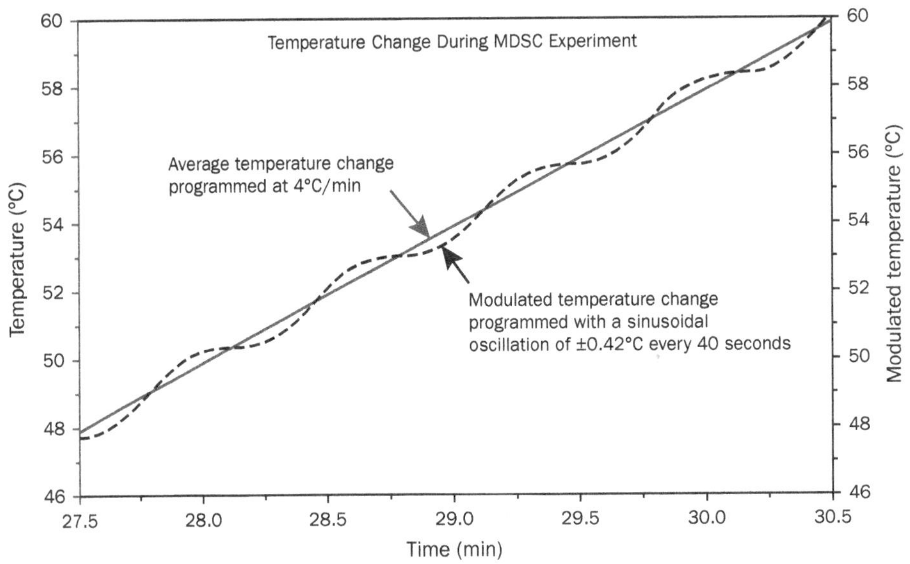

FIGURE 8.9 Schematic showing the heating rate of a conventional DSC with respect to a MDSC.[15] (Courtesy TA Instruments Technological Papers.)

the heat capacity component and (2) nonreversing heat flow or the kinetic component, which are defined as follows:

$$\text{Reversing Heat Flow} = \text{Heat Capacity} \times (\text{Average Heating Rate})$$

$$\text{Nonreversing Heat Flow} = \text{Total Heat Flow} - \text{Reversing Heat Flow}$$

The reversing heat flow provides information on the following:

- Heat capacity
- Glass transition: defined as the temperature (T_g) at which the transition occurs from stiff and brittle polymer chains system to a soft and more ductile polymer chain system. Below T_g, the polymer chains are unable to move and are locked in a specific position. However, on increasing the temperature above T_g, the polymer chains relax and are able to rotate more freely.
- Melting: defined as the loosening and breaking of intermolecular bonds after absorbing sufficient energy at high enough temperature, known as *melting temperature* T_m.
- Nonreversing heat flow: provides information on the kinetic processes such as enthalpy transitions and crystallization, which is the heat evolved or absorbed when a mole of substance crystallizes from its saturated solution (Fig. 8.10).

8.4 Scanning Transmission Electron Microscopy

Let us reconsider the example given in Sec. 8.1; in short, you synthesized Au NPs using a recipe given by a colleague. After synthesizing, you observe the LSPR peak at 520 nm;

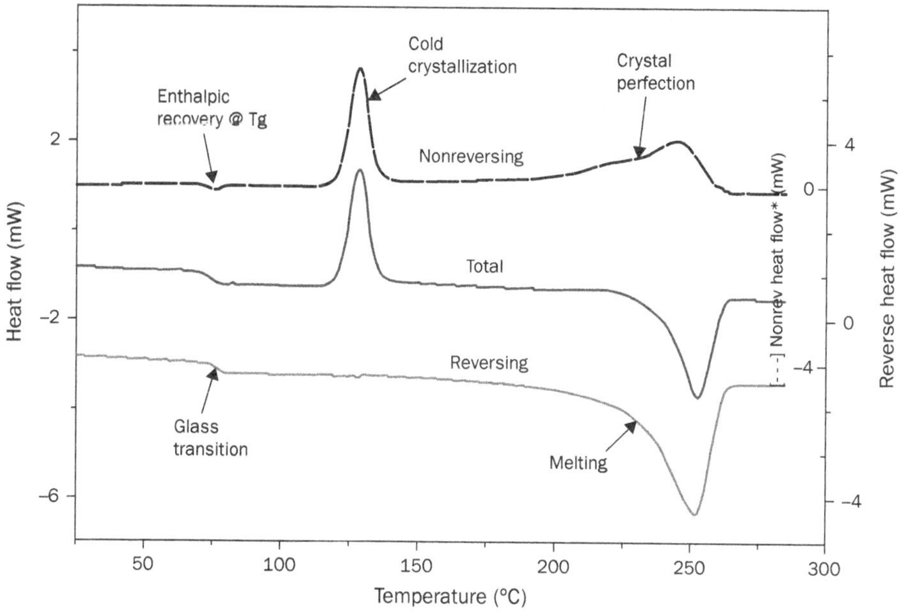

Figure 8.10 Schematic representing the reversing and the nonreversing heat flows. (Courtesy TA Instruments Technological Papers.)[15]

however, your colleague also doesn't know what shape it is supposed to be. In such and many other cases, microscopy can be used to determine the shape and proving the adage: a picture (image in this case) speaks a thousand words. This microscopy technique is based on the interaction between electrons and the sample on which the electrons are focused. When a beam of electrons is transmitted through a sample to form an image, the technique is called transmission electron microscopy (TEM). The technique of using a beam of electrons to scan an object to render its image is called scanning electron microscopy (SEM). S(T)EM is the technique which combines the functionalities of both TEM and SEM and can be used in SEM or TEM mode. In S(T)EM, a focused beam of electrons is passed through a thin sample; the beam is focused over a narrow region to scan in a raster pattern. The electrons can be scattered or transmitted, and these are captured by several detectors that are available. These interactions generate a series of signal streams which are correlated to generate a virtual picture of the scanned object. The ability to generate signals from various sources such as secondary electrons, scattered electrons, x rays, and electron energy loss is the single most advantage of S(T)EM.

Electrons may be scattered elastically due to repulsion from nuclei or inelastically due to scattering from other atoms of the sample. The inelastic scattering has a scattering angle less than 90° while the elastic scattering, also known as back scattering (as the electrons are directed back out of the sample), has an angle more than 90°. These are depicted in Fig. 8.11.

X rays are generated due to interactions of the electrons with the elements present in the sample. This is useful in identifying the elemental composition of the sample. When the electrons are transmitted, some of their energy is lost and this can be used to identify the type of chemical bonding, valency, etc. As a result of these different types of interactions, S(T)EM can be used in different modes: bright field mode and annular dark field mode, (Imaging) electron energy loss spectroscopy, and energy dispersive x-ray spectroscopy (EDX).

The different types of scattering electrons are shown in Fig. 8.12.

When electrons interact with the sample, depending on the escape depth (region in which electrons are scattered elastically), it results in secondary electron emission and

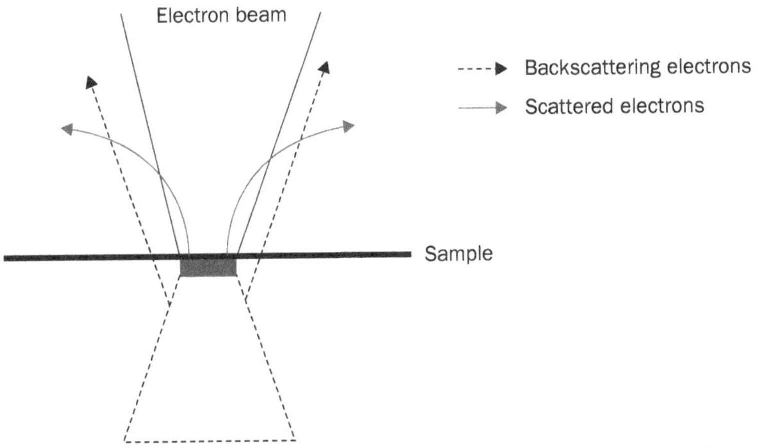

FIGURE 8.11 Generation of scattering and back-scattering electrons. (Credits: NTNU Nanolab.)

back scattering of electrons. The secondary electrons give an image with a good resolution. However, backscattering depends on the sample; the heavier the metal, the higher is the backscattering. Therefore, images from backscattered electrons provide contrasts between the different elements in the sample; however, the resolution will be lower, as represented in Fig. 8.13.

When the electrons are transmitted through the thin sample, depending on the operating conditions and the sample, some of them are scattered. The electrons that are scattered at higher angle are detected by the dark field detector, while the transmitted

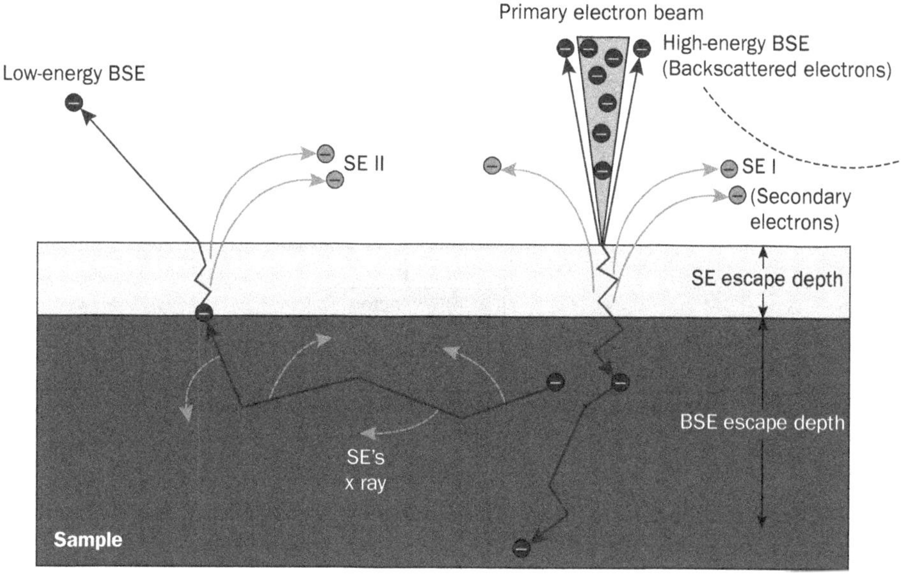

FIGURE 8.12 Interaction of electron beam with the sample. (Credits: NTNU Nanolab.)

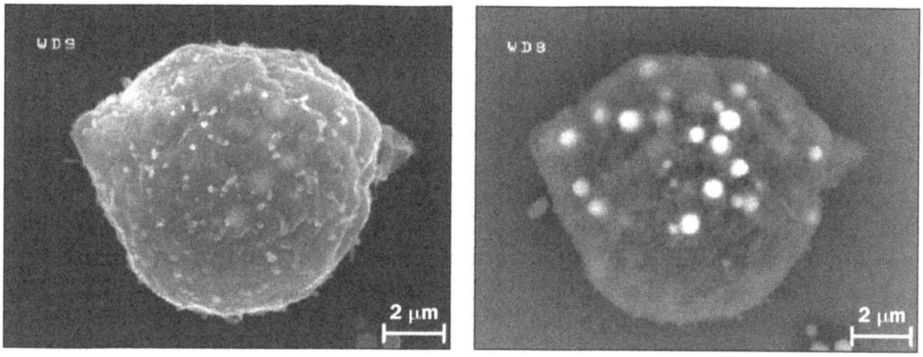

FIGURE 8.13 Images from secondary electron (SE) and backscattered electron (BSE) modes. (Credits: NTNU Nanolab.)

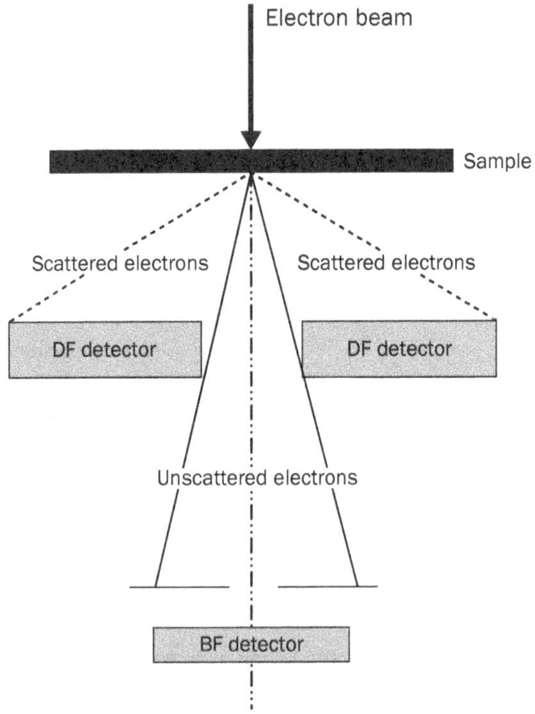

FIGURE 8.14 Setup for transmission mode. (Credits: NTNU Nanolab.)

electrons are detected by the bright field detector as the detectors are placed in the path of the beam (unlike annular dark field detector) (Fig. 8.14).

When operated under SEM mode, secondary and backscattered electrons are also used to form the image. However, in the TEM mode only transmission electrons are used, resulting in a high-resolution image (Fig. 8.15). Figure 8.15 shows tetrahexahedral Au NPs imaged using SEM and TEM modes.

S(T)EM is a versatile instrument; the different imaging options give insight into the sizes and shapes of NPs. It is also possible to get a first and rather quick insight into the elemental composition of the NPs using EDX.

Consider a situation where you synthesized core-shell Fe@Au NPs (growing Au atop Fe NPs as seeds). UV-Vis will give an insight into the LSPR, which is at 525 nm in this case. Had you synthesized Au NPs using a standard synthesis protocol, you could have guessed the resulting NPs to be spherical. However, in the case of Fe@Au NPs, the resulting NPs could be Fe@Au core-shell structure or could be Janus shaped or dumbbell shaped with Fe and Au at either ends. Using DLS you measure the size of Fe seeds and the final product, and you see an increase in size, but this still doesn't confirm the actual shape. However, using S(T)EM, you find Janus-like structure, and when opting SE and BSE modes along with EDX you can determine where Fe and Au are present in the NPs. Thus, electron microscopy may be used in combination with other solution-based characterization techniques, such as DLS and UV-Vis spectrometry, to obtain a vivid overview of physicochemical properties of NPs in solution.

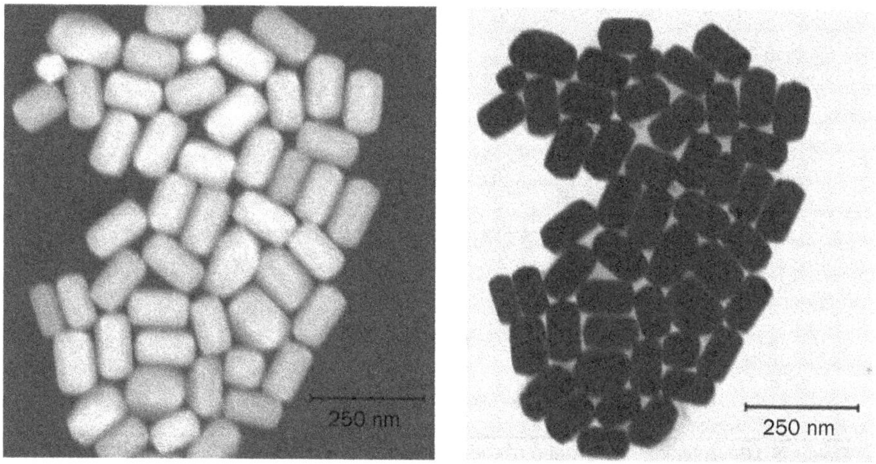

Figure 8.15 Images of SEM and TEM modes. (Image courtesy: Jibin Antony—Master Student, Department of Chemical Engineering, NTNU.)

It should, however, be remembered that images obtained from microscopic techniques should be representative of the whole sample and should contain several particles for statistical significance. Although a quick insight into the shape of the synthesized NPs is easily available from microscopy, other size measurement techniques (such as DLS for spherical particles) should be used as complementary techniques to ascertain hydrodynamic sizes, polydispersity indices, width of the size distribution, among others. Further, since most microscopic techniques are used on dried samples, which are irradiated with electron beams, it is almost impossible to detect the functional groups present on the surface of the NPs. Surface characterization of metal NPs may be carried out using techniques such as Fourier transform infrared (FTIR) spectroscopy, Raman spectroscopy, x-ray photoelectron spectroscopy (XPS), and others. FTIR spectroscopy and XPS are the final topics of discussion in this chapter.

8.5 Fourier Transform Infrared Spectroscopy

Fourier transform infrared (FTIR) spectroscopy is an extremely powerful characterization technique that allows for the identification and quantitation of chemical constituents in a material sample. It identifies the type of chemical bonds existing in the sample and utilizes Fourier transform (FT) for signal analysis like few other spectroscopic techniques such as nuclear magnetic resonance (NMR) spectroscopy and electron spin resonance (ESR) spectroscopy.

The total internal energy of a molecule can be depicted as:

$$U = E_R + E_V + E_E \tag{8.61}$$

where E_R, E_V, and E_E are the rotational, vibrational, and electronic energy levels, respectively. In IR spectroscopy, the electromagnetic waves interact with the molecular vibrations where the molecule can be excited to a higher energy state by absorbing the IR radiation. The interaction is greatly enhanced if the photon energy coincides with the

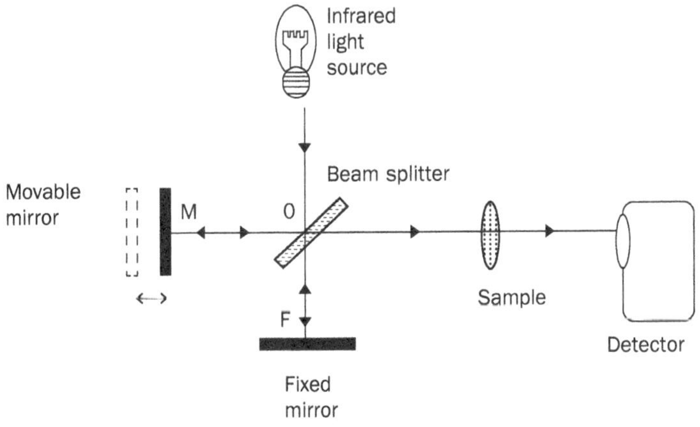

FIGURE 8.16 A typical schematic of the Michelson interferometer showing the essential components.[16]

vibrational energy levels of the molecule and this is a unique characteristic of individual molecules. Thus, IR spectroscopy is a very powerful technique providing fingerprint information on the chemical composition of the sample.

A drawback of IR spectroscopy is that it is time-consuming, as the material is scanned with one individual wavelength at a time. This considerably increases the total scanning time with a slight increase in scanning range. FTIR spectroscopy overcomes this issue because it allows the user to pass multiple wavelengths through the sample at once instead of one wavelength at a time. A FT is used to interpret the data collected from all the wavelengths passing through the sample at once.

FTIR spectroscopy measures interferograms with an optical device known as *interferometer*, which is based on Michelson interferometer (Fig. 8.16).

In an interferometer, the beam splitter splits the IR beam coming from the source into two—one-half of the IR light reflected to the movable mirror while the other half transmitted to the fixed mirror. The beams from the mirrors are reflected back to the beam splitter where each of the two beams is again spilt in two. One-half goes back to the source and is lost, whereas the other half passes through the sample and the non-absorbed light continues to the detector. Therefore, the detector records two beams that are made to interfere with each other, generating an optical phase difference between the two depending on the position of the movable mirror. This optical phase difference pattern is referred to as the interferogram that encodes spectral information about the sample. The difference between the two beams of infrared light shows which wavelengths of light are absorbed by the sample. A sample will absorb infrared light that matches the frequency of the vibration of the bonds of the atoms that make up the material. The magnitude of the difference between the reference beam and the beam from the sample shows how much of that bond is present that allows the exact chemical makeup of the sample to be determined. The interferogram of a laser light is a pure cosine wave function with a constant amplitude (Fig. 8.17).

The derivation of the working equation for a Michelson interferometer is adapted from *Handbook of Applied Solid State Spectroscopy*.[17]

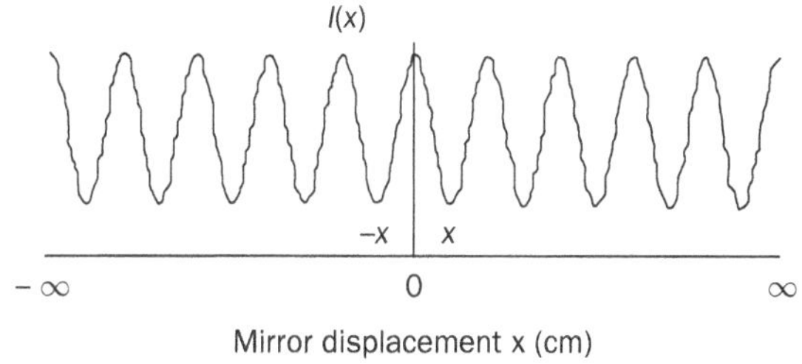

FIGURE 8.17 Output from a Michelson interferometer as a function of mirror displacement *x* shown as an interferogram *I*(*x*) for a monochromatic infrared source.[16]

The amplitude of the wave (traveling in the *z* direction) incident on the beam splitter is given as:

$$E(z,v)dv = E_0(v)e^{i(\omega t-2\pi v z_1)}dv \tag{8.62}$$

where $E_0(v)$ is the maximum amplitude of the beam at $z = 0$.

Let the distances travelled by the reflected and the transmitted beams before recombining be z_1 and z_2 respectively having *r* and *t* as the reflected and transmitted beam coefficients. Then the amplitude of the recombined beam (E_R) can be represented as:

$$E_R[z_1,z_2,v]dv = rtE_0(v)[e^{i(\omega t-2\pi v z_1)} + e^{i(\omega t-2\pi v z_2)}]dv \tag{8.63}$$

Therefore, the intensity of the recombined wave for a fixed spectral range *dv* can be written as:

$$I[z_1,z_2,v]dv - E_R[z_1,z_2,v]E_R^*[z_1,z_2,v]dv = 2E_0^2(v)|rt|^2[1+\cos 2\pi(z_1-z_2)v]dv \tag{8.64}$$

On integrating Eq. (8.64) we get the total intensity for the whole spectrum at any path difference $x = (z_1 - z_2)$ as:

$$I_R(x) = 2|rt|^2\int_0^\infty E_0^2(v)dv + 2|rt|^2\int_0^\infty E_0^2(v)\cos(2\pi xv)dv \tag{8.65}$$

When applying FT to Eq. (8.65), the intensity is converted into spectrum resulting in:

$$E_0^2(v) = (1/\pi|rt|^2)\int_0^\infty\left[(x) - \frac{1}{2}I_R(0)\right]\cos(2\pi vx)dx \tag{8.66}$$

and the resulting spectrum ($S(v)$) as:

$$S(v) \propto E_0^2(v) = CONSTANT\int_0^\infty\left[I_R(x) - \frac{1}{2}I_R(0)\right]\cos(2\pi vx)dx \tag{8.67}$$

The interferogram is Fourier transformed using a computer to convert the space domain into the wave number domain. As the mirror cannot be moved infinitely, the finite mirror displacement makes it impossible to calculate FT integral equation with

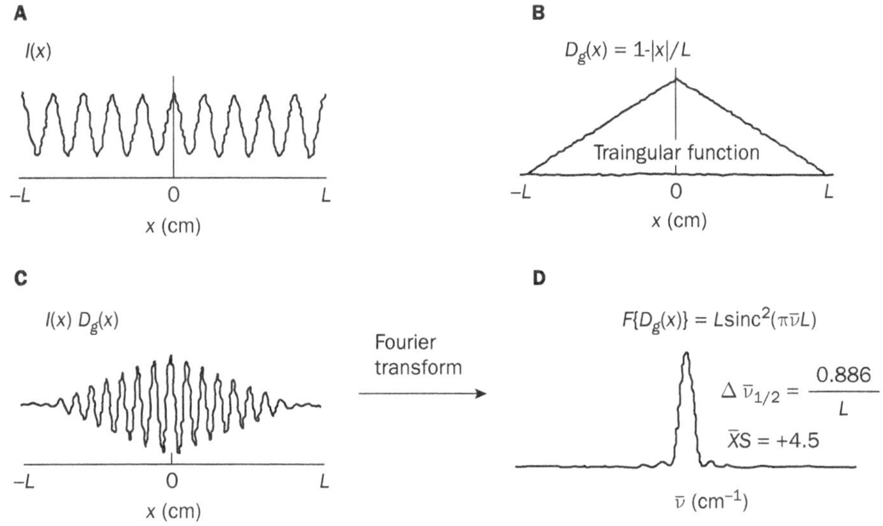

FIGURE 8.18 Triangular truncation function being used to achieve a final Fourier transform spectrum.[16]

infinite limits. Therefore, it is necessary to correctively modify the FT integral by the method of apodization that removes the false side lobes of the spectrum. In order to achieve this, the interferograms are usually multiplied by a function prior to FT. There are numerous apodization functions that can be used, namely, linear, triangular, exponential, Gaussian truncation functions, among others. Figure 8.18 shows the use of triangular truncation function to achieve the final FT spectrum.

8.6 X-ray Photoelectron Spectroscopy

X-ray photoelectron spectroscopy (XPS) is one of the most advanced surface analysis tools that gives information about the surface up to a depth of about 10 nm. It is one of the high-end characterization techniques, increasingly becoming more important than ever, when it comes to capturing information about the energy distributions of x-ray irradiated surfaces. The spectral data capture information about the valence electrons as well as the core electrons. Our discussion here will be limited toward understanding the fundamental principle of operation and a subset of applications where XPS characterization plays a pivotal role. For further details, you may refer to an excellent description by Chusuei and Goodman.[18]

You may be aware of photoelectric emission, which is the emission of electrons (photoelectrons) when light of a certain wavelength is incident upon a metal surface. The principle of operation of XPS involves similar photoelectric interaction between monochromatic x rays and a sample leading to emission of electrons with discrete kinetic energies. The sample in this case can be solid or gaseous and the emitted photoelectrons contain chemical information about the surface analyte. The working principle of XPS is based on energy balance, as depicted in the following equation:

$$h\nu = E_k + \phi + E_b \tag{8.68}$$

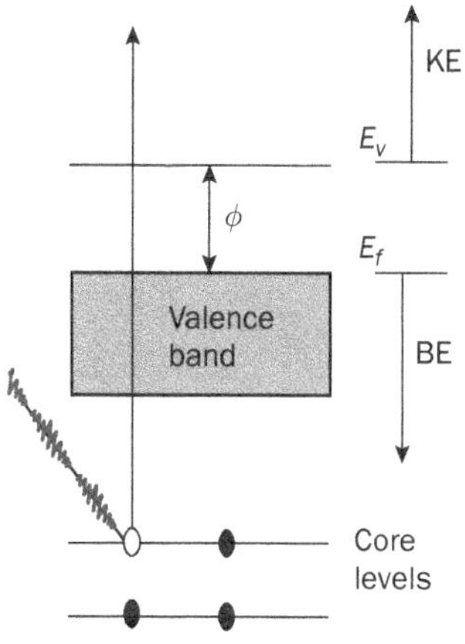

FIGURE 8.19 Energy transitions in x-ray photoelectron emission.

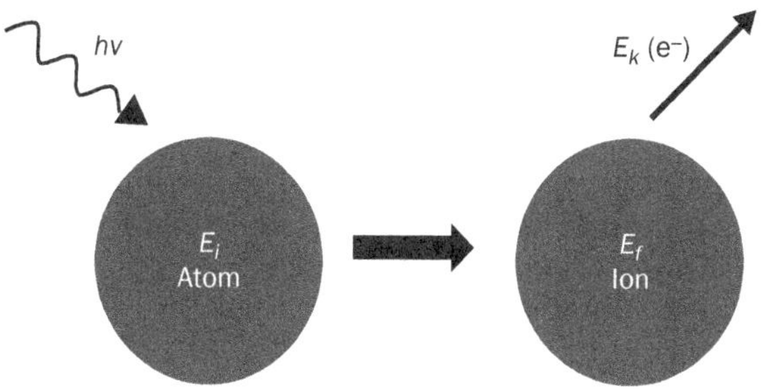

FIGURE 8.20 Electronic configuration of an atom before and after impingement from a photon.

where $h\nu$ is the incident photon energy, h is the Planck's constant, and ν is the frequency of incident x-ray photon. E_k refers to the kinetic energy of the emitted electron, ϕ is the metal work function, and E_b is the electron binding energy. The energy transition process is shown in Fig. 8.19.

In the photoemission process, the incident x-ray photon knocks out a core-level electron whose kinetic energy is recorded by an electron energy analyzer. The energy required to remove an electron from the Fermi level (a reference point with which binding energies of photoelectrons are measured) to vacuum level is known as the *metal work function* (ϕ). Let us take a closer look into the meaning of the binding energy, E_b since it contains information about the chemical state of the surface. To understand the concept, let us consider the following situation (Fig. 8.20).

As the photon hits the atom having a total energy of E_i, an electron is emitted with kinetic energy $E_k(e^-)$, resulting in an ion with a total energy of E_f. Therefore, an energy balance gives:

$$h\nu + E_i = E_k(e^-) + E_f \tag{8.69}$$

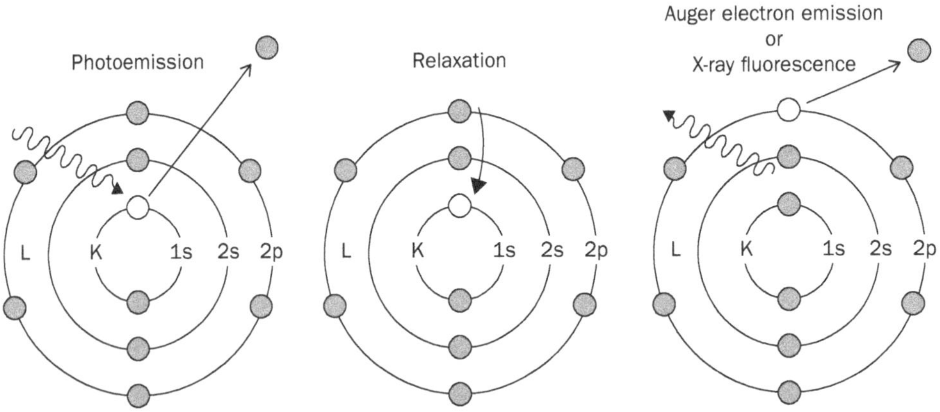

FIGURE 8.21 Different modes of electron emission from incident x-rays or electrons.

Rearranging the terms, we can obtain a physical meaning of the binding energy:

$$h\nu - E_k(e^-) = E_f - E_i = E_b \tag{8.70}$$

Thus, binding energy is the energy of the exiting photoelectron produced by the photoionization process of core-level electrons and contains discrete chemical information. This photoionization is, however, accompanied by secondary emission processes—either a photon emission resulting in x-ray fluorescence or an emission of a valence-level electron called an *Auger electron*. These processes are depicted in Fig. 8.21.

These two secondary emission processes are competitive in nature; however, since XPS utilizes a low photon energy (Mg K_α at $h\nu = 1253.6$ eV or Al K_α at $h\nu = 1486.6$ eV), contributions from x-ray fluorescence are negligible. However, Auger electron emission is characterized by the relaxation of an upper level valence electron into the vacant core-level state after the initial photoionization. This results in ejection of another electron in the valence level. The Auger electrons are independent of the photon energy and their peaks commonly appear in XPS spectra.

The XPS spectra are also affected by inelastic scattering at high binding energies. This results in higher background in the high binding energy side of the XPS peak due to partial loss of kinetic energies of the ejected electrons while they travel through the solid on their way to the detector. High background observed at about 900 eV extending up to the bombardment photon excitation energy is due to bremsstrahlung radiation generated from the x-ray source. Considering all these effects, typical XPS spectra could look like as shown in Fig. 8.22.

Overall, an XPS can do a myriad of things based on the principle of measurement of binding energies which carry information about the chemical states. Surface analysis, ion etching, depth profiling, imaging, and angle-resolved XPS are some of the various techniques that can be tried with such a characterization tool. In principle, an operating XPS looks similar to Fig. 8.23. It consists of three important parts—a load lock (LL), where the sample is introduced; a sample transfer chamber (STC) for parking multiple

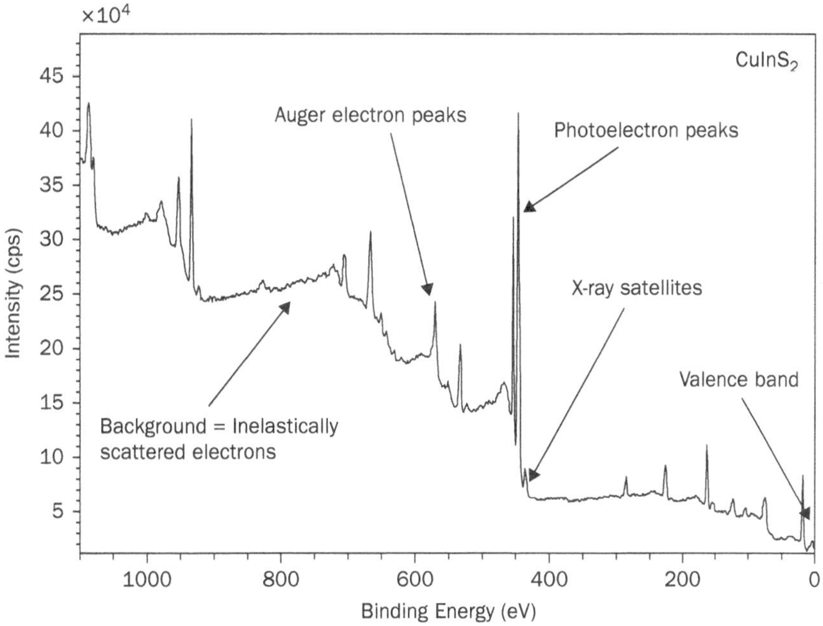

Figure 8.22 Representative XPS spectra showing all possible peaks for an arbitrary sample.

Figure 8.23 XPS instrument available at NTNU premises for characterization.

samples before doing analysis, and a sample analysis chamber (SAC) equipped with a monochromatic x-ray source and an electron detector that measures the kinetic energy.

A quick analysis can be carried out to know about the various elements present in the sample by obtaining survey spectra. Detailed analysis to know about the chemical states of the constituting elements will need high-resolution spectra. As an example, Fig. 8.24 shows the survey and high-resolution spectra of core-shell Fe@Au nanoparticles synthesized using XPS facility at NTNU and analyzed as follows. A hybrid lens (electrostatic and magnetic) mode was employed with an analysis area of approximately 300 μm × 700 μm. Survey spectra were obtained over the range of 0–1100 eV binding energy with an analyzer pass energy of 160 eV. High-resolution spectra of Fe 2p and Au 4f were obtained with an analyzer pass energy of 20 eV. XPS data were processed with Casa XPS software (Casa Software Ltd., UK).

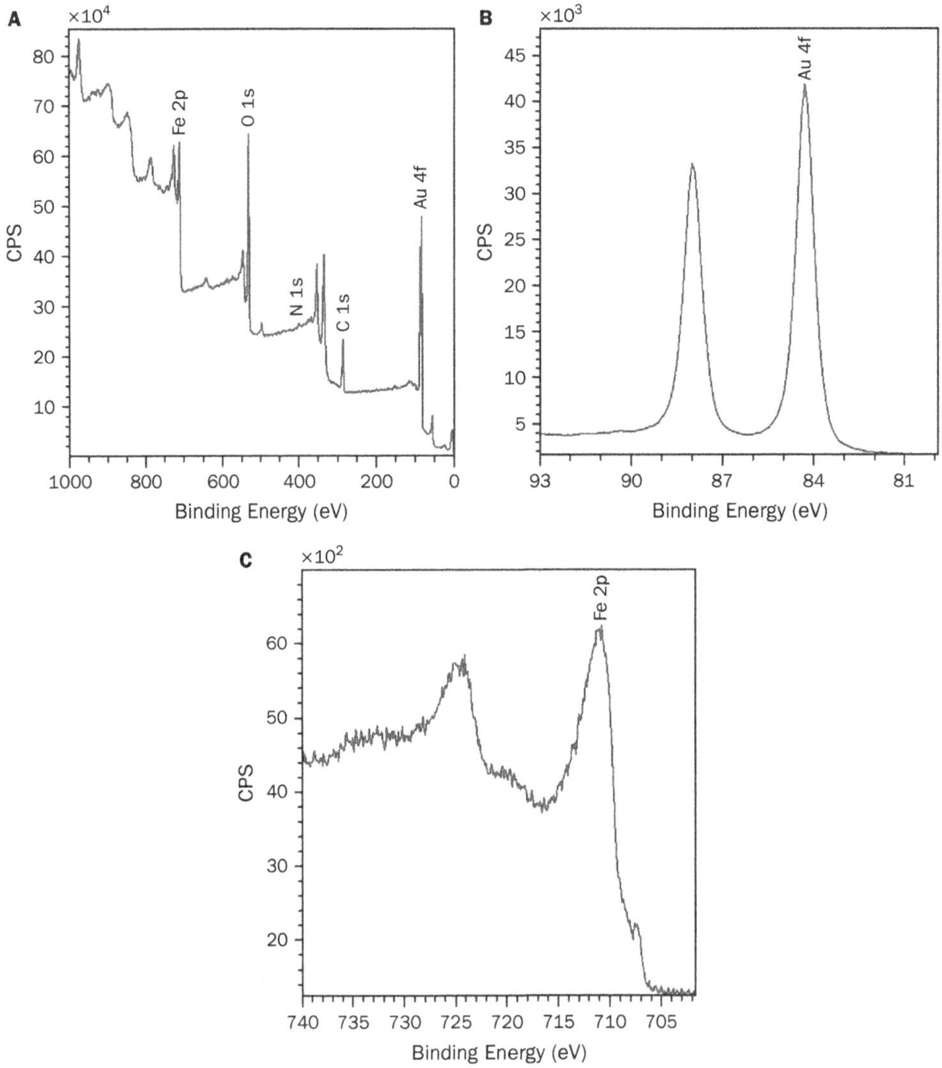

FIGURE 8.24 (A) Survey spectra for core-shell Fe@Au NPs synthesized in our labs using a solution-based approach. High-resolution spectra for (B) Au and (C) Fe, respectively.

This chapter surely has enabled you to get a detailed understanding about the physicochemical properties of the synthesized NPs—both solution properties as well as properties in dry state. From simple visualization of the NPs to detailed understanding of the functional groups present on the surface, different characterization techniques discussed in this chapter will enable you to select nanomaterials for intended applications, some of which will be discussed in Chaps. 9 to 11.

8.7 Study Questions

8.7.1 Concept-Based Questions

8.1 Au NPs coated with sodium citrate were synthesized using a modified Turkevich method. Then they were also coated with PEG to use them for biological applications.

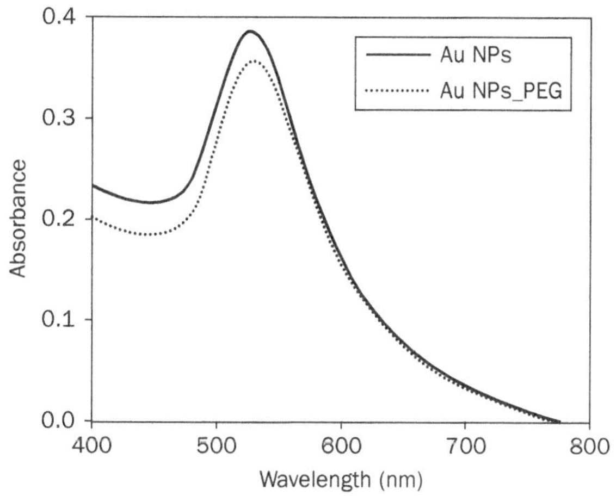

a. These Au NPs show strong absorption around 525 nm as evident from the figure above. Explain this phenomenon.

b. Using the same phenomenon or otherwise, explain why the maximum in wavelength shifts to higher wavelengths upon coating with PEG (shifts from 527 nm to 530 nm).

8.2 In order to calculate the size of these NPs in solution, you plan to use DLS. List the assumptions underlying the principle. Using these assumptions and fundamental principles, derive the governing equation for measurement of size using DLS.

8.3 A master student has collected sufficient data using the DLS setup to come up with the following plot.

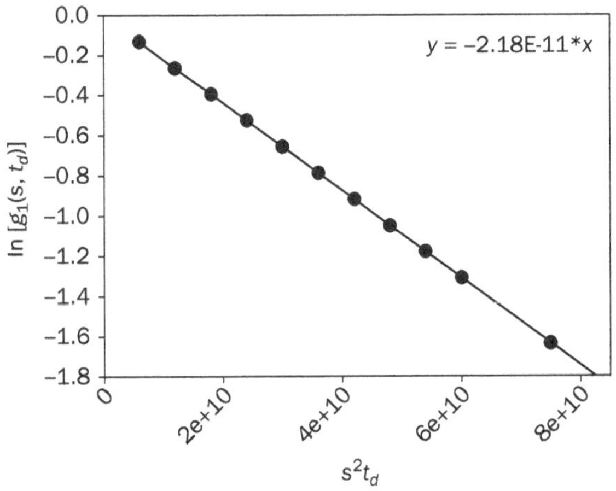

From his note book, you have found out that the experiments were performed at 298 K ($\eta = 0.001$ kg/m-s). Using the equation derived in question 1.2, calculate the size of the Au NPs, assuming the particle distribution to be monodisperse (Boltzmann constant is 1.38×10^{-23} J/K). Assume SI units.

Siegert relation:

$$\ln g_1(s,t_d) = -\overline{D}s^2t_d + \sigma^2 s^4 \frac{t_d^2}{2!} + \cdots$$

For dilute dispersions:

$$\frac{C(s,t_d)}{[\bar{i}(s)]^2} = g_2(s,t_d) = 1 + \xi\left|g_1(s,t_d)\right|^2$$

8.4 Set up the governing equation for x-ray photoelectron spectroscopy. What different information may be obtained by performing XPS measurements on samples?

8.5 Fe@Au core-shell NPs were synthesized using a modified method in Ugelstad Laboratory. The NPs were characterized using DLS and were found to have a size of 55 ± 3 nm. Since these particles were supposed to be used in dry form, a drop of the solution was put on carbon-coated copper grids and observed under S(T) EM. The NPs showed an average size of 30 ± 2 nm. Compare and contrast the two techniques used for determining the particle sizes. Which one would you trust more?

8.6 What is the fundamental principle for operation of DSC? Mention the different modes in which a DSC can function. Set up the final governing equations that are widely used for operation in the two modes.

8.7 What is an auto-correlation function? How is it used for data analysis in a DLS?

8.8 Mention the significant difference in the working principles of IR and FTIR spectroscopies.

8.7.2 Research-Based Questions

8.9 Using a bottom-up synthesis method, one researcher in our lab synthesized citrate-stabilized Fe@Ag core-shell NPs. Since the DLS setup in our lab is a black box, he re-measured the size using an in-house DLS in Dr. Sandvig's lab. In this case, the technician provided him with the following data set.

$s^2t_d * 10^{-9}$ (m^{-2}s)	2.30	4.60	6.90	9.19	11.50
$\frac{\ln[g_1(s,t_d)]}{s^2t_d} * 10^{12}$ (m^{2}s^{-1})	−9.86	−9.72	−9.58	−9.54	−9.31
$s^2t_d * 10^{-9}$ (m^{-2}s)	13.80	16.10	18.40	20.70	23.00
$\frac{\ln[g_1(s,t_d)]}{s^2t_d} * 10^{12}$ (m^{2}s^{-1})	−9.03	−9.03	−8.90	−8.68	−8.62

a. From his notebook, we have found out that the experiments were performed at 298 K ($\eta = 0.001$kg/m-s). Calculate the size of the NPs (Boltzmann constant is 1.38×10^{-23} J/K).

b. Comment on the particle size distribution. What kind of scattering is taken into consideration for DLS measurements?

c. Explain if you can measure the size of non-spherical particles using this technique.

Siegert relation:

$$\frac{C(s,t_d)}{[\bar{i}(s)]^2} = g_2(s,t_d) = 1 + \xi\left|g_1(s,t_d)\right|^2$$

For dilute dispersions:

$$\ln g_1(s,t_d) = -\bar{D}s^2t_d + \sigma^2 s^4 \frac{t_d^2}{2!} + \cdots$$

8.10 Hu et al.[19] synthesized nanostructured lipid (NSL) nanocarriers and loaded them with a model drug, nimodipine. A modified solvent diffusion method was developed to overcome the disadvantages of low drug encapsulation by the conventional solvent method. The nanocarriers from both the systems were loaded with nimodipine and characterized using DSC. The results are shown by the curves below.

Curve (a): NSL prepared by conventional solvent diffusion method after washing with 0.2 wt% SDS solution.

Curve (b): NSL prepared by conventional solvent diffusion method before washing with 0.2 wt% SDS solution.

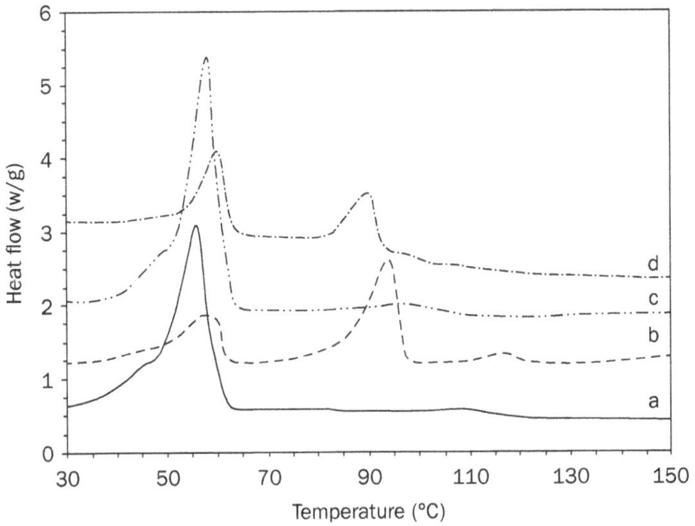

Curve (c): NSL prepared by solvent diffusion method in a drug-saturated aqueous system after washing with 0.2 wt% SDS solution.

Curve (d): NSL prepared by solvent diffusion method in a drug-saturated aqueous system before washing with 0.2 wt% SDS solution.

DSC was used to demonstrate higher drug encapsulation efficiencies (E.E.) in lipid nanocarrier systems using nimodipine as the model drug, as shown in the figure above. The peaks between 50°C and 60°C and 90°C and 100°C indicate the presence of NSL and nimodipine, respectively.

a. Comment on the drug efficiency from the DSC studies by comparing the curves.

b. It can be seen that the drug peak vanishes after washing with SDS solution. Does it reveal anything about the location of the drug in the NSL nanocarrier system?

c. What other characterization technique would you prefer over DSC in order to confirm the encapsulation of drug in the NSL nanocarrier device?

8.11 A student working at Ugelstad Lab synthesized a batch of gold nanospheres and gold nanorods. The average size of nanospheres and nanorods was measured using DLS and the values were 35 ± 3 nm and 12 ± 2 nm, respectively. However, when looked under S(T)EM the nanospheres have an average size of 23 ± 5 nm and nanorods have a length of 47 ± 4 nm and a width of 13 ± 2 nm.

Comment on the discrepancy that you observe and the reason behind it. If the student decides to publish the data, which size should be used?

8.12 After reading about crystallization and about gold nanorods, you are intrigued by their growth processes. You just got access to UV-Vis spectrometer and the lab assistant informs you that you can analyze samples at higher temperature as the equipment has a temperature control. You come up with a plan to synthesize Au NRs using seed-mediated growth. You keep the UV-Vis temperature at the reaction temperature and withdraw samples from the growth solution at various

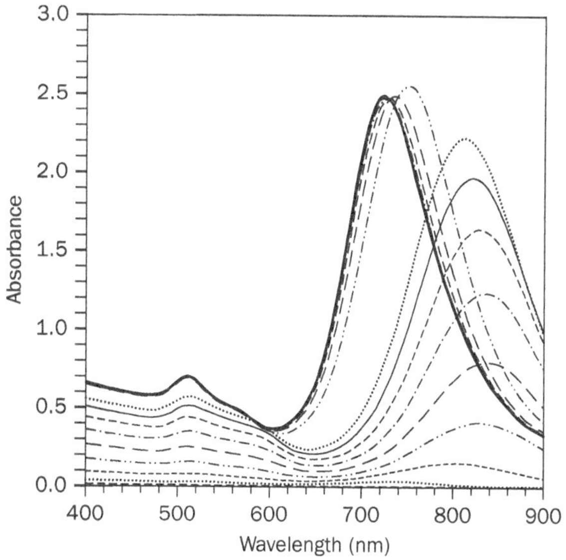

time points and check their LSPR peaks at different time points to spot differences. You continue this until the end of the reaction (8 hours) and plot absorbance vs wavelength and obtain a plot as shown in the figure above.

Initially you see no peaks and later two peaks are visible. The longitudinal peaks grow in absorbance values and also show blue shift with increasing time. Based on the knowledge from Chaps. 2 and 3 and Beer Lambert's law, comment on the growth of the NRs.

You also notice a change in the LSPR peak of the longitudinal axis, while the LSPR peak of transverse axis is almost the same throughout the growth process. Explain the reason for this phenomenon.

After the UV-Vis measurements you are left with some more samples and you decide to look at these samples using S(T)EM. (Images have been obtained using NTNU Nanolab Electron Microscope.)

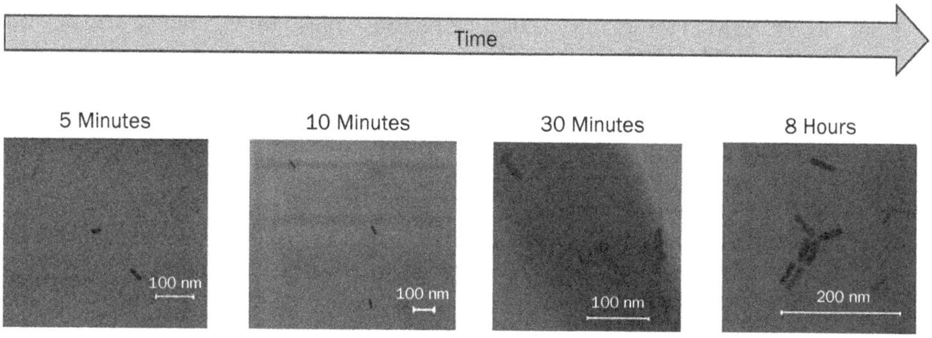

You observe fully grown NRs as early as 5 minutes, and the size of NRs obtained after 30 minutes is equal to the size of NRs obtained after the completion of the reaction (8 hours). Comment on the growth and growth mechanism based on these observations.

8.13 You are fascinated by the simplicity of the modified Turkevich method and decide to synthesize a batch of Au NPs using 10 mL of 1.5 mM chloroauric acid and 10 mL of 10 mM citrate as per the following equation:

$$2AuCl_4^- + Ctr^{3-} + 2H_2O \rightarrow 2Au^0 + 3H_2O + 3CO_2 \uparrow + 8Cl^- + 3H^+$$

The size and electrophoretic mobility of the particles were measured to be 32 nm and -4.23×10^{-4} cm^2/V-s, respectively.

The following data are provided to you by your supervisor:

Dielectric constant = 78.5

Permittivity in free space $(\varepsilon_0) = 8.854 \times 10^{-12}$ F/m

Viscosity $(\eta) = 8.9 \times 10^{-4}$ kg/m-s 0.01 cP

Boltzmann constant $(k_B) = 1.38 \times 10^{-23}$ J/K

Avogadro number $(N_A) = 6.023 \times 10^{23}$ molecules/mole

Elementary charge $(e) = 1.602 \times 10^{-19}$ C

Your supervisor wants your inputs about the experiments and asks you to calculate the zeta potential of the NPs using the data given above. He is of the opinion that the NPs are agglomerated. Based on your calculations, would you agree with your supervisor? In either case, give a clear explanation.

8.8 References

1. Hornyak, G. L.; Tibbals, H. F.; Dutta, J.; Moore, J. J., *Introduction to Nanoscience and Nanotechnology*. Boca Raton, FL: CRC press, 2008.
2. Petryayeva, E.; Krull, U. J., Localized surface plasmon resonance: nanostructures, bioassays and biosensing—a review. *Analytica Chimica Acta*. **2011,** *706*(1), 8–24.
3. Zhang, J. Z.; Noguez, C., Plasmonic optical properties and applications of metal nanostructures. *Plasmonics*. **2008,** *3*(4), 127–150.
4. Willets, K. A.; Van Duyne, R. P., Localized surface plasmon resonance spectroscopy and sensing. *Annual Review of Physical Chemistry*. **2007,** *58*, 267–297.
5. Mayer, K. M.; Hafner, J. H., Localized surface plasmon resonance sensors. *Chemical Reviews*. **2011,** *111*(6), 3828–3857.
6. Link, S.; El-Sayed, M. A., Shape and size dependence of radiative, non-radiative and photothermal properties of gold nanocrystals. *International Reviews in Physical Chemistry*. **2000,** *19*(3), 409–453.
7. Mulvaney, P., Surface plasmon spectroscopy of nanosized metal particles. *Langmuir*. **1996,** *12*(3), 788–800.
8. Okamoto, H.; Imura, K., Near-field imaging of optical field and plasmon wavefunctions in metal nanoparticles. *Journal of Materials Chemistry*. **2006,** *16*(40), 3920–3928.

9. Stetefeld, J.; McKenna, S. A.; Patel, T. R., Dynamic light scattering: a practical guide and applications in biomedical sciences. *Biophysical Reviews.* **2016,** *8*(4), 409–427.
10. Frisken, B. J., Revisiting the method of cumulants for the analysis of dynamic light-scattering data. *Applied Optics.* **2001,** *40*(24), 4087–4091.
11. Koppel, D. E., Analysis of macromolecular polydispersity in intensity correlation spectroscopy: the method of cumulants. *Journal of Chemical Physics.* **1972,** *57*(11), 4814–4820.
12. Thomas, J. C., The determination of log normal particle size distributions by dynamic light scattering. *Journal of Colloid and Interface Science.* **1987,** *117*(1), 187–192.
13. Bhadeshia, H., Differential scanning calorimetry. Research paper. Department of Materials Science and Metallurgy, University of Cambridge, London, 2002.
14. Mraw, S. C., Mathematical treatment of heat flow in differential scanning calorimetry and differential thermal analysis instruments. *Review of Scientific Instruments.* **1982,** *53*(2), 228–231.
15. Thomas, L. C., Modulated DSC® Paper# 1: Why Modulated DSC®? An Overview and Summary of Advantages and Disadvantages Relative to Traditional DSC. *TA Instruments Technol. Papers,* **2005.**
16. Markovich, R. J.; Pidgeon, C., Introduction to Fourier transform infrared spectroscopy and applications in the pharmaceutical sciences. *Pharmaceutical Research.* **1991,** *8*(6), 663–675.
17. Jaggi, N.; Vij, D. R., Fourier transform infrared spectroscopy. In: *Handbook of Applied Solid State Spectroscopy*, Vij, D. R., ed. New York, NY: Springer, 2006, pp. 411–450.
18. Chusuei, C. C.; Goodman, D. W., X-ray photoelectron spectroscopy. In: *Encyclopedia of Physical Science and Technology*. 3rd ed. Meyers, R. A., ed. New York, NY: Academic Press, 2003, pp. 921–938.
19. Hu, F.-Q.; Zhang, Y.; Du, Y.-Z.; Yuan, H., Nimodipine loaded lipid nanospheres prepared by solvent diffusion method in a drug saturated aqueous system. *International Journal of Pharmaceutics.* **2008,** *348*(1–2), 146–152.

CHAPTER 9

Applications in Catalysis

9.1 Introduction

Catalysis is a key technology in our modern society. Through speeding up and controlling the selectivity of chemical reactions, we can produce a range of materials that we rely on, convert energy from one form to another, as well as clean and remove emissions. Catalysis plays a key role in biology, through the action of enzymes, which are very selective catalysts but with a limited range of operational conditions. The earliest human application or invention of "technical" catalysis we know of dates back to around 2800 BC, when Babylonians made and stored soap-like materials in clay vessels.[1] The material was made by boiling fat in the presence of ash. This method was widely used in later civilizations, e.g., in Egypt and in the Roman Empire.

Industrial catalysis is an important business at present. The value of the global catalyst market is expected to reach $34.3 billion by 2024, according to a report by Grand View Research, Inc. The numbers get even more impressive when looking at the value of the products and processes that rely on catalysis. Currently, about two-thirds of the top commodity chemicals are produced in catalytic processes (ammonia, nitric acid, methanol, etc.). For $1 used on a catalyst, the value of the corresponding products amounts to $200–1000. This means that in an industrialized country such as the United States, the value of goods produced catalytically amounts to 15–30 percent of the GDP (gross domestic product).[2]

By definition, a catalyst enhances the rate of a chemical reaction, without being consumed or altered by the same reaction. Catalysts can have many forms. In this context, our interest is only in *heterogeneous catalysts,* solid materials converting reacting species in the gas or liquid phase. These materials rely on nanosized structures in order to function, for example, nanosized pores that provide high surface area and nanosized metal particles that provide unique active sites for reactions. The role of the active site is to activate the reactants and provide a reaction path with a lowered energy barrier, the activation energy.

In the 19th century, catalysis was only an interesting phenomenon with a few practical applications. Modern industrial catalysis was developed in early 20th century, with discoveries such as the Ostwald process[3] for the production of nitric acid (Nobel Prize in Chemistry in 1909) and the Haber–Bosch process of ammonia synthesis (Haber was controversially awarded the Nobel Prize in Chemistry in 1918), and a range of other catalysts and processes that have changed and shaped society. As an example, the French chemist Paul Sabatier (Nobel Prize in Chemistry in 1912) developed catalysts for hydrogenation, consisting of "finely divided metals," which has applications in food preparation (fat hydrogenation to produce margarine), oil refining, and petrochemistry.

This was an early reference to nanotechnology, some 60 years before the term was coined (see, e.g., information on the homepages of the Swedish Nobel Institute, www.nobel.se).[4] A side-reaction, the formation of carbon, also called *coke* or *soot*, from hydrocarbons or synthesis gas is a considerable problem for these catalysts. The carbon is an unwanted by-product that deactivates and can destroy the catalyst. Only with the invention of the high-resolution electron microscope (developed in the 1930s, commercially available from 1939) it was possible to observe this carbon and learn that it actually could form nanostructures. Today we differentiate between carbon nanotubes (CNT), with hollow (tubular) morphology made from concentric graphene layer(s), and carbon nanofibers (CNF) where the orientation of the graphene layers influences the properties of the material. (For further details refer to, e.g., Hornyak et al.,[4] Serp and Figueiredo,[5] Boehm, 1997.[6])

The purpose of this chapter is to provide a brief introduction to catalysis and examples of nanomaterials used in catalysis, all with emphasis on the materials. The reader is referred to textbooks on catalysis (see Sec. 9.5 for a list of books) for a detailed treatise of heterogeneous catalysis.

9.1.1 Brief Introduction to Heterogeneous Catalysis

Heterogeneous catalysis is probably the oldest industrial application for which the nanostructure of the material is deliberately utilized. The nanostructure in question can be the porous system (nanoporous materials, with large surface areas and pore structures with specific properties) allowing transport of very specific structural and chemical properties. The other main type of material, often used in combination with porous materials, contains nanosized particles with specific catalytic properties. The role of the porous material is then to be a support, or carrier of the small particles. The catalytic properties of the nanoparticles can be enhanced by their small size. The particles can be metallic, or even can be metals in chemical states other than metallic, e.g., as oxides or sulfides. A key parameter for these materials is the *dispersion*—the fraction of the material available on the surface (interface to the fluid phase). Table 9.1 gives examples of important catalyst systems and some of their most important applications.

Heterogeneous catalysts have applications in a number of areas, and Fig. 9.1 illustrates some of the classes of materials used, with examples of active phases and reactions where these catalysts are used. As a technical and scientific subject, catalysis involves the understanding of thermodynamics, kinetics, electronic interactions, (nano) structures, reactor design, and process design and development. As such, it is an interdisciplinary subject, where contributions from materials science, chemistry, physics, and chemical engineering are necessary in order to understand and develop catalytic solutions.

By definition, a catalyst enhances the reaction rate of a chemical reaction without being consumed. This is explained in terms of an alternative reaction pathway, also underlining the fact that the overall reaction thermodynamics, e.g., the heat of reaction and reaction free energies (determining the reaction equilibrium), is unchanged by introducing a catalyst. The temperature dependence of the reaction rate is usually described by the Arrhenius equation:

$$k = Ae^{\left(-\frac{E_{act}}{RT}\right)} \tag{9.1}$$

Type of Material	Form/Structure	Typical Active Phases	Examples of Applications
Metal	Bulk, porous/ skeletal, supported	Metallic Pt, Ni, Co, Ir, Ru, Cu, Pd, Au, Ag, Fe Multi-metallic systems	Hydrogenation, dehydrogenation, oxidation, methanol synthesis, Fischer–Tropsch synthesis, ammonia synthesis, exhaust gas cleaning
Metal oxides	Bulk, supported, pellets, extruded shapes	Oxides of V, Al, Si, Ti, Cu, Bi, Ni, Mn, Co, mixed oxides	Oxidation, dehydration, isomerization, alkylation, cracking
Metal sulfides	Supported	Sulfides of Mo, Ni, W	Hydrogenation, hydrotreating (hydrodesulfurization, hydrodenitrogenation), hydrodearomatization, hydrodeoxygenation
Metal carbides	Bulk, supported	Carbides of Fe, Mo, W	Hydrogenation, Fischer–Tropsch synthesis
Other forms	Supported, anchored	Chlorides, nitrides, enzymes, metal complexes	Polymerization, hydrogenation, alkylation, organic transformations, biological reactions

Table 9.1 Examples of Catalytic Materials

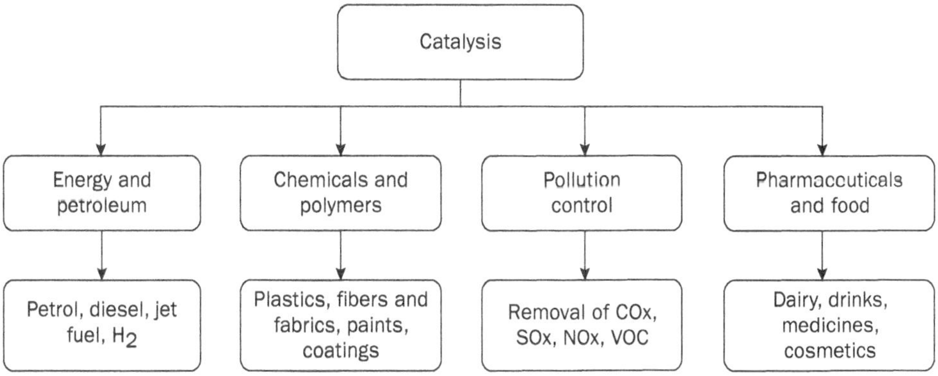

Figure 9.1 Overview of applications of catalysis.

where k is the temperature-dependent rate constant in the rate equation (the relationship between the reaction rate and the concentrations or pressures of the components involved), A is the pre-exponential factor, often called frequency factor. A and k have the same units, for a first-order reaction this is [s^{-1}], giving rise to the term frequency factor for A. E_{act} is the activation energy [J/mol], R is the gas constant [J/K·mol], and T is the temperature [K]. The rate constant is the temperature-dependent term in the rate equation, and the activation energy represents the energy barrier toward reaction. The action of the catalyst is to allow a new reaction path, with a lower energy barrier.

When we discuss the properties of catalysts we rely on some important definitions:

Activity: The activity characterizes the ability to which a given substance can act as a catalyst, and is usually defined as the specific reaction rate, e.g., as [mol/g cat, time].

Selectivity: The selectivity is the ability to convert the reactants into the desired product (instead of side products) (amount of desired product/amount of reactant converted).

Lifetime: All heterogeneous catalysts lose some activity with time; this phenomenon is called deactivation. The lifetime is the time during which a certain (minimum) activity is maintained.

Heterogeneous catalysis is a surface phenomenon typically involving a solid phase, the catalyst, and a fluid, either a gas or a liquid. The catalytic process can be divided into specific steps: The reactants are in the fluid bulk phase—they need to be transported to the surface, where they are adsorbed and react. Subsequently, the products desorb from the surface and are transported to the bulk phase. This is a complex series of events, and any step can at given conditions be slow and dominate the rate of the process.

9.1.2 Adsorption

The key step in catalysis is adsorption on the surface. You have already been introduced to the topic of adsorption in Chap. 5. Adsorption is by definition the preferential accumulation of material on a surface of a material due to surface forces, to be distinguished from absorption, which is a bulk phenomenon. The material adsorbed is called the *adsorbate.* The amount adsorbed is proportional to the surface of the material, whereas the amount absorbed relates to the volume of the material in question. This distinction is not always easy to make, e.g., for materials that are microporous, such as zeolites.

Adsorption is divided into physical adsorption (physisorption) and chemical adsorption (chemisorption). Physisorption involves weaker bonds, typically involving van der Waals forces, while chemisorption is a chemical process akin to bond forming. Almost all adsorption processes are exothermic. Since the process takes place spontaneously we must have $\Delta G < 0$, and ΔS is negative (since molecules going from gas phase to the adsorbed state lose degrees of freedom), and since

$$\Delta G = \Delta H - T\Delta S \tag{9.2}$$

we must have a negative ΔH of a certain size. The enthalpy of adsorption, also called the heat of adsorption, is a key parameter describing the nature of the adsorption process, and in case of chemisorption processes it defines the strength of the adsorption bond. Table 9.2 illustrates the main differences between the two types of adsorption.

The relationship between the conditions and the amount adsorbed is usually described in terms of isotherms (the amount adsorbed at equilibrium at a constant temperature as a function of the pressure of the gas) or isobars (the amount adsorbed as a function of temperature at constant pressure). Isotherms or isobars can be measured using suitable instruments and are reported for a number of systems. The term *isotherm* is also used to describe mathematical models describing such phenomena. For example, a monolayer situation (an adsorption that leads to a monolayer or saturation of a surface) is usually described by the Langmuir isotherm (already discussed in detail

	Physisorption	Chemisorption
Type of attachment	Weak van der Waals forces	Strong chemical bonding
Specificity	Not specific	Highly specific
Reversibility	Reversible	Irreversible (but usually adsorbates can be removed by extended pumping or heating)
Temperature dependency	No appreciable activation energy, can occur at low temperature	Can be an activated process, rate increases with *T* (but so does also rate of desorption)
Heat of adsorption	Low heat of adsorption (10–40 kJ/mol)	Heat of adsorption typically higher (80–200 kJ/mol)

TABLE 9.2 Physisorption and Chemisorption Characteristics

in Chap. 5) shown in another form (widely used for catalysis applications) in Eq. (9.3), describing molecular adsorption of a simple molecule on sites with constant heat of adsorption:

$$\theta_A = \frac{K_A p_A}{1 + K_A p_A} \tag{9.3}$$

where θ_A is the fractional surface coverage, K_A is the equilibrium constant for adsorption, and P_A is the partial pressure of the component *A*. This relationship was originally derived from the kinetics of adsorption,[7] but can also be established from thermodynamics, in both cases assuming an equilibrium between adsorption and desorption. There are many such models or isotherms that describe the amount adsorbed, e.g., the Freundlich isotherm, Temkin isotherm, Henry's law, and the Brunauer–Emmet–Teller (BET) equation.

The BET equation describes multilayer adsorption, and is commonly used to characterize the total surface area of samples [Eq. (9.4)]:

$$\frac{p}{V(p_0 - p)} = \frac{1}{V_m C} + \frac{C-1}{V_m C} \cdot \frac{p}{p_0} \tag{9.4}$$

where V is the volume of gas adsorbed, V_m is the volume adsorbed equivalent to a monolayer of gas coverage, p is the pressure, p_0 is the saturated vapor pressure of the gas, and C is a constant. Equation (9.4) can be linearized, and V_m can be determined from a plot of V_m vs. p/p_0, typically for p/p_0 in the range 0.05–0.3. The gas used is often N_2, adsorbed at its boiling point (–196°C), and the surface area of the sample can then be calculated from the cross-sectional area of closely packed N_2 molecules on the surface, 0.162 nm^2/molecule. The measurements are carried out using liquid N_2 as a coolant in dedicated instruments. When the pressure is extended toward $p/p_0 = 1$, condensation occurs, starting in narrow pores. Figure 9.2 shows a classification of isotherms, also indicating situations where hysteresis can occur.[8] The shape of the isotherms can be used as an indicator of the structure of the material investigated. As an example, type I

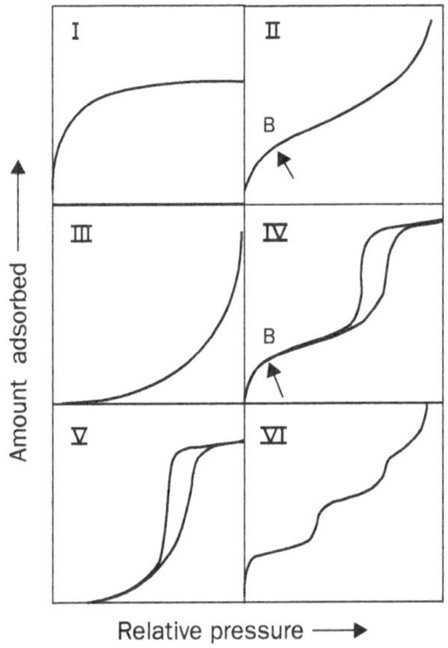

Figure 9.2 Classification of isotherms. (Reproduced with permission from Sing.[8])

is the result of adsorption on a microporous material such as zeolites or activated carbons, the maximum level reflecting the micropore volume of the material. Typical mesoporous materials used as catalysts or supports are often of type IV, and the hysteresis loop condensation behavior can be used to characterize the pore size distribution [see, e.g., Sing (2001)[9] for details and a further description]. The point B marked in the case of isotherm types II and IV indicates the beginning of the almost linear middle section of the isotherm. This is often taken to indicate the stage at which monolayer coverage is complete and multilayer adsorption is about to begin.

The pore system of a material is important because it influences the transport of material from the bulk phase (gas or liquid) to the active phase (e.g., metal sites) where the reaction occurs. The transport in the porous system is inevitably by diffusion, and the diffusion rate is very dependent on the pore size and shape. A hierarchical pore structure, where macropores facilitate transport into the material, and meso- and micropores branching (such as branches on a tree) provide the large surface area necessary for the catalytic activity. In large pores, diffusion is by bulk diffusion, but if the pore size is smaller than the mean free path of the molecules the transport mechanism is Knudsen diffusion. This is commonly the case in mesopores. The diffusion in micropores (e.g., in zeolites) is more difficult to classify, several barriers for transport are present,[10, 11] and the rate of diffusive transport can be very slow. Configurational diffusion occurs when the diameter of the diffusing molecule is in the same range as the channel or pore diameter in the solid. This is the case for many molecules in zeolites.[12]

The distance molecules have to travel by diffusion is dependent on the size of the catalyst particles. Heterogeneous catalysts can have a range of physical structures, ranging from fine powder (for fluidized beds or slurry reactors), to spheres and extrudates (typically in the size range 1–30 mm), to macroscopic shapes (monoliths with fairly large open cross sections), all designed to give the necessary contact and residence time for the reacting fluid, with a suitable pressure drop.

Often, physical adsorption is presumed to precede chemisorption. Following adsorption on the active phase of the catalyst, adsorbed reactants may move on the surface (by diffusion), and in some cases dissociate before they undergo reaction and form products that desorb. The description of the mechanisms of catalytic reactions is beyond the scope of this chapter, but the reader is directed toward the vast literature in the field, starting with the textbooks listed in Sec. 9.5.

9.2 Nanoporous Materials—Catalyst Supports and Active Materials

9.2.1 Oxide Supports

As outlined above, most heterogeneous catalysts are nanoporous materials. An important application of porous materials is as catalyst supports. Catalyst support makes up a large fraction of the catalyst and provides the physical structure and surface area for deposition of the active phase. In some cases, the support itself has catalytic activity and takes part in the reaction mechanism, e.g., by providing acid sites for a bifunctional mechanism where supported metal particles carry out hydrogenation and dehydrogenation reactions. Desired properties of a support are as follows:

- High surface area
- Desirable surface chemistry
  - Anchor metal particles and provide high dispersion
  - Affect the functionality, e.g., SMSI (strong metal support interaction)
  - Specific catalytic functionalities (acidity, oxygen storage, reactant activation, etc.)
- Suitable porosity
- Stability
  - Thermal
  - Mechanical
  - Chemical
- Processability and shaping

The most common materials used are oxides with high melting points (e.g., alumina—Al_2O_3, and silica—SiO_2, or mixed oxides). Some oxides also have catalytic properties, and can be catalysts for dehydration reactions (acid sites) or oxidation reactions (typically semiconducting oxides). Other supports can be clays or carbon materials, the latter often used in electrochemical applications because of their electrical conductivity. Table 9.3 gives some examples of supports and the active phases and industrial applications they are used in.

9.2.2 Alumina and Silica

High surface area supports are by nature unstable; the aluminas are an example. They are produced by stepwise thermal dehydration of aluminum hydroxides.

Support	Active Phase	Process
γ-Al_2O_3 (gamma-alumina)	Pt, PtRe, PtSn	Catalytic reforming (gasoline production)
	Cu-ZnO	Methanol synthesis
	Ni, Pt, Pd	Hydrogenation
	PtSn, Cr_2O_3	Dehydrogenation
	$CuCl_2$	Oxychlorination
	CoMoS, NiMoS	Hydrodesulfurization and hydrotreating
	Co	Fischer–Tropsch synthesis
α-Al_2O_3 (alpha-alumina)	Ni	Steam reforming
	Ag	Epoxidation
SiO_2 (silica)	Cr_2O_3	Polymerization
	H_3PO_4	Hydration
	V_2O_5	Oxidation
TiO_2 (titania)	V_2O_5	Oxidation
Carbon	Pd, Pt	Hydrogenation

Table 9.3 Examples of Catalytic Supports for Industrial Processes

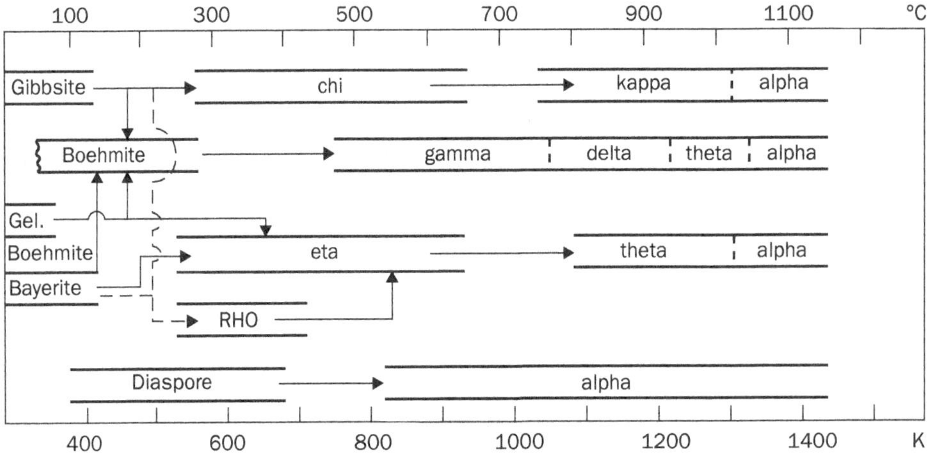

Figure 9.3 Transition aluminas. (Reprinted with permission from Santos et al.[13])

The end-product (which is stable, and not strictly a transition alumina) is α-Al_2O_3. Figure 9.3 shows the possible transitions from aluminum hydroxide compounds to alumina with different structures. Different hydroxides can be the starting point.

Gibbsite ($Al(OH)_3$), boehmite [γ-AlO(OH)], bayerite ($Al(OH)_3$), or diaspore [α-AlO(OH)] gives rise to different alumina structures when heated and dehydrated, but upon further heating only α-Al_2O_3 is stable. The high porosity and large surface areas are only available in the transition aluminas, but as long as these are not exposed to higher temperatures they are remarkably stable. The most used material is γ-Al_2O_3,

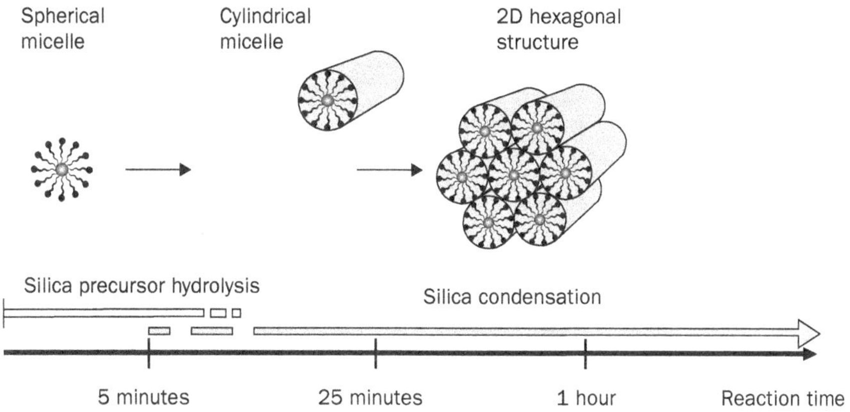

FIGURE 9.4 Initial stages of ordering during preparation of mesoporous silica SBA-15. (Reprinted with permission from Zholobenko et al.[16])

which has a spinel-like structure, typically exposing a surface area of 100–300 m^2/g, linked to a mesoporous structure with pores in the range 5–15 nm. Pores are classified by International Union of Pure and Applied Chemistry (IUPAC) according to their pore sizes as micropores (with diameters <2 nm), mesopores (2–50 nm), and macropores (>50 nm). The final material in the transition series is α-Al_2O_3, a crystalline material with low surface area (typically below 10 m^2/g), which can be used as a support for high temperature reactions, such as the steam re-forming of methane.

Silica, SiO_2, is less used than alumina, but also has important catalytic applications. Silica is amorphous, with a tunable pore structure and high surface area (typically in the range 50–500 m^2/g). Its limitation is its lack of stability toward sintering in the presence of steam at high temperatures.[14] There are many methods to prepare silica. In the sol-gel method, silicon alkoxides (e.g., tetraethylorthosilicate, or TEOS) are polymerized to form a gel that can be dried to form silica.

The most common process of this kind is the Stöber process.[15] If TEOS is reacted in the presence of a template made of micellar rods the resulting silica contains pores with regular size and arrangement. A nice example is the material family called *SBA* (*Santa Barbara Amorphous*), where the silica structure itself is amorphous, but the pore structure is regular.

The initial steps of the preparation according to the cooperative self-assembly mechanism are outlined in Fig. 9.4.[16] This involves formation of spherical micelles of the block co-polymer used as a template in the first step. This is followed by the formation of organic-inorganic hybrid cylindrical micelles. These then organize into a hexagonal structure, and the density of silica increases as silicate species condensate in the void between the micelle cylinders. This structure is maintained as the material is heated and the organic part removed. Figure 9.5 shows a catalyst made this way.

The support material contains straight pores with a diameter estimated to be about 6 nm, containing Pt particles of approximately the same size. This silica material was made using a triblock copolymer (EG–PG–EG) as the template for the support preparation, and then depositing platinum using impregnation of a Pt salt. This particular sample material had a surface area of 860 m^2/g.[17]

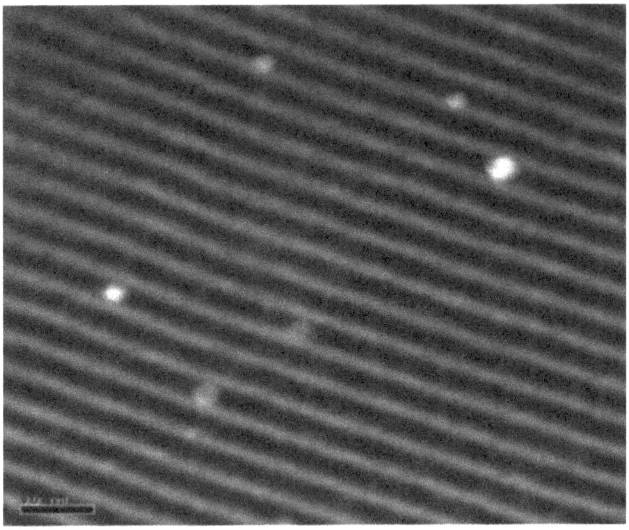

Figure 9.5 STEM image of mesoporous silica SBA-15 with Pt nanoparticles. Dark channels are pores, lighter lines are pore walls made of SiO_2, and bright spots are Pt particles located inside the pore structure. The size bar represents 20 nm. (Reproduced with permission from Chytil et al.[17])

9.2.3 Zeolites and Zeotype Materials

Zeolites are crystalline oxides composed of oxygen tetrahedra connected by metal atoms, Si and Al. These tetrahedra are linked by common oxygen atoms to form octahedra, sodalite cages, which are building blocks for different structures. Some zeolites are naturally occurring, but a wide range of structures are synthesized using hydrothermal synthesis techniques. By varying the chemical composition and choice of structure-directing agent (template), a range of pore structures and pore sizes can be obtained. Zeolites are classified according to the size of the cavities and the size of the pore openings.

This is controlled by the number of O-atoms forming the pore opening. Table 9.4 gives some examples of dimensions and structures of common zeolites used in catalysis and gas cleaning. A comprehensive source for details of structures and dimensions of zeolites is the "Database of Zeolite Structures" (www.iza-structure.org/databases/).

Figure 9.6 shows an example of how the pore size determines the chemical reactions inside the material, in this case the isomerization of *n*-hexane over two catalysts: a zeolite with 10-ring channels (ZSM-5), compared to a zeolite with 12-ring channels (BEA). The smaller channels do not allow the formation of the larger molecules.[18] This phenomenon is termed *shape selectivity*, and is further grouped as reactant selectivity (reactants are too large to enter the pores), product selectivity (products are too large to exit cavities in the zeolite), and transition state selectivity (the cavity is not large enough to accommodate the transition state of a reaction).

The *n*-hexane isomerization reaction is acid catalyzed. The acidity of zeolites is linked with their content of aluminum. Zeolites are mainly composed of Si and O, but they also contain some aluminum. Replacing Si^{4+} with Al^{3+} ions slightly distorts the geometry of the framework, but also introduces a charge imbalance to the structure. This must be

Zeolite	Pore Structure	Pore Opening	
		Number of O-atoms	Dimensions (nm)
Zeolite A	3-D	8	0.41
ZSM-5	3-D	10	0.54 × 0.56
Mordenite	2-D	12	0.74
Faujasite, Y	3-D	12	0.74

Table 9.4 Examples of Zeolite Structures

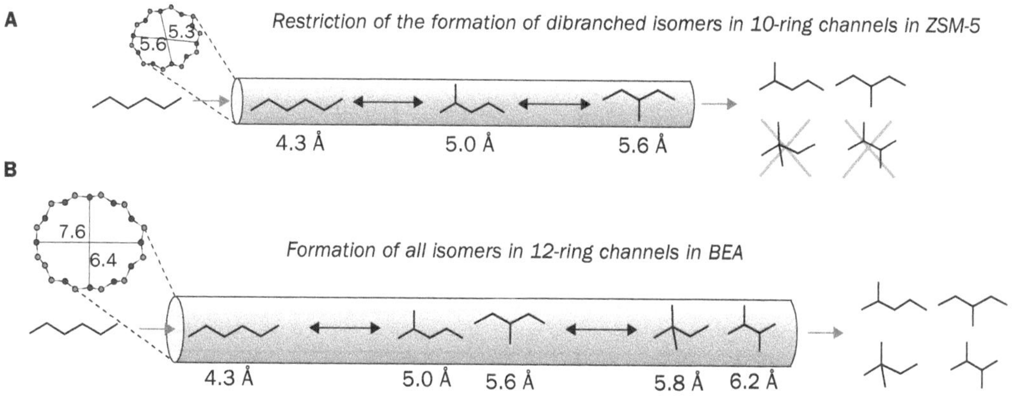

Figure 9.6 Illustration of shape selectivity in zeolite channels in isomerization of *n*-hexane. (Reproduced with permission from Sazama et al.[18])

balanced by a cation—commonly Na^+ is used as part of the synthesis. This cation can be ion-exchanged, e.g., for a proton (typically with NH_4^+, which is subsequently decomposed to form NH_3 and a proton); this proton provides an acid site on the surface. Figure 9.7 illustrates a model for this acidity, showing how the acidic proton is located.

Thus, one important application area of zeolites is as solid acid catalysts, as illustrated in Fig. 9.6. The number and the strength of the acid sites are tunable, and the Si:Al ratio of a zeolite is an important measure of this. It is also possible to exchange other metal cations into a zeolite, giving very specific active sites for a range of reactions. As mentioned above, diffusion rates in the micropores of zeolites are low. The strategy used to overcome this and ensure that fast reactions are not limited by slow diffusion rates is to introduce larger pores, mesopores as well as macropores. There are several ways of doing this, such as de-alumination by steaming, de-silication, and special synthesis routes involving pore-forming templates.[20, 21]

9.3 Carbon Materials

Carbon materials come in many forms and have wide applications as catalyst supports and as catalysts. Figure 9.8 illustrates possible carbon allotropes based on structure and hybridization.

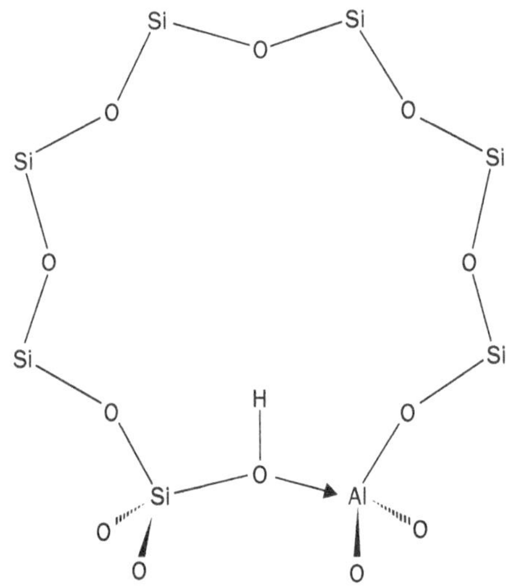

FIGURE 9.7 Illustration of acidity in zeolites. (Reproduced with permission from Busca.[19])

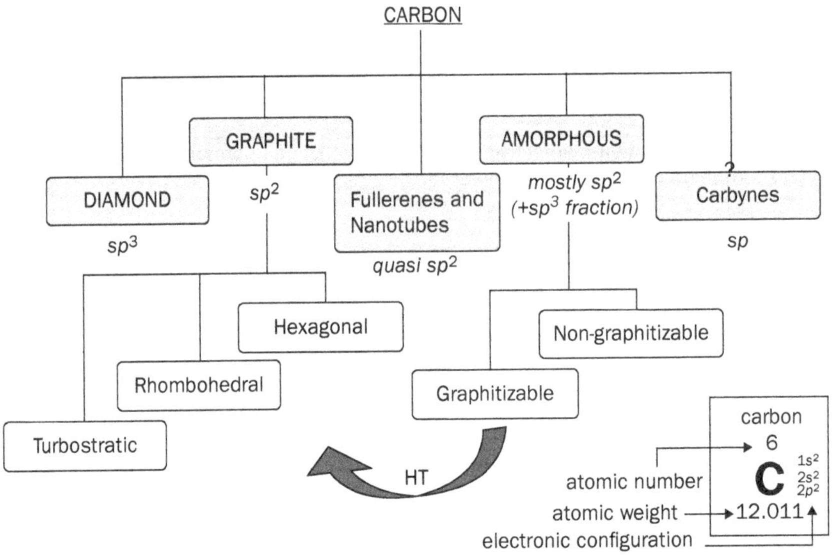

FIGURE 9.8 Carbon allotropes. (Reprinted with permission from Lam and Luong.[22])

Porous carbon materials are important catalysts and supports, based on their properties such as high surface area and porosity, ranging from micropores (<2 nm), to mesopores (2–50 nm), to macropores (>50 nm). Their electric conductivity is utilized in applications as supports for electrode catalysts in fuel cells and electrolyzers. Activated carbon is a catalyst for several reactions, and also has important applications as a sorbent, e.g., in gas and liquid cleaning. Activated carbon may be made from

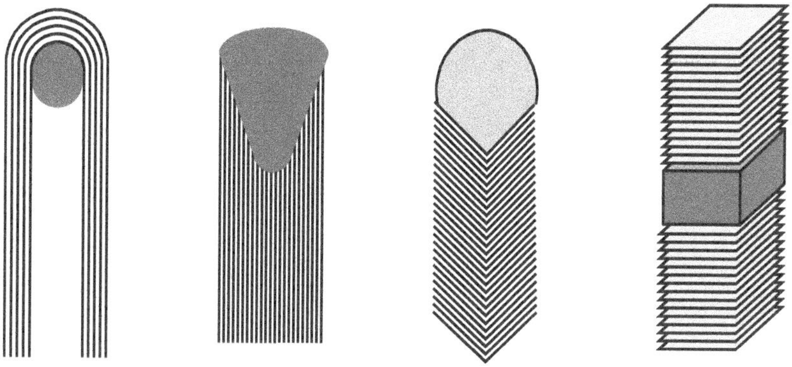

Figure 9.9 Different carbon nanostructures produced by catalytic chemical vapor deposition: (A) multi-walled carbon nanotubes; (B) ribbon-carbon nanofibers (r-CNFs); (C) fishbone-carbon nanofibers (h-CNFs); and (D) platelet-carbon nanofibers (p-CNFs). (Adapted from Serp and Figueiredo.[24])

any carbon-containing solid, such as biomass (rice husks, coconut shells, wood, sawdust, etc.) or coal. Sawdust is activated with a suitable chemical, e.g., $ZnCl_2$, followed by pyrolysis at around 700°C. This gives a porous, disordered graphitic material with many surface oxide groups. Carbon black has a lower surface area, and is made by pyrolysis of hydrocarbons (natural gas or petroleum fractions).[23] Carbon nanofibers and nanotubes are made by catalytic or thermal decomposition of hydrocarbons. By choice of catalyst, source of hydrocarbon, and reaction conditions, it is possible to control the structure of these materials.

Figure 9.9 shows idealized structures of different types of nanostructured carbon materials. Carbon nanotubes (CNTs) can be multiwalled (Fig. 9.9A) or single-walled, the latter basically consisting of a rolled-up graphene sheet with a large aspect ratio. The properties of these materials depend on the direction of the graphene structure with respect to the axis around which they are rolled. Multi-walled CNTs have several graphene layers, with a distance between the layers close to that of graphite (3.4 Å). Carbon nanofibers (CNFs), shown in Fig. 9.9B–D, can have different alignments of the graphene sheets with respect to the length axis of the material, and the properties can be tuned through the preparation route. Graphene sheets are also investigated as catalyst supports with unique properties.[24] CNFs can be grown catalytically by chemical vapor deposition (CVD) over (supported) metal catalysts. Figure 9.10 shows an example of CNF grown over Fe catalysts from a mixture of CO and H_2 (synthesis gas). Figure 9.10A shows a platelet arrangement, whereas Fig. 9.10B shows a tubular arrangement of the fiber.[25] The metal particle is attached to the end of a carbon fiber. This is because of the growth mechanism for CNFs. The carbon-containing feedstock, in this case of CO, adsorbs and dissociates on a metal particle, typically attached to a catalyst support. When carbon is formed on the surface, it can under suitable conditions self-organize into graphene sheets on the surface, and these sheets then grow away from the particle and form fibers. The shape and structure of the fiber are then a consequence of the catalytic chemistry (how fast the carbon is formed and transported on the surface) and of the shape and size of the catalytic particle. Figure 9.10 indicates that the shape of the

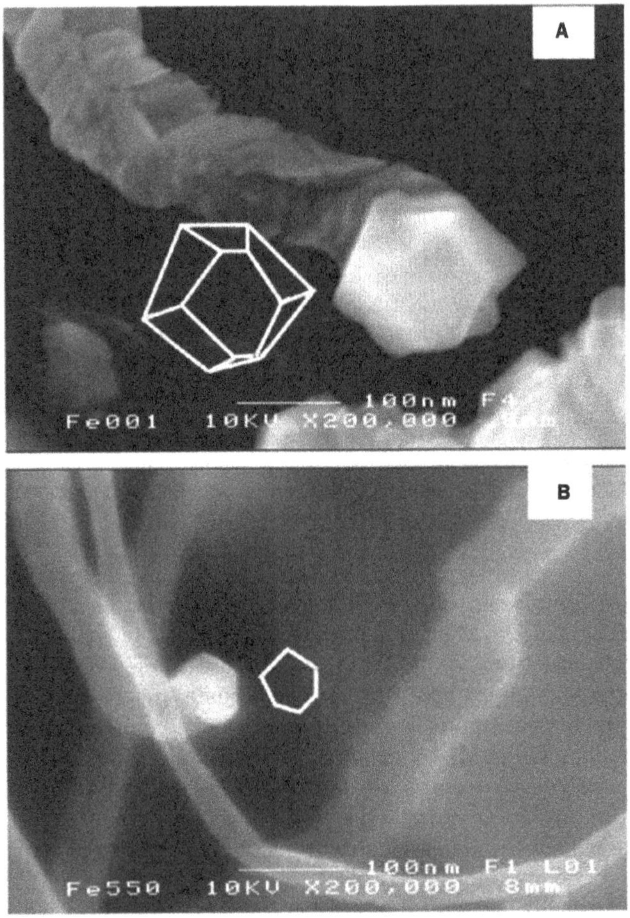

Figure 9.10 Highly magnified HR-SEM photographs of Fe catalysts on top of a (A) platelet CNF and (B) tubular CNF. (Reproduced with permission from Tanaka et al.[25])

Fe particles in this case can give rise to either a platelet CNF or a tubular arrangement. Fe, Ni, and Co are among the metals that have this ability, and the carbon-containing reactant can be CO (synthesis gas), methane, ethane, ethene, or other hydrocarbons. CNFs have many applications, e.g., as catalyst supports, as filters and membranes, and as electrode materials (cathodes/anodes in electrolyzers and fuel cells), as well as in applications utilizing high mechanical strength of CNFs.

9.4 Metal Particles

A range of metals are catalytically active, and metal catalysts can have many forms. Some catalysts are bulk metals, e.g., metal wires or solid porous metals providing some surface area of the metal. But the most important metal catalysts are supported metals, where the metal is *dispersed* on the surface of a *support* or *carrier* with a large surface area, e.g., Al_2O_3, as discussed earlier in this chapter. The available surface area of the

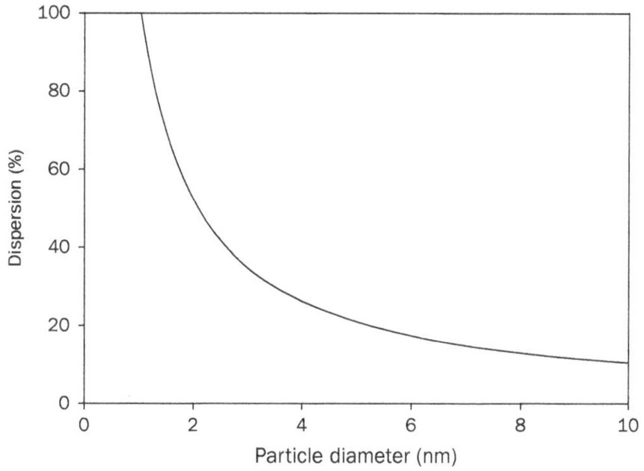

FIGURE 9.11 Dispersion as a function of particle diameter for small Cu particles.

metal increases dramatically when distributed in small particles. This is in some cases necessary in order to reduce the amount of metal used because the material is expensive and/or rare. This is very often the case for platinum and other noble metals. But dividing the metal into nanoparticles can also be necessary in order for the metal in question to have the desired catalytic properties, a situation that is relevant for metallic gold catalysts. The size of metal nanoparticles is often reported and related to the dispersion of the metal. The dispersion (often noted as *D*) is the fraction of the metal atoms in a sample that is available on the surface of the material [Eq. (9.5)]:

$$D = \frac{N_S}{N_T} \tag{9.5}$$

where N_S is the number of atoms on the surface and N_T is the total number of atoms of the metal in question in the sample. The dispersion is related to the particle size and the particle size distribution, the shape of the particle, and also the material properties, e.g., the crystalline structure. If we know the properties of the material, this relationship can be calculated. Figure 9.11 shows the result of such a calculation for Cu (FCC metal, with spherical particles exposing 100, 110, and 111 planes in equal measures). Other metals show similar relationships.

It is important to have information on the particle size or dispersion of a material. This can be characterized using a range of techniques. Some methods measure the particle size directly, such as electron microscopy, by directly observing the individual particles. Other methods measure average properties and relate them to particle size. An example of this approach is x-ray diffraction (XRD)–line-broadening analysis. Chemisorption of probe molecules provides a direct estimate of the dispersion as long as the stoichiometry of the chemisorption process is known. Figure 9.12 illustrates a common situation, where hydrogen gas is chemisorbed on a transition metal catalyst, in this case platinum. A 1:1 stoichiometry between H atoms and surface Pt atoms is often

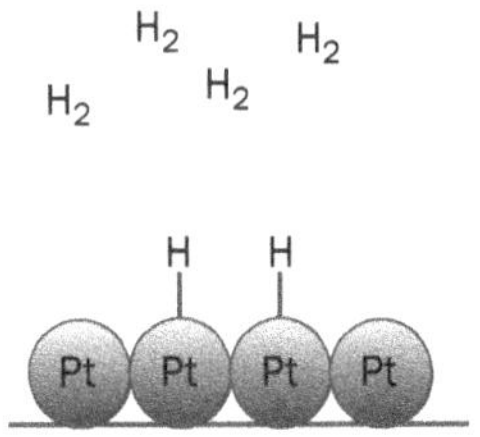

FIGURE 9.12 Chemisorption of H_2 on Pt assuming 1:1 stoichiometry.

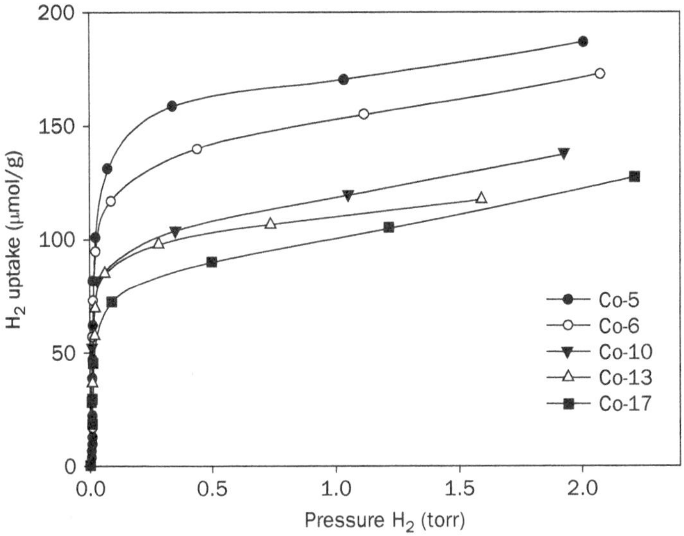

FIGURE 9.13 H_2 adsorption isotherms for a series of Co-Al_2O_3 catalysts with different particle sizes. Adsorption at 313 K. (Reprinted with permission from Patanou et al.[26])

assumed, and many studies show that this is close to the real situation. The surface is titrated with H-atoms until saturated, and a monolayer is formed.

Figure 9.13 illustrates a real-life situation, where hydrogen gas is chemisorbed on a series of supported cobalt catalysts with different particle sizes.[26] The numbers reflect an estimated particle size of each sample in nanometer, and the smallest sample, with approximately 5-nm particles, adsorbs the most hydrogen, and the amount adsorbed declines with increasing particle size as expected. It is worth noting that these adsorption processes do not follow the Langmuir formalism, in that the adsorbed amount does not level off with increasing pressure. In practice, the monolayer amount is estimated by extrapolating the linear part of the curve back to zero pressure. A real supported metal particle is seldom spherical, so the models we use to tie the link between different measurable properties are not accurate. The degree of interaction with the support can have a strong influence on the shape of the metal particles. The part of the metal particle that is bonded to the support is of course not available for reaction or chemisorption, but this fraction is not easy to assess.

A wide range of metals are active, and are used as catalysts. Metals are the most important active materials—about two-thirds of known catalytic reactions involve

Metal	Reaction
Cu	Methanol synthesis, water gas shift, hydrogenation
Ag	Oxidation of methanol to formaldehyde, oxidation of ethylene to ethylene oxide
Au	Water gas shift, acetylene hydrochlorination
Ni	Hydrogenation, steam reforming
Pd	Oxidation of hydrocarbons, hydrogenation, dehydrogenation
Pt	Hydrogenation, dehydrogenation, oxidation of CO and hydrocarbons, reduction of NO_x
Co	Hydrogenation, Fischer–Tropsch synthesis
Fe	Ammonia synthesis, Fischer–Tropsch synthesis

Table 9.5 Catalytic Metals and Their Reactions

metal catalysts. The catalytic properties of metals are similar to other chemical properties and follow trends across the periodic table. Metals in the same group have similar properties and often catalyze the same reactions. Most applications of catalysis use transition metals with d-electrons. In Table 9.5 some catalytic metals and some of the reactions they are used in are listed. Next, we will discuss some metal catalysts that are used as supports.

Example 1: Hydrogenation of CO over Metal Catalysts

The hydrogenation of CO is an industrially important reaction. By choosing from a range of catalysts, different products can be made. The mixture of CO and hydrogen is called synthesis gas, or syngas, and can be made from any carbon-containing feedstock (coal, oil, natural gas, or biomass). Syngas reactions are important for converting energy-rich feedstock to desired products. Using Cu as catalyst, and with some CO_2 in the gas, we can make methanol; using Ni as catalyst, the main product is methane; using Co and Fe and Ru as catalysts, hydrocarbons (alkanes and alkenes) can be produced. CO can also react with H_2O to produce H_2 (and CO_2) in the water gas shift reaction; this forms the basis for the very important ammonia synthesis.

Supported Co catalysts are used when the goal is to make hydrocarbons for fuel purposes. The activity of these catalysts was observed to be proportional with the available Co surface area, i.e., an optimal catalyst had as large a Co surface as possible, corresponding to very small Co particles. It has been found that below a certain Co particle size, the activity is smaller, and declines with reducing particle size (Fig. 9.14). There is also a particle size effect on the selectivity of this reaction. A suggested explanation for this behavior is that the smaller particles have a larger fraction of surface atoms with lower coordination numbers, e.g., corner atoms, compared to terrace atoms. These atoms adsorb CO stronger, and CO dissociation, a necessary step in the process, is hindered. So when the nanoparticles become too small, they are no longer suited to perform the desired reaction.[27] The most common form of this effect is that atoms with low coordination numbers lead to higher catalytic reaction rates, but in this case the effect is the opposite, probably because the bond between CO or carbon on the surface is so strong that further reaction is hindered. This is often described as the Sabatier principle; the adsorption strength should be just right—not too weak or too strong [see, e.g., Che (2013)].[28] The next example is of small particles which are more active, as they provide highly uncoordinated atoms.

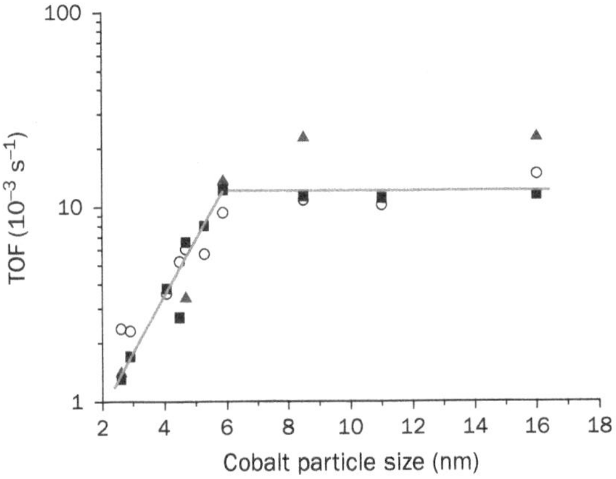

FIGURE 9.14 Relationship between cobalt particle size and TOF found at H_2/CO = 2:1 bar, 220°C (■) and 35 bar, 210°C (▲); and H_2/CO = 10:1.85 bar, 210°C (○). (Reprinted with permission from Den Breejen et al.[27])

Example 2: Gold Catalysis

Gold was traditionally considered chemically inert and not suited as a heterogeneous catalyst. But based on some new discoveries, supported gold nanoparticles are now recognized as important catalysts, e.g., in CO oxidation at low temperature and in the water-gas shift reaction (both important reactions in producing pure hydrogen gas for fuel cells). For CO oxidation, the catalyst preparation route is the key to activity. CO oxidation is a very important reaction in hydrogen fuel cells. Polymer electrolyte membrane (PEM) fuel cells need hydrogen gas with a high purity; especially the presence of CO is critical in such cases. Hydrogen can be produced on-site, e.g., in a vehicle, but the gas needs to be cleaned for all traces of carbon monoxide down to a concentration of below 10 ppm. One way of removing CO is to oxidize it to CO_2, and several catalysts can do this. However, many of the catalysts also oxidize H_2 to water, which here represents a loss and is unwanted. To be able to do the CO oxidation at very low temperature, where the reactivity of hydrogen over the same catalysts is negligible, is an advantage. Figure 9.15 shows the seminal work on CO oxidation activity over gold catalysts reported by Haruta and coworkers.[29,30]

The catalytic activity is shown as CO conversion as a function of temperature for a range of materials. A co-precipitated Au/α-Fe_2O_3 prepared in a specific manner was by far most active, converting CO at much lower temperatures than any other sample. Later it was shown that very high activity also can be achieved with other supports such as TiO_2. Figure 9.16 shows a model of such a material, and a scanning tunneling microscope (STM) image showing how the activity is correlated with the size of the gold clusters—in this case supported on an ideal TiO_2 (110) surface.[31] Very specific size and shape of gold clusters are necessary in order to achieve high catalytic activities.

Figure 9.17 provides a general explanation, linking the catalytic activity to the availability of uncoordinated gold atoms. The left panel of the figure shows the general trend of catalytic activity as a function of gold particle size over a range of catalytic supports. The solid line in the figure represents 1/(particle diameter to the power of 3), which in a simplified model of gold particles can be assumed to be proportional

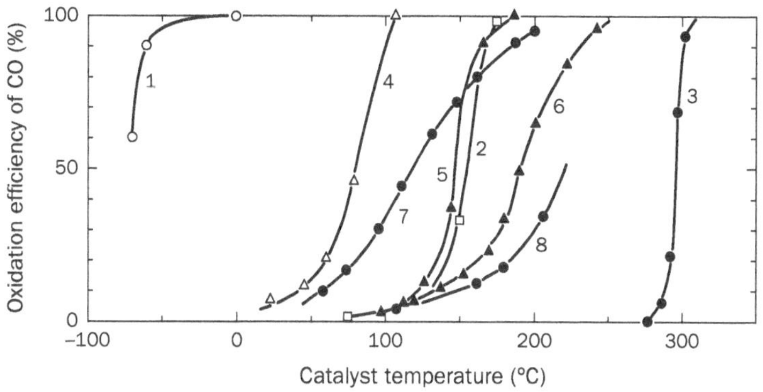

Figure 9.15 CO oxidation activity reported as CO conversion over different catalysts as a function of temperature. Catalysts: (1) Au/α-Fe_2O_3 (Au/Fe = 1/19, co-precipitation, 400°C), (2) 0.5 wt% Pd/γ-Al_2O_3 (impregnation, 300°C), (3) fine Au powder, (4) Co_3O_4 (carbonate, 400 °C), (5) NiO (hydrate, 200°C), (6) α-Fe_2O_3 (hydrate, 400°C), (7) 5 wt% Au/α-Fe_2O_3 (impregnation, 200°C), and (8) 5 wt% Au/γ-Al_2O_3 (impregnation, 200°C). (Reproduced with permission from Haruta et al.[30])

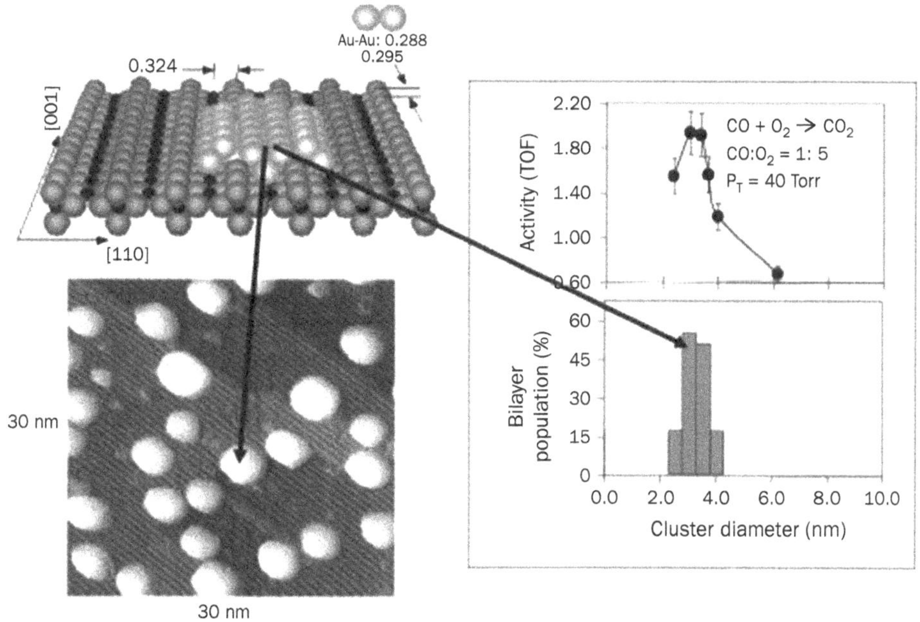

Figure 9.16 An STM image (bottom left) and schematic (top left) of Au bilayer clusters on TiO_2 (1 1 0) whose population is dominant (bottom right) for a Au/TiO_2 catalyst most active for CO oxidation (top right). The activity (at 350 K) is expressed as turnover frequency (TOF) in units of CO_2 molecules produced per Au site per second. (Reproduced with permission from Choudhary and Goodman.[31])

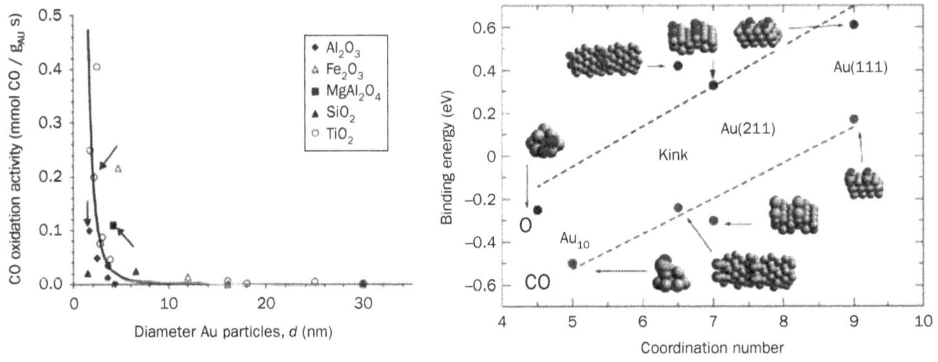

FIGURE 9.17 Left panel: Catalytic activity for CO oxidation over supported gold nanoparticles at 273 K. Right panel: CO binding energy as a function of coordination number for the gold atoms. (Reproduced with permission from Lopez et al.[32])

to the number of corner atoms in small particles. Both experimentally and from theoretical calculations (the density functional theory calculations, a very useful method in predicting and explaining catalytic materials and reactions) it is demonstrated that the binding energy of the reactants, in this case CO and O_2, is also strongly dependent on the coordination number; this is illustrated in the right panel of Fig. 9.17. Here the binding energies of the reactants need to be high in order to activate the reactants, which is possible only in small particles (or perhaps even single atoms of gold).[32]

9.5 Summary and Further Reading

In this chapter we have tried to introduce the main features of heterogeneous catalysis and the importance of nanostructures and nanomaterials. The whole field revolves around the manipulation of nanostructures, both through their physical properties (surface area, pore system, pore sizes) and through their chemical properties that are highly dependent on sizes in the nano-range. For the interested reader, we recommend some textbooks in catalysis that give a more comprehensive treatise of the field.

Brief introductions to catalysis:

- M. Bowker: *The Basis and Applications of Heterogeneous Catalysis*, Oxford University Press, Oxford, 1998.
- J. T. Richardson: *Principles of Catalyst Development*, Plenum Press, New York, 1989.

More comprehensive textbooks:

- I. Chorkendorff, J. W. Niemandtsverdriet: *Concepts of Modern Catalysis and Kinetics* (3rd ed.), Wiley-VCH, Weinheim, 2017.
- J. M. Thomas, W. J. Thomas: *Principles and Practice of Heterogeneous Catalysis* (2nd ed.), Wiley-VCH, Weinheim, 2015.
- G. A. Somorjai, Y. Li: *Introduction to Surface Chemistry and Catalysis* (2nd ed.), Wiley, Hoboken, NJ, 2010.

Handbooks:

- C. H. Bartholomew, R. J. Farrauto: *Fundamentals of Industrial Catalytic Processes* (2nd ed.), Wiley Interscience, Hoboken, NJ, 2006.
- *Handbook of Catalysis* (2nd ed.), Edited by G. Ertl, H. Knözinger, F. Schüth, J. Weitkamp (8 volumes), Wiley, Weinheim, 2008.

9.6 Study Questions

9.6.1 Concept-Based Questions

9.1 What is a zeolite, and how can a synthetic zeolite be prepared?

9.2 Illustrate and describe briefly three types of selectivity in reactions occurring in zeolites.

9.3 Carbon nanofibers (CNFs) are promising catalyst materials. List three nanostructures of CNFs.

9.4 Mention the main steps in catalyst synthesis and activation. Which are the key phenomena in the formation of solid catalysts from precursors?

9.5 Describe how carbon nanofibers (CNFs) can be grown using catalysts—typical temperature, gaseous atmosphere, and catalyst particle composition.

9.6 Compare the two types of adsorption. How can you characterize adsorption?

9.7 You have prepared a 1 wt% Pt/Al_2O_3 catalyst for catalytic combustion, and you would like to estimate the dispersion based on chemisorption. Pt is fcc and the lattice constant is 3.920 Å.

a. Suggest a probe molecule and a reasonable adsorption stoichiometry (number of surface atoms per probe molecule).

b. You measure an uptake of your probe molecule of 0.7 μmol/g cat. Estimate the dispersion.

c. Suggest a reasonable area per surface Pt atom, and estimate the specific surface area of the metal (metal surface area per gram of metal).

d. Assuming that the Pt particles are spherical, estimate the diameter.

9.6.2 Research-Based Questions

9.8 Co/Al_2O_3 (cobalt particles deposited on a high-surface area γ-Al_2O_3 support) is a catalyst used for the production of hydrocarbons from synthesis gas, termed the *Fischer–Tropsch process*:

$$n\mathrm{CO} + 2n\mathrm{H_2} \rightarrow (-\mathrm{CH_2}-)_n + n\mathrm{H_2O}$$

where n is the chain length of the formed aliphatic hydrocarbons (a mix of alkenes and alkanes). The catalyst contains 15 wt% cobalt, with a dispersion $D = 10\%$.

a. You have the support, Al_2O_3 as a mesoporous powder, with surface area 200 m^2/g, and cobalt as $Co(NO_3)_2$, a salt soluble in water at moderate to low pH.

Suggest and describe briefly how you can deposit a cobalt-species on the surface of the support.

b. Which steps following the deposition are needed to obtain metallic cobalt particles?

c. Given that a surface cobalt atom on average occupies 0.068 nm^2, and assuming that the cobalt particles are metallic and spherical, what is the average particle diameter d of the cobalt particles?

Data: Cobalt molecular weight: 59

Cobalt metallic density: 8.9 g/cm^3

Avogadro number: 6.22×10^{23} mol^{-1}

9.9 $NiCo_2O_4$ is a spinel-like material which is catalytically active for the complete oxidation (combustion) of hydrocarbons (like methane), as an alternative to noble metal-based (Pd) catalysts to prevent engine emissions.

a. Why is high specific surface area of this material beneficial?

b. Explain how the specific surface area can be measured by adsorption.

c. What are the typical conditions at which this is done (p, T, atmosphere/probe molecule)?

d. Explain the BET equation and how it is used to extract the specific surface area.

e. What are the assumptions for the applicability of this equation?

f. The catalyst—synthesized by co-precipitation by one of this year's Master's students—provided the following BET result. Calculate the surface area of this catalyst, using the constants given:

T(STP):	273.15 K
P(STP):	1 bar = 100 kPa = 0.986 atm
R (universal gas constant):	8.33×10^{-5} m^3-bar/(K-mol)
N_A (Avogadro's number):	6.02×10^{23} mol^{-1}
A_0 (area occupied by a single probe molecule):	0.164 nm^2

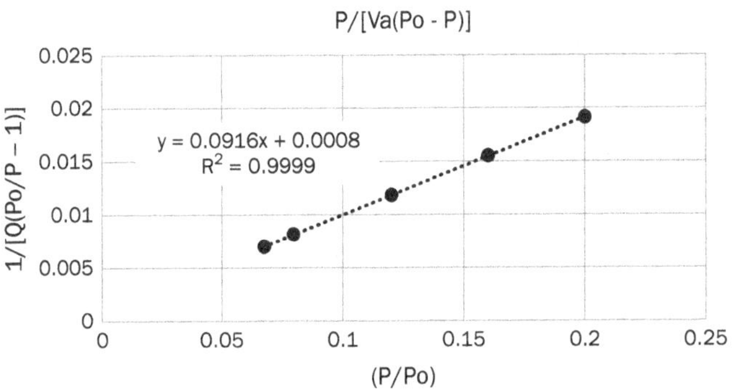

9.10 Answer the following questions based on the research work of Hammer et al.[33]

a. Describe the catalysts studied in the paper. Which are the active phases? Which are the support materials?
b. Which reaction has been studied over the catalyst(s)? What is this reaction applied for? In this particular case, is the reaction more of a model/test reaction to elucidate the catalyst properties, or is it studied under industrially relevant conditions?
c. What are the starting materials for preparation of the different phases present in the catalysts? Try to describe briefly the synthesis steps.
d. Was the BET equation applied? What was the result?
e. Was any metal particle size measured and for which metal? If so, which measurement technique was applied and what was the result?
f. Could a direct correlation be established between the metal dispersion and the turnover frequency (TOF)? How was the dispersion estimated? Was chemisorption applied? Why is reporting activity in terms of TOF important/relevant in this case?

9.7 References

1. Mitchell, J. A., Heterogeneous catalysis. *Journal of Chemical Education.* **1932,** *9*(2), 261.
2. Morbidelli, M.; Gavriilidis, A.; Varma, A., *Catalyst Design.* Cambridge, UK: Cambridge University Press, 2001.
3. Bartholomew, C. H.; Farrauto, R. J., *Fundamentals of Industrial Catalytic Processes.* Hoboken, NJ: John Wiley & Sons, 2011.
4. Hornyak, G. L.; Tibbals, H. F.; Dutta, J.; Moore, J. J., *Introduction to Nanoscience and Nanotechnology.* Boca Raton, FL: CRC Press, 2008.
5. Serp, P.; Figueiredo, J. L., *Carbon Materials for Catalysis.* Hoboken, NJ: John Wiley & Sons: 2009.
6. Boehm, H., The first observation of carbon nanotubes. *Carbon.* **1997,** *35*(4), 581–584.
7. Langmuir, I., The constitution and fundamental properties of solids and liquids. Part I. Solids. *Journal of the American Chemical Society.* **1916,** *38*(11), 2221–2295.
8. Sing, K. S., Reporting physisorption data for gas/solid systems with special reference to the determination of surface area and porosity (Recommendations 1984). *Pure and Applied Chemistry.* **1985,** *57*(4), 603–619.
9. Sing, K., The use of nitrogen adsorption for the characterisation of porous materials. *Colloids and Surfaces A: Physicochemical and Engineering Aspects.* **2001,** *187*, 3–9.
10. Farooq, S.; Qinglin, H.; Karimi, I., Identification of transport mechanism in adsorbent micropores from column dynamics. *Industrial & Engineering Chemistry Research.* **2002,** *41*(5), 1098–1106.
11. Masuda, T., Peculiar diffusion mechanisms within micropores of zeolite catalysts. *Journal of the Japan Petroleum Institute.* **2003,** *46*(5), 281–294.
12. Xiao, J.; Wei, J., Diffusion mechanism of hydrocarbons in zeolites—I. Theory. *Chemical Engineering Science.* **1992,** *47*(5), 1123–1141.
13. Santos, P. S.; Santos, H. S.; Toledo, S., Standard transition aluminas. Electron microscopy studies. *Materials Research.* **2000,** *3*(4), 104–114.

14. Leboda, R.; Tertykh, V.; Sidorchuk, V.; Skubiszewska-Zięba, J., Peculiarities of steam-phase hydrothermal modification of pyrogenic silicium dioxide. *Colloids and Surfaces A: Physicochemical and Engineering Aspects.* **1998,** *135*(1–3), 253–265.
15. Stöber, W.; Fink, A.; Bohn, E., Controlled growth of monodisperse silica spheres in the micron size range. *Journal of Colloid and Interface Science.* **1968,** *26*(1), 62–69.
16. Zholobenko, V. L.; Khodakov, A. Y.; Impéror-Clerc, M.; Durand, D.; Grillo, I., Initial stages of SBA-15 synthesis: an overview. *Advances in Colloid and Interface Science.* **2008,** *142*(1–2), 67–74.
17. Chytil, S.; Glomm, W. R.; Kvande, I.; Zhao, T.; Walmsley, J. C.; Blekkan, E. A., Platinum incorporated into the SBA-15 mesostructure via deposition-precipitation method: Pt nanoparticle size estimation and catalytic testing. *Topics in Catalysis.* **2007,** *45*(1–4), 93–99.
18. Sazama, P.; Pastvova, J.; Kaucky, D.; Moravkova, J.; Rathousky, J.; Jakubec, I.; Sadovska, G., Does hierarchical structure affect the shape selectivity of zeolites? Example of transformation of n-hexane in hydroisomerization. *Journal of Catalysis.* **2018,** *364*, 262–270.
19. Busca, G., Acidity and basicity of zeolites: a fundamental approach. *Microporous and Mesoporous Materials* **2017,** *254*, 3–16.
20. Abildstrøm, J. O.; Kegnæs, M.; Hytoft, G.; Mielby, J.; Kegnæs, S., Synthesis of mesoporous zeolite catalysts by in situ formation of carbon template over nickel nanoparticles. *Microporous and Mesoporous Materials.* **2016,** *225*, 232–237.
21. Zhu, J.; Meng, X.; Xiao, F., Mesoporous zeolites as efficient catalysts for oil refining and natural gas conversion. *Frontiers of Chemical Science and Engineering.* **2013,** *7*(2), 233–248.
22. Lam, E.; Luong, J. H., Carbon materials as catalyst supports and catalysts in the transformation of biomass to fuels and chemicals. *ACS Catalysis.* **2014,** *4*(10), 3393–3410.
23. Auer, E.; Freund, A.; Pietsch, J.; Tacke, T., Carbons as supports for industrial precious metal catalysts. *Applied Catalysis A: General* **1998,** *173*(2), 259–271.
24. Serp, P.; Machado, B., *Nanostructured Carbon Materials for Catalysis.* Cambridge, UK: Royal Society of Chemistry, 2015.
25. Tanaka, A.; Yoon, S.-H.; Mochida, I., Preparation of highly crystalline nanofibers on Fe and Fe–Ni catalysts with a variety of graphene plane alignments. *Carbon.* **2004,** *42*(3), 591–597.
26. Patanou, E.; Tveten, E. Z.; Chen, D.; Holmen, A.; Blekkan, E. A., Microcalorimetric studies of H_2 and CO on Co/γ-Al_2O_3 catalysts for Fischer–Tropsch synthesis. *Catalysis Today* **2013,** *214*, 19–24.
27. Den Breejen, J.; Radstake, P.; Bezemer, G.; Bitter, J.; Frøseth, V.; Holmen, A.; Jong, K. D., On the origin of the cobalt particle size effects in Fischer–Tropsch catalysis. *Journal of the American Chemical Society* **2009,** *131*(20), 7197–7203.
28. Che, M., Nobel Prize in Chemistry 1912 to Sabatier: organic chemistry or catalysis? *Catalysis Today.* **2013,** *218*, 162–171.
29. Haruta, M.; Kobayashi, T.; Sano, H.; Yamada, N., Novel gold catalysts for the oxidation of carbon monoxide at a temperature far below 0 C. *Chemistry Letters.* **1987,** *16*(2), 405–408.
30. Haruta, M.; Yamada, N.; Kobayashi, T.; Iijima, S., Gold catalysts prepared by coprecipitation for low-temperature oxidation of hydrogen and of carbon monoxide. *Journal of Catalysis.* **1989,** *115*(2), 301–309.

31. Choudhary, T.; Goodman, D., Catalytically active gold: the role of cluster morphology. *Applied Catalysis A: General.* **2005,** *291*(1–2), 32–36.
32. Lopez, N.; Janssens, T.; Clausen, B.; Xu, Y.; Mavrikakis, M.; Bligaard, T.; Nørskov, J. K., On the origin of the catalytic activity of gold nanoparticles for low-temperature CO oxidation. *Journal of Catalysis.* **2004,** *223*(1), 232–235.
33. Hammer, N.; Kvande, I.; Xu, X.; Gunnarsson, V.; Tøtdal, B.; Chen, D.; Rønning, M., Au–TiO_2 catalysts on carbon nanofibres prepared by deposition-precipitation and from colloid solutions. *Catalysis Today.* **2007,** *123*(1), 245–256.

CHAPTER 10

Applications in Drug Delivery and Biomedicine

Unique properties of nanomaterials stemming from size, shape, surface functionality, among others, discussed in Chaps. 5 and 7, offer many advantages in biomedicine. Their applications range from drug delivery, to imaging, to biosensing, covering both therapeutic and diagnostic functions. Of late, nanomaterials have become essential candidates for diagnosing, treating, and/or monitoring vital diseases such as cancer, Parkinson's, Alzheimer's, and so on. The surface tunability of nanoparticles (NPs) in addition to their size-dependent properties make them interesting candidates. You may be aware of our body's defense mechanisms continuously working to eliminate unwanted foreign particles from entering our bodies. A few simple examples may be when you sneeze to get rid of dust particles or when the body temperature shoots up (a condition commonly known as *fever*) to combat a virus or bacteria. The body's defense mechanisms indeed make administration of nanomaterials containing active cargo (like drugs) a challenging task. There are various hindrances that need to be overcome before the required therapeutic function may be delivered at the desired location in the body. Such challenges include long circulation of the nanocarrier inside the body, prevention of capture by the body's defense mechanisms, precise delivery of the nanocarrier at the diseased site, and timely rupture of the carrier to release the active material, to mention a few. By now, you might have guessed that drug delivery is indeed a complicated topic and requires advanced knowledge both in fabrication of the nanocarriers as well as in their applications downstream.

In this chapter, the focus is on the development of drug delivery systems including loading and encapsulation of drugs, biodistribution of the nanocarrier containing the drug, and subsequent release of the drug from these systems at a desired location. Apart from this, following a brief introduction to magnetic resonance imaging (MRI), NP systems that find wide applications in MRI are discussed. Finally, fundamental concepts of biosensing and the applications of some of the nanomaterials covered in this book are also discussed.

Section 10.1 of this chapter deals with the applications of nanomaterials in drug delivery, Sec. 10.2 deals with the applications in MRI, and Sec. 10.3 deals with biosensing.

10.1 Drug Delivery

One of the most common drugs taken orally worldwide is paracetamol. A lesser known fact is that only a very small fraction (10–30 percent) of paracetamol is absorbed from its site of administration and reaches, in an unchanged form, the systemic circulation. You might be wondering what happens to the remaining amount. Most of it binds to the proteins present in blood, a process known as *opsonization*, and is converted into other toxic or non-toxic products.

Thus, for a desired therapeutic effect, better carrier systems capable of protecting the drug from the circulation environment are required. However, the protective carrier should not initiate immunological responses in the system leading to capture by the body's defense mechanisms. Further, once the carrier with its contents has reached the target site, it should be able to deliver the drug in a rather easy fashion, without affecting the healthy cells. So, a lot of these situations may be linked to how the drug is loaded into the nanocarrier, what are the plausible interactions that the drug might have with the nanocarrier, and what are the various mechanisms that may influence both loading and release. In case of drug delivery for targeted therapies, clinically important drugs are incorporated in the NP system and administered for their release at a targeted diseased site yielding desired theranostic effect. Biodistribution of these nanocarriers, which is an important step in drug delivery, may be achieved by passive and/or active targeting.[1]

10.1.1 Passive Targeting

In a healthy person, endothelial cells line the walls of blood vessels as tight junctions thereby resisting entry of macromolecules. However, in case of pathological conditions such as tumors, infarcts, and inflammations, these vessels become more permeable to larger molecules (i.e., Enhanced Permeability). Molecules ranging in size from 10 to 500 nm can leave the vascular bed and accumulate inside the interstitial space (i.e., Retention).

This enhanced permeability may be exploited by NPs to deliver drugs to the diseased site. Several pathological conditions are often characterized by growing tumors. When a tumor grows abnormally, the demand to provide oxygen and essential nutrients to tumor cells becomes inevitable. New small blood vessels are formed to meet the ever-increasing nutrient demand of the tumor. This chaos and imbalance due to angiogenesis results in disordered leaky blood vessels with gaps at the junctions of endothelial cells. When drug carriers are administered intravenously, circulating NPs tend to accumulate at the tumor site through these spaces in endothelial lining of blood vessels.[2] This process by which NPs can extravasate and accumulate in such interstitial areas is known as Enhanced Permeability and Retention (EPR), the term coined by Maeda et al.[3] Figure 10.1 shows how nanocarriers loaded with drugs can extravasate through leaky vasculature passively into the interstitial space.

One of the requirements for drug delivery via passive targeting is long blood-circulation of the carrier system, so that there is sufficient level of accumulation in the target. Different approaches to functionalize NPs (discussed in Chaps. 5 and 7) may be recalled here to understand how these might aid in longer circulation times.

Development of nanomaterials with therapeutic possibilities largely depends on their size, shape, surface chemistry, and modifications. These properties would influence the circulation duration, targeting, and biodistribution at the intended site. Some of the factors that influence passive targeting are discussed here.

Size

In passive targeting, as the NPs are transported passively through diffusion, their size plays an important role in extravasation. Any active biomolecule injected inside the human body will have to overcome barriers constituting the body's defense mechanisms

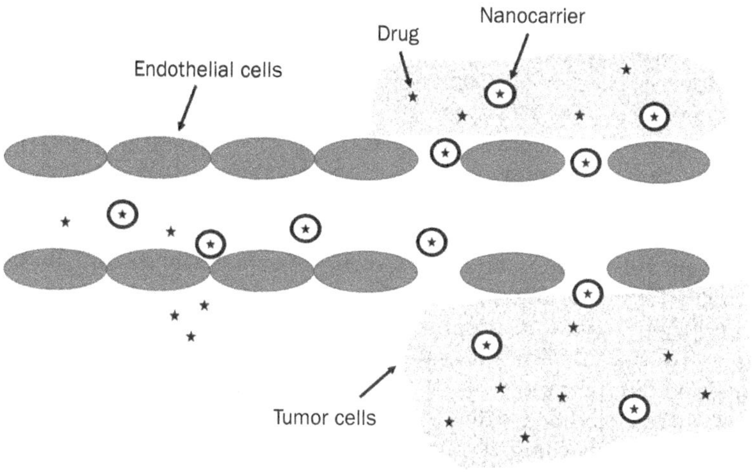

Figure 10.1 Schematics of Enhanced Permeability and Retention (EPR) effect or "passive" targeting. Drug-loaded nanocarriers cannot extravasate through normal endothelium. Only small molecules of free drug can traverse across normal endothelium to a certain extent in both directions. Nanocarriers loaded with drugs can extravasate through leaky vasculature around the tumor cells and accumulate in such areas creating high local drug concentrations.

to achieve its functions. Circulating NPs having size less than 2 nm diameter are subjected to renal clearance via glomerular filtration.[4] However, nanocarriers bigger than 200 nm are easily removed via phagocytosis. According to Liechty's opinion, the favorable size range for tumor targeting ranges from 40 to 200 nm.[5] The type of NPs used for tumor targeting also has an effect on the resulting efficacy of the system. Liposomes having size greater than 200 nm cannot be taken up by the tumor cells. However, polyethylene glycol (PEG)-coated liposomes have reportedly shown prolonged circulation times with better anticancer activity between 50 and 150 nm size range, at the expense of increased loading efficiencies.[6,7] Thus, it is important to note that the size of the NPs is just one of the important factors that influences the mode of uptake.

Shape

Like size, shape is also an important property of NPs that influences their uptake by cells. By now, you already have a good grasp of how NPs of different sizes and shapes can be synthesized and functionalized for desired applications (refer to Chaps. 5–7). Cellular uptake mechanisms are greatly influenced by shape and curvature of the NPs when they interact with cellular membranes.

Spherical NPs have been in use for nanomedicine research mostly because of their easy synthesis processes. However, non-spherical NPs have also shown promise with their high drug loading abilities, longer circulation, and enhanced cellular uptake.[8] Several factors need to be taken into consideration at different levels (organ, tissue, and cell) to realistically investigate the interaction of NPs of different shapes with cells.

Surface Modifications

The ideal nanomaterial designed for biomedical applications must mimic the natural aesthetics of the human body. The degree of retention of NPs in circulation and subsequent uptake by the cells greatly depend on their surface charge. To a greater extent, highly cationic NPs are more rapidly cleared from circulation than highly anionic NPs. However, neutral NPs, and those with a slight negative charge, show significantly prolonged circulating half-lives.[9]

One of the most common ways to enhance circulation lifetimes of nanocarriers is to surface modify them by using cloaking polymers such as PEG. This creates a hydrophilic environment around the nanocarrier, preventing opsonization of plasma proteins and clearance by the reticuloendothelial system (RES). Increasing molecular weight of PEG might also increase the residence of NPs in blood, endowing them with increased specificity for tumor targeting.[10]

Endocytosis of the nanocarriers inside the cells is also determined by nature and charge of surface modifications of NPs. The interaction and collision of NPs with cell membranes are prompted by the activation energy. In order to facilitate endocytosis, NPs should be modified with slight cationic charge that would enable the NPs to interact with negatively charged cell membranes.[11]

It should, however, be noted that in several pathological situations the integrity of vascular endothelium remains unaffected and there is no opportunity for EPR. Although the EPR effect can be an effective tool in targeting nanocarriers to tumor microenvironment, practical insights about extent of their uptake by solid tumors are still poorly understood. To overcome the limitations of passive targeting, active targeting can be a more robust method in order to selectively target the diseased cells.

10.1.2 Active Targeting

Active targeting is another way of targeting nanocarriers loaded with active cargo to the desired site of action. This is primarily achieved through receptor-mediated endocytosis by the use of ligands, peptides, or other small molecules attached to nanocarriers. The ligands recognize the diseased cell receptors expressed on the cell surfaces. By receptor-mediated endocytosis, subcellular sites of interest can also be targeted yielding the desired effect, which can help reduce toxicity-related effects on healthy cells.

The binding affinity between cell receptors and NP ligands determines their specificity, which is required to induce endocytosis. Upon entering into the cell, nanocarriers can release the drug to produce its therapeutic effect. Similar approaches can also be used for delivering genes and other biomolecules of interest. It is important to note that both conjugating ligand and receptor must be in close proximity (<0.5 nm) in order to acquire significant interaction.[12]

Active targeting can bypass the blood–brain barrier and can also help overcome multiple drug resistance in cancer therapies. Monoclonal antibodies, aptamers, and targeting ligands (e.g., folic acid, transferrin, etc.) are some of the prominent biorecognition elements used for actively targeting diseased cells. Active targeting has been employed largely to target tumors.

Overexpression of certain receptors occurs on the surface of cells, e.g., folic acid, in a variety of cancer types (such as ovarian and cervical carcinoma). Folate receptors are commonly bound to the surface of nanocarriers containing anticancer drugs (such as doxorubicin, DOX) for specific targeting. These formulations have shown increased uptake by the cells when compared with non-functionalized nanocarriers.

Although it might seem that active targeting is a robust strategy for cancer targeting, the reality is farfetched. The design of NP system for cancer targeting depends on several factors including malignancy, hemodynamics of the tumor microenvironment,

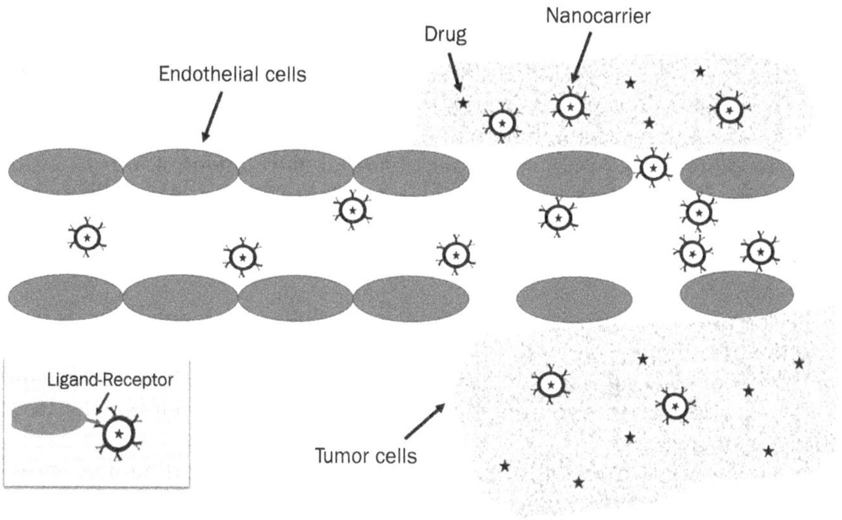

FIGURE 10.2 Schematics showing active drug targeting by use of ligands on the surfaces of nanocarriers. Ligands recognize specific receptors expressed on the surfaces of tumor cells and/or angiogenic endothelial cells.

location of tumor, tumor stage, tumor type, host organ, and hostility. Distinguishing diseased cells from healthy ones can be a major challenge to overcome in passive targeting, which may lead to its rapid clearance by physiological processes and hence possible toxicity issues. However, in active targeting, intracellular processes can also hinder the delivery of drug inside the cell, e.g., endosomal escape. Similarly, NP systems targeting only overexpressed entities of interest, without encountering any other biological barrier, can be a challenging task. Moreover, different types of tumors show distinct characteristics such as gene expression, proliferation, and aggressiveness. Sometimes these characteristics show variability in different regions within the same tumor. Therefore, a single approach or targeting strategy for different types of heterogenic cancers is not a realistic way of encountering them. More research work is required to tap the potential, quality, and efficacy of nano systems in order to produce them commercially.

In addition, for a nanocarrier to cause a therapeutic effect, it should be sufficiently loaded with an active material that should be first and foremost protected during circulation and released upon reaching the target. This brings us to the discussion about loading and encapsulation of active cargo in different nanocarrier systems.

10.1.3 Loading and Encapsulation

Development of NPs with theranostic abilities relies highly on loading and encapsulation strategies. The synthesis and surface chemistry, structure of drug molecule, desired application, and intended circulation time play a vital role in encapsulation methods. These factors also contribute to subsequent drug release from NP system at the targeted site.

We will discuss here two important methods by which a drug may be loaded into the nanocarriers. This classification is based on at what point of the synthesis protocol the drug is loaded into the nanocarriers. The first method is via incorporation of the drug at the time of NP synthesis, commonly referred to as *in situ* loading. The second method is drug absorption after the synthesis of NPs, i.e., post-synthesis loading.

Based on the interactions (between the drug and nanocarrier) responsible for loading, two distinctions may be made—physical and chemical entrapment. Physical loading relies on non-specific binding through van der Waals forces, hydrogen bonding, and/or other weak forces. However, chemical entrapment of a drug inside NPs happens through specific bonding such as covalent or ionic. Although stronger binding is preferable for better encapsulation, it may hinder an efficient release at the target site. Hence, a trade-off needs to be considered while designing nanomaterials for drug delivery.

One of the commonly used strategies employed for loading drugs in hydrogels is via the breathing-in method. In this method, lyophilized hydrogels are imbibed with a concentrated drug solution. Similar to water absorption in a sponge, the hydrogels swell up with the drug solution by filling the porous architecture of hydrogels with the drug solution.

As can be deciphered from the above discussions, loading and encapsulation of drugs in nanocarriers present several challenges. Some strategies that have been employed to load different drugs in inorganic NPs (specifically, magnetic NPs), liposomes, and polymeric micelles are described next. The choice of the nanocarrier and the drug depends largely on the application at hand, hence it is not possible to enlist features from a general perspective. The aim, however, is to show that a wide range of applications in drug delivery may be catered through efficient fabrication of the nanomaterials.

Magnetic NPs

Magnetic NPs (MNPs) have been widely studied for drug delivery applications, since these can be targeted to the diseased site using magnetic fields or tuned to control release of the drug using an external magnetic field. Ideally, MNPs contain three fundamental parts—core, a charged surface, and functional outer coating, which are designed as per requirements of the application. As MNPs tend to agglomerate and oxidize quickly, stabilizers are used to protect their surface. Further functionalization with polymers (e.g., PEG) or other cargo molecules increases their functions, stability, and biocompatibility.

Drug loading onto the surface of MNPs can be achieved by attachment of several functional groups such as amines, carboxylic acid, biotin, antibodies, etc. This depends on the type of drug to be loaded, which shows affinity toward a specific functional group attached with MNPs. MNPs have been functionalized with carboxylic groups and amines to attach tumor-targeting biomolecules including peptides and/or antibodies. Anticancer drugs such as doxorubicin (DOX), epirubicin, paclitaxel, etc. have also showed good binding affinity toward MNPs in developing targeting nanocarriers.[13]

Liposomes

Liposomes have been widely used in drug delivery applications owing to their ability to protect drugs from degradation, good intended site delivery, and transport cargo molecules for targeted therapies. Drugs are mixed directly into the lipid solution to produce drug-loaded liposomes. Their steric stability must be considered because they can cause unspecific interactions with cell membranes and can become vulnerable to clearance by physiological processes. Certain modifications are done on their surfaces for attaining stability and to make them more biocompatible, e.g., PEGylation.

Drug loading in liposomes can be done either actively or passively. In active loading, the drugs are loaded into the interior of already synthesized liposomes. However, in case of passive loading, the drugs are mixed in the lipid solution to produce drug-loaded liposomes. Alternatively, active and passive loadings may be referred to as post-synthesis and *in situ* loading, respectively. The loading of drug in liposomes is influenced by many factors, including pH, transmembrane gradient, and so on. Controlled release of drug from liposomes is dependent on the type of synthesis method, surface modifications, and application type. Release mechanisms also rely on type of stimuli applied around lipid membranes, e.g., pH-dependent neutralization (charged functional groups), pH-dependent hydrolysis (non-charged components), thiolysis (disulfide bonds break), and temperature-dependent release.[9, 14-18] Several drug types have been tested for encapsulation in liposomes in order to evaluate their efficacy.

There are quite a number of formulations of liposomes that are commercially available and can be used for treating a variety of pathological conditions, e.g., tumors, fungal, bacterial, and viral infections. DOX, one of the widely used anticancer drugs, has been shown to be effectively encapsulated in liposomes. Liposomes have shown to have decreased DOX toxicity and increased retention time, although the values may vary based on applications and intended performances. A commercially available liposome suspension *Lipo-dox* is formulated by loading DOX into PEGylated liposomes. It is administered intravenously and used to combat different cancer conditions such as Kaposi's sarcoma, ovarian/breast cancer, etc.[19]

Polymeric Micelles

As briefly described in Chap. 6, polymeric micelles are core-shell nanomaterials formed from amphiphilic block copolymers. They offer structural stability, which makes them well suited for drug-delivery applications. The type of intramolecular forces acting among the polymer chains determines the type of micelle formation. Polymeric micelles can be categorized as amphiphilic micelles formed via hydrophobic assembly, poly–ion complex micelles, and micelles stemming from metal complexation. The hydrophobically assembled micelles have hydrophobic and hydrophilic domains and therefore self-assemble when exposed to aqueous medium–forming core-shell structures. Drug of choice can be encapsulated inside these hydrophobic cores.[20]

Drugs tend to be physically entrapped inside the core of micelles or chemically conjugated onto the surface. Drug loading into the core is achieved through dialysis, solid dispersion, and oil in water (o/w) emulsion using organic solvents. Incorporation of anticancer drugs within polymeric micelles has shown significant inhibition of tumor growth. Moreover, reports suggest higher blood circulation time and higher drug accumulation in tumors for micellar systems.[21,22]

Not only cancer drugs but also antibiotics have been loaded in polymeric micelles. Slow and controlled release of drugs from micelles have been reported in several studies.[23-25] Release of the drug depends on several factors including the diffusion efficiency and drug affinity with the core, loaded amount of the drug, and length of core which forms a polymer segment.[26]

Surface of micelles are usually modified for drug delivery using hydrophilic moieties. PEG is a common hydrophilic polymer that offers longer circulation time and biocompatibility. Other cargo molecules can also be attached to the surface, e.g., targeting ligands, receptors, antibodies, etc., as they offer higher chances of cellular uptake and stability as compared to non-modified micelles.[20,27]

Although we have discussed a few categories of nanomaterials that are widely used for drug delivery applications this chapter, there exist several other alternatives that span from polymeric NPs to inorganic NPs and several combinations. Note that loading of drugs to nanocarriers is not an easy task, since this not only depends on the properties of the drug and the nanocarrier but also dictates the release of the drug. Next, we will discuss how a loaded drug may be released at the site of action. The discussion here will be limited to polymer-based drug delivery systems.

10.1.4 Drug Release and Mathematical Models

Release of the drug at the site of action is a key event that ultimately defines the efficiency of the drug delivery system. Drug solubility, its diffusion, and erodible nature of the carrier system are key factors that influence the release process. Loading efficiency of the drug and NP size are also important as smaller particles diffuse easily than larger ones. Moreover, if the drug is weakly bound to the NP system and/or drug is evenly distributed on NP surface, a quick release (burst) of the drug can occur.[28] Coating of NPs with the use of polymers can also help regulate the controlled release process.

Mathematical models for drug release are intended to estimate rates of drug release and nature of the drug diffusion from the drug delivery processes. In order to optimize the release kinetics and affirmation of the processes associated with drug transport, mathematical models can provide significant understanding. Additionally, comparison of the models with the data from experimental work can provide better understanding of release kinetics. Mathematical models can facilitate better understanding of the effects of different parameters involved in the drug delivery systems such as shape, composition, and size of the systems involved on the rate of drug release. More efficient drug formulations can be achieved by using mathematical models which can estimate the drug release rate and can cut the cost and time required for the formulation of drug delivery systems.

Various models, semi-empirical, empirical, and mechanistic, have been employed to understand the drug release from drug delivery frameworks.[29,30] If you have a chemical engineering background, you may have already spotted that Fick's law governs the mechanistic models, where the concept of diffusivity is applied. These models can give details about the degradation, erosion, and diffusion of the drug. Drug delivery systems may usually be categorized as having Fickian or non-Fickian diffusion-based release of the drug. Diffusion is considered to be of Fickian nature when the relaxation time t_r is greater than the solvent diffusion time t_d. However, in case of non-Fickian diffusion, these times are comparable, i.e., $t_r \approx t_d$.

The basic aim to perform drug release modeling is to get detailed insight into the release phenomena occurring for different drug delivery systems based on different materials. However, as there are various factors which interplay during the course of drug release, inconsistencies exist between the experimental data and the mathematical models. Furthermore, models which are currently available do not efficiently explain the dynamic nature of drug release process from complex drug delivery devices. Some of the relevant models are described next.

Mathematical Models Based on Diffusion

A drug release system purely dependent on diffusion control can primarily be distinguished as (i) reservoir systems, having a cache of drug surrounded by a membrane that controls the diffusion of the drug through it, and (ii) monolithic systems that have the drug incorporated in the polymer matrix as a *one-block* system rendering no membrane barrier (Fig. 10.3).[31]

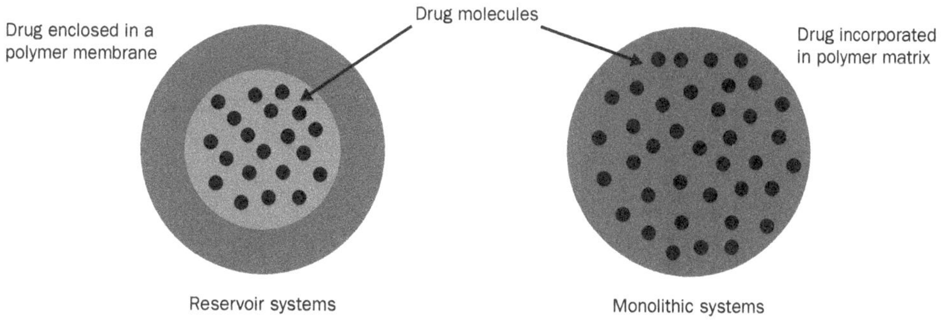

Figure 10.3 Schematic showing the reservoir and monolithic systems.

Reservoir Systems Two cases may arise in this particular scenario, one having drug concentration below the drug solubility in the device, whereas the other having drug concentration above the drug solubility, resulting in a saturated system. In the former case, the drug concentration constantly decreases with drug release through the membrane and the release follows a first-order release kinetics. Thus the cumulative drug release, M_t, can be represented, irrespective of the geometry, as:

$$\frac{dM_t}{dt} = \frac{ADKc_t}{l} = \frac{ADK}{l}\frac{M_0 - M_t}{V} \tag{10.1}$$

where M_t denotes the absolute cumulative amount of drug released over time t, M_0 is the initial amount of drug inside the matrix, A is the total surface area of the device, D represents the diffusion coefficient of the drug within the membrane, K is the partition coefficient of the drug between the membrane and the reservoir, c_t is the concentration of the drug in the release medium at time t, l is the thickness of the membrane, and V is the volume of the drug reservoir.

In the latter case, the reservoir maintains a constant drug concentration owing to the dissolution of the drug crystals/aggregates inside the device. In this case, the system follows a zero-order release kinetics represented by:

$$\frac{dM_t}{dt} = \frac{AJ_{\text{lim}}}{l} = \frac{ADKc_s}{l} \tag{10.2}$$

where J_{lim} is the membrane limiting flux and c_s is the solubility of the drug in the reservoir.

Figure 10.4 shows the cumulative drug release curves as a function of time for both kinds of reservoir systems discussed in this section.

Monolithic Systems Contrary to the previous system, the system geometry has a major impact on the drug release kinetics when it comes to monolithic systems. However, similar to its previous counterpart, two cases might arise depending on the concentration of drug inside the NP matrix. These two scenarios are the monolithic dispersions and monolithic solutions, where the drug concentration is higher and lower than the drug solubility inside the device, respectively. Let us first discuss the case of monolithic solutions.

The diffusion of drug from the system follows Fick's second law, with appropriate boundary conditions to respective geometries.[31]

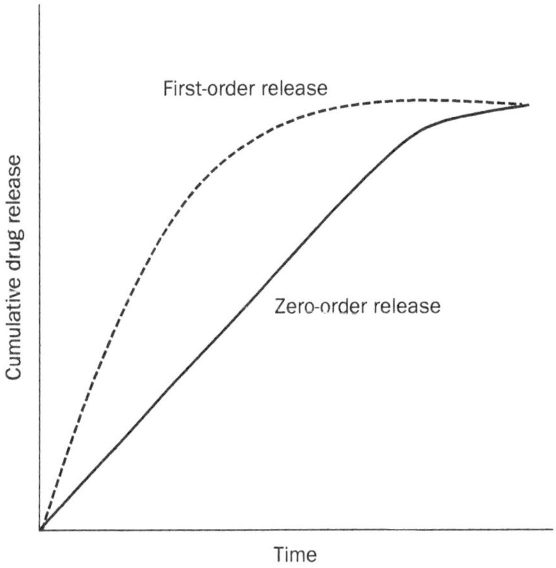

FIGURE 10.4 Curves showing the cumulative drug release dependence on time.

Fick's second law of diffusion where the diffusion coefficient is independent of concentration is defined as:

$$\frac{\partial c}{\partial t} = D\frac{\partial^2 c}{\partial x^2} \tag{10.3}$$

In the case of thin-film geometry, the following boundary conditions can be applied:[32]

$$t = 0 \quad -\frac{L}{2} < x < \frac{L}{2} \quad c = c_0 \tag{10.4}$$

$$t > 0 \quad x = \pm\frac{L}{2} \quad c = 0 \tag{10.5}$$

where L is the thickness of the film.

These result in the following equations:

$$\frac{M_t}{M_\infty} = 1 - \frac{8}{\pi^2}\sum_{n=0}^{\infty}\frac{1}{(2n+1)^2}\exp\left(\frac{-D(2n+1)^2 \pi^2 t}{L^2}\right) \tag{10.6}$$

$$\frac{M_t}{M_\infty} = 4\sqrt{\frac{Dt}{\pi L^2}} \quad \text{for } 0 \le \frac{M_t}{M_\infty} \le 0.6 \tag{10.7}$$

$$\frac{M_t}{M_\infty} = 1 - \frac{8}{\pi^2}\exp\left(\frac{-\pi^2 Dt}{L^2}\right) \quad \text{for } 0.4 \le \frac{M_t}{M_\infty} \le 1.0 \tag{10.8}$$

where M_∞ is the absolute cumulative drug released at infinity and M_t is the absolute cumulative drug released over time t.

Spheres	$$\frac{M_t}{M_\infty} = 1 - \frac{6}{\pi^2}\sum_{n=1}^{\infty}\frac{1}{n^2}\exp\left(\frac{-Dn^2\pi^2 t}{R^2}\right)$$ Where R is the radius of the sphere
Cylinder	$$\frac{M_t}{M_\infty} = 1 - \frac{32}{\pi^2}\sum_{n=1}^{\infty}\frac{1}{q_n^2}\exp\left(\frac{-q_n^2}{R^2}Dt\right) \times \sum_{p=0}^{\infty}\frac{1}{(2p+1)^2}\exp\left(-\frac{(2p+1)^2\pi^2}{H^2}Dt\right)$$ Where H is the height of the cylinder

Table 10.1 Drug Release Equations for Different Geometries in the Case of Monolithic Solution Type of Systems

The drug release equations for other geometries are presented in Table 10.1.

The situation becomes more complex for monolithic dispersions. Takeru Higuchi describes the model for the simplest thin-film geometry having negligible edge effects (i.e., considering an infinitely large film) with the equation:

$$\frac{M_t}{A} = \sqrt{D(2c_0 - c_s)c_s t} \tag{10.9}$$

where D is the drug diffusivity in the carrier, and c_0 and c_s are the initial drug concentration and the drug solubility within the polymer, respectively.

An interesting conclusion that can be inferred for thin-film release geometry is that whether we follow the Fickian diffusion model or Higuchi model, both show a $t^{1/2}$-time dependence of drug transport.

Mathematical Models Based on Polymer Swelling

The models described in the above sections do not incorporate polymer swelling and assume no significant changes in the polymer matrix that is not applicable for all polymer-based NP systems. Swelling creates substantial changes to the release geometry by increasing the length of diffusion pathways, resulting in decreased drug concentration gradients that might decrease the release rate. Additionally, the mobility of macromolecules greatly increases causing an increase in the release rates. Therefore, the drug release kinetics might increase or decrease depending on whether the diffusion distance or mobility of the macromolecules dominates. For this reason, semi-empirical power law was developed, which has been widely used for understanding drug release mechanism from polymer-based systems. The law incorporates both the diffusion-controlled as well as the relaxation-controlled drug release mechanisms, represented by the following equation:[32]

$$\frac{M_t}{M_\infty} = k_1\sqrt{t} + k_2 t \tag{10.10}$$

where k_1 and k_2 are constants.

That can be further simplified by combining the two constants as depicted by Eq. (10.11), also referred to as the famous *Peppas equation:*

$$\frac{M_t}{M_\infty} = kt^n \tag{10.11}$$

where the value of coefficient, n, dictates the release mechanism for different particle geometries, as stated in Table 10.2.

Mathematical Models Based on Degradation

Now we will discuss systems where the polymeric matrix degrades due to either dissolution or erosion.

Polymer Dissolution As the polymer swells indefinitely, it may result in polymer dissolution at the surface, which is explained by Narasimhan and Peppas.[33,34] Above a certain critical water concentration, C_{crit}, the polymer chains at the surface start to disentangle and diffuse into the bulk solvent surrounding the NP, resulting in polymer dissolution (Fig. 10.5).

In this case, the polymer mass loss is characterized by polymer dissolution rate constant, k_{diss}, and is represented as:

$$M_{pt} = M_{p0} - k_{diss} A_t t \tag{10.12}$$

where M_{pt} and M_{p0} are the dry polymer matrix mass at time t and $t = 0$, respectively, and A_t is the surface area of the device.

The release model developed by Narasimhan and Peppas can be depicted as follows:[33–35]

$$\frac{M_t}{M_\infty} = \frac{\upsilon_{d,eq} + \upsilon_d^*}{l}\left(\sqrt{2At + Bt}\right) \tag{10.13}$$

where $\upsilon_{d,eq}$ and υ_d^* are the equilibrium and the characteristic drug concentrations respectively, and A and B are variables.

Thin Film	Sphere	Cylinder	Drug Release Mechanism
0.5	0.45	0.43	Fickian diffusion
$0.5 < n < 1$	$0.45 < n < 0.89$	$0.43 < n < 0.85$	Anomalous transport
1	0.89	0.85	Case-II transport/polymer swelling
Higher than 1			Super case-II transport

TABLE 10.2 Values of n for Various Geometries Showing Different Drug Release Mechanisms

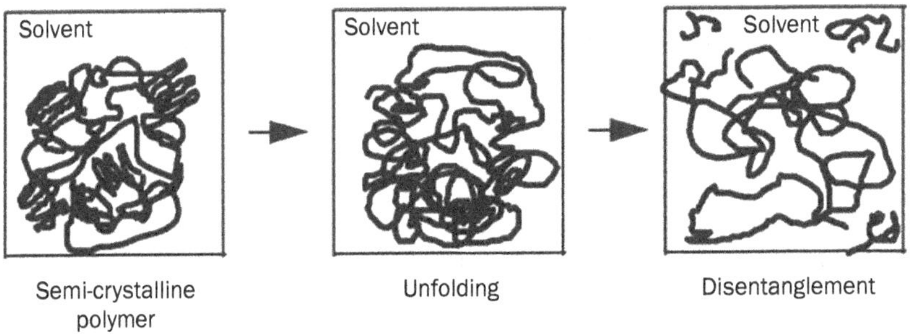

FIGURE 10.5 Schematic showing the disentanglement and dissolution of polymeric chains.[33]

Polymer Erosion Erosion may be defined as the process of material loss from the polymer bulk that is further sub-categorized as surface (heterogeneous) and bulk (homogeneous) erosion. In the former case, the polymer chain cleavage is much faster than the penetration of water into the matrix, whereas it is opposite in the case of homogeneous erosion. The rapid water penetration in the polymer matrix results in faster wetting, and degradation occurs throughout the polymer network resulting in bulk erosion.

A semi-empirical model was proposed by Hopfenberg, where rate-limiting step is being quantified from the zero-order surface release.[28,36] The equation that can be applicable for the case of spheres, slabs, and cylinders is represented as:[31]

$$\frac{M_t}{M_\infty} = 1 - \left(1 - \frac{k_0 t}{c_0 a}\right)^n \tag{10.14}$$

where k_0 is the rate constant, c_0 is the uniform initial drug concentration within the system, a is equal to the radius of the sphere or cylinder or half-thickness in case of slab, n is the shape factor; $n = 3$ for spheres, $n = 2$ for cylinders, $n = 1$ for slab geometry.

This model has a shortcoming that it does not give good correlation with the experimental data in case of a cylindrical tablet.[37] The improved model is developed, which has considered the axial and radial factors for erosion for the case of drug release from the erodible material matrix and is depicted as:

$$\frac{M_t}{M_\infty} = 1 - \left(1 - \frac{k_a t}{C_0 a_0}\right)^2 \left(1 - \frac{2k_b t}{C_0 b_0}\right) \tag{10.15}$$

where k_a and k_b are the radial and axial erosion rate constants, respectively, a_0 is the initial radius of the tablet, and b_0 is the initial thickness of the tablet.

A variety of release models have been developed in order to study the release kinetics, but there is not a single model that can describe the release behavior from each polymer type. It is because of the fact that the release of drug is dependent upon several key factors including type and synthesis of the delivery system, release mechanisms (e.g., stimuli responsive), physicochemical properties, and so on.

So far we have discussed modes of drug delivery and how different nanomaterials find applications in this field. Further, we have discussed how loaded drugs may be released from polymer-based systems. Next topic of discussion is another important area of application of nanomaterials in biomedical fields—magnetic resonance imaging (MRI).

10.2 Magnetic Resonance Imaging

Imaging in biology is indispensable to analyze and diagnose biological processes at organ, cellular, and molecular levels. Additionally, in the field of nanomedicine, it is widely used to study NP "*in vivo* behavior." One of the imaging modalities extensively applied for this purpose is MRI. MRI provides an unsurpassed soft-tissue contrast and high spatial resolutions, down to about 30 μm in animals and typically 0.5 mm in humans. MRI is a widely used clinical imaging modality to diagnose, for example, cancer, joints, and Parkinson's disease, among others.[38,39] MRI is based on nuclear magnetic resonance (NMR) principles. In the subsequent sections, MRI working principles will be discussed along with NP-based contrast agents (CAs) for biomedical applications with a few examples from the literature.

Before discussing NPs for MRI, let us discuss different types of magnetic materials that can be used for imaging purposes. Electron spin and orbital moments result in magnetic dipoles in materials, which finally determine the magnetic properties of the materials as a whole. Based on the response shown by materials toward external magnetic field B_0, they may be classified as diamagnets, paramagnets, ferromagnets, ferrimagnets, or antiferromagnets. Further, their magnetic properties can be described in terms of three parameters—coercivity, saturation magnetization, and remnant magnetization. Coercivity is a measure of the resistance of a magnetic material to changes in magnetization. Mathematically, it is measured as the external magnetic field required to reduce the magnetization to zero. Saturation magnetization is the maximum value of magnetization that the material can reach with increasing external field. Remnant magnetization is a measure of residual magnetization in the material when there is no applied field.

An interesting phenomenon that appears at nanoscale is superparamagnetism. The behavior of such materials is that of paramagnetic, but their saturation magnetizations are typical of ferromagnetic materials. Magnetization can randomly flip direction under the influence of temperature, and the time between two flips is called *Neel relaxation time.* A superparamagnetic material shows zero coercivity and remanence, meaning there is no hysteresis between magnetization and applied magnetic field. However, superparamagnetism is strictly a time-dependent property that depends strongly on the relaxation time of the material and on the measuring time of the technique chosen. Although any ferro or ferromagnetic material undergoes a transition to a paramagnetic state above its Curie temperature, superparamagnetism occurs below the Curie temperature of the material. Further, superparamagnetism may be destroyed by cooling—this temperature being commonly known as *blocking temperature.*[40]

10.2.1 Fundamentals of Magnetic Resonance Imaging

MRI is one of the high-resolution noninvasive imaging techniques used in clinical science today. In most MRI applications, the MRI signal originates from water protons. When placed in a magnetic field, an equilibrium magnetization of the protons in the direction of the magnetic field arises. This equilibrium can be excited by the application of an additional magnetic field, called a *radiofrequency (RF) pulse.* The RF pulse will result in rotation of the magnetization equilibrium vector away from the B_0 direction. After the RF pulse, excited protons return to their original states through a process called *relaxation.* Two independent relaxation processes occur: relatively slow longitudinal or T_1 relaxation, which describes the return to the equilibrium magnetization along the B_0 field, and the relatively fast transverse or T_2 relaxation, which results in a loss of magnetization in the x-y plane perpendicular to the direction of B_0 (z-direction). These magnetic fluctuations induce electrical voltages in receiver coils of the MR scanner, which are translated to signal intensities. Shorter T_1 results in faster recovery of the net magnetization along B_0 after an RF pulse and higher signal intensity. Shorter T_2 results in a faster loss of magnetization in the x-y plane and lower signal intensity. As such, signal intensity in MRI is primarily dependent on water proton density, and T_1 and T_2 relaxation. Since water content of different tissues is variable, and also T_1 and T_2 relaxation significantly differ depending on the specific physiochemical properties of the tissue, this allows for contrast generation between virtually all biological tissues. By applying suitably combined magnetic gradients and RF pulses in concert, called *MRI sequences,* images can be weighted to predominantly contain contrast based on proton density, T_1, or T_2.

The use of CAs in MRI to increase the contrast in target tissues further increases the diagnostic power of MRI. All conventional MRI CAs alter image contrast by decreasing T_1 and T_2 relaxation times. CAs predominantly shortening T_1 increase the signal intensity in their surrounding and are referred to as *positive contrast agents*. CAs predominantly shortening T_2 decrease signal intensity and are referred to as *negative contrast agents*. The efficiency by which CAs alter T_1 and T_2 relaxation times is called T_1 relaxivity (r_1) and T_2 relaxivity (r_2), respectively. Contrast agents can be successfully used to generate positive contrast in T_1-weighted images when r_2/r_1 ratios are in the range of 1–5. If this ratio is higher than 5, the contrast agent is more suitable for generating T_2 contrast in T_2-weighted images.

Note that for efficient T_1 shortening, close interaction between protons and T_1 agents is the prerequisite, which can be hampered by the thickness of magnetic NP coating[41] or cellular internalization.

10.2.2 Nanoparticle Contrast Agents for Magnetic Resonance Imaging

NP-based CAs also broaden spectrum of functions for MRI. Because of their multifunctional aspects of high payload ability, penetration, soft tissue contrast, and high spatial resolution, they have become effective noninvasive imaging modality in medical diagnostics. Although clinically approved, superparamagnetic iron oxide (SPIO) NPs typically have a high r_2/r_1 ratio and result in negative contrast, which can complicate distinction between NP accumulation and lesions. Nevertheless, they are famous for MRI use among researchers owing to their high sensitivity and biological tolerance.[42]

Gadolinium (Gd) ions are extensively applied in the clinic as T_1 CA-producing positive image contrast. Because of its electronic configuration (Gd possesses seven unpaired electrons in f orbital, which gives S electronic state), it provides a large-moment magnetization and produces high T_1 relaxation. Toxicity with the use of Gd-based CAs has been reported, but chelation with suitable ligands can help reduce its toxicity.[43] Nevertheless, these toxicity concerns stimulated the development of iron oxide–based NPs with low r_2/r_1 ratio as a safer alternative positive contrast agent. In a most recent study carried out, Wei et al.[44] suggested the use of core-shell zwitterion-coated small SPIO NPs (3-nm core with 1-nm ultrathin hydrophilic coat) as positive CAs for MRI. They have suggested that these small core-shell NPs have the potential to address the significant need for Gd-free CAs having strong T_1 contrast.

Imaging agents can be loaded into NPs in a controlled concentration and can be targeted to the organ of interest. Surface modifications of such NPs loaded with imaging agent can equip them with site-specific targeting and can also increase their circulation time in blood.[45] MRI CAs have the ability to image brain, heart, and vasculature and also possess the ability to detect tumors.[46,47] Magnetic iron oxide (MIO) NPs can be used as MRI CAs. They have the ability to produce heat when exposed to fluctuating magnetic field. This property can be exploited in targeted cancer therapies as to cause heat to release theranostic agents at tumor site.[48] With an aim to perform a dual function McDonagh et al. have functionalized manganese oxide NPs with L-DOPA (drug for Parkinson's diseases) for MRI and drug delivery. These NPs were degraded in biological medium releasing Mn^{2+} ions along with L-DOPA. The manganese oxide generates negative contrast and Mn ions released upon degradation generate positive contrast. This resulted in a contrast switch that is accompanied by the release of L-DOPA, potentially allowing to study drug release in real time *in vivo*.[49]

NPs provide many functions which can enhance CA use in MRI for diagnosis, monitoring, and prognosis of medically important diseases. One very promising application of iron oxides is patient selection for nanotherapy. The use of nanomedicine to treat solid tumors is dependent on the EPR effect (explained in Sec. 10.1.1). Since the EPR effect is highly heterogeneous, both within tumors as well as from tumor to tumor, some patients will not respond to nanotherapy, while others will. However, currently no clinical tools to assess the EPR effect in individual patients are established. Miller et al. demonstrated that iron oxide accumulation in tumors in mice, as assessed with MRI, correlated with nanomedicine tumor accumulation and therapeutic response.[50] Subsequent work in the clinic has shown promising evidence of correlation between iron oxide MR imaging and tumor response in patients.[51]

These studies provide a compelling application of MRI CA to resolve a pressing clinical issue for nanomedicine. Another intriguing field of application of nanomaterials—biosensing—is discussed next.

10.3 Biosensing

A potent and innovative analytical device involving a biological sensing moiety capable of detecting and analyzing biomolecules is termed as *biosensor.* These novel devices find wide applications in drug discovery, biomedicine, diagnosis, food safety, and environmental monitoring, to name a few. Clarks and Lyons developed the first biosensor to measure the glucose levels in biological samples relying on the technique of electrochemical detection of oxygen or hydrogen peroxide via immobilized glucose oxidase electrode. The performance of a biosensor is defined in terms of two important properties—detection limit and sensitivity. While the former gives an indication of how much low amounts can be effectively sensed, the latter quantifies how small changes in the amount can be significantly sensed. Significant advances have been made in the field of biosensing, evolving from the classical electrochemical to optical, polymer, glass, silica, and nanomaterials, improving the detection limits, as well as system sensitivity along the way.

Generally, biosensing strategies are categorized into labeled and label-free assays (Fig. 10.6). The label-free biosensors detect the presence of an analyte directly through biochemical reactions on the surface of the transducer that converts the biochemical signal into an electrical output. While in the case of labeled assay, the analyte is sandwiched in between the capture agents and detector agents. The capture agents are immobilized on the surface of the solid, such as electrodes, glass, NPs, etc., whereas the detector agents are associated with signaling tags such as fluorophores, enzymes, NPs, etc.

Nanomaterials offer great advantages in development of biosensing devices owing to their highly specific surface area, enabling them to immobilize increased amounts of bioreceptor units on their surfaces thereby detecting substances down to a single molecule. This trait of NPs was utilized in the label-free detection of the potentially deadly hepatitis B virus (HBV) using glassy carbon surface modified with AuNPs and poly(allylamine)-branched ferrocene (PAA-Fc) (Fig. 10.7).[52] The AuNPs provide the appropriate surface for antigen binding and create the favorable microenvironment. The attachment of HBV antigen is detected through the decrease in the amphoteric activity of the sensor resulting from the binding of the antigen to the antibody with detection limits being as low as 40 pg/mL.

Besides having high surface functionality, Au and Ag NPs possess LSPR characteristics in the visible light spectrum which are used for signal amplification labels for biorecognition

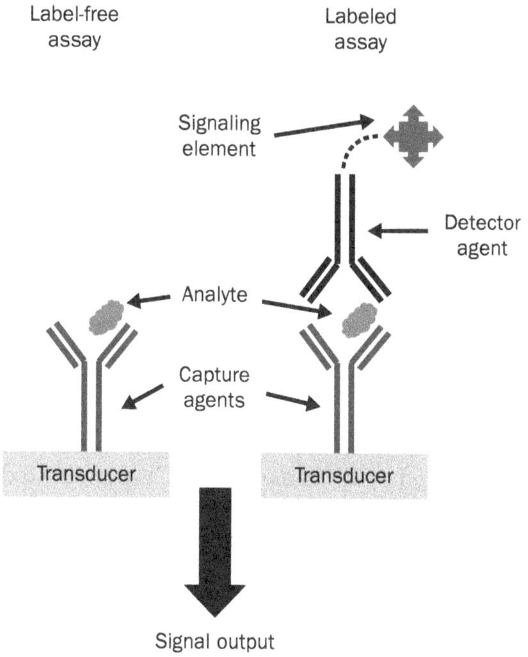

Figure 10.6 Schematic representing label-free and labeled assay used in biosensing.

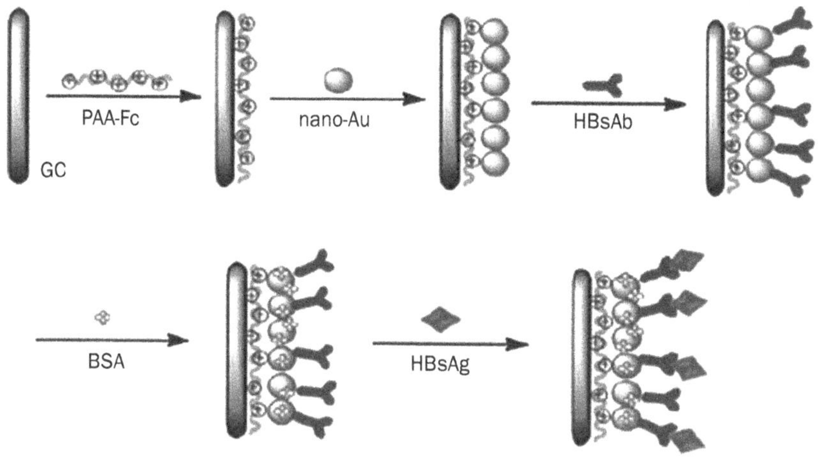

Figure 10.7 PAA-Fc modified AuNPs for detection of hepatitis B virus.[52]

processes. This phenomenon has been utilized as a part of an integrated microfluidic chip for the detection of HIV and syphilis from a needle-prick amount of blood.

Magnetic NPs allow for solution phase detection as compared to the diffusion-dependent surface structure–based detection as discussed previously, resulting in comparatively faster assay times. Another advantage of using MNPs is that they are superior at dephasing nuclear spins of neighboring water protons, allowing for a signal amplification in NMR-based sensors by causing a decrease in the spin-spin relaxation time.[53] This property has been exploited for the detection of clinical pathogens using bacteria's 16S rRNA molecule as the target marker, and for dextran-coated iron oxide NPs (21nm in size) using a μNMR device.

Optrode-based optic fiber biosensors, which combine both the optical and electrical properties into an analytical device to measure the concentrations of a chemical or a group of chemicals in a sample, have been used in the industry for a few decades.[54] However, the advent of nanotechnology has enabled these devices to get smaller and measure the analytes inside a single living cell. Tapered fibers with a tip diameter of around 20–80 nm are synthesized by pulling optical fibers from an initial diameter of a few microns using a modified micropipette puller. Nanobiosensors synthesized using this technique were utilized by Kopelman and coworkers[55,56] to monitor intercellular pH inside living cells. Similar devices were used to detect nitric oxide (NO) content using fluorescent-labeled cytochrome C (Cyt C) immobilized on the sensor tip. Cyt C undergoes conformational changes in the presence of NO; hence, changes in the energy transfer between Cyt C and the fluorescent dye are collated with the NO concentrations.

This chapter gives insights into biomedical applications of NPs, such as drug delivery, imaging, biosensing, and diagnostics, to name a few. With newer synthesis routes being developed and innovative NPs being discovered, their potential uses in the medical industry are also increasing. Therefore, NPs are there to stay and aid in the improvement of the biomedical research and facilities.

10.4 Study Questions

10.4.1 Concept-Based Questions

10.1 Mention two different ways by which a drug can reach its site of action.

10.2 Discuss some of the qualities that a nanocarrier should possess for it to be used in drug delivery applications.

10.3 What is the main role of PEG for nanocarriers used in drug delivery?

10.4 What are the factors that control release for a system that follows Higuchi rate model?

10.5 Graphically show the differences between a zero-order and first-order equations for diffusion-based rate models in relation to drug release. Using rate equations, explain the nature of the curves.

10.6 Describe the fundamental working principle for MRI. Differentiate between positive and negative contrast agents.

10.7 What are the different ways of weighting MRI images?

10.8 State the differences between labeled and label-free assays for biosensing.

10.4.2 Research-Based Questions

10.9 Hybrid Fe@Au, coated with pNIPAm-based hydrogels, was synthesized to investigate the loading and release of a hydrophilic drug, cytochrome C (Cyt C).

UV-Vis spectroscopy was used to determine loading efficiency of the hydrogels using a standard calibration curve. The samples showed high loading of about 80%. The data for the release study are shown in the table below.

t (h)	% Release of Cyt C
0.00	0.00
0.50	0.08
1.00	2.87
2.00	5.50
3.00	7.93
4.00	11.08
5.00	12.67
6.00	26.93
7.00	28.56
8.00	28.79
9.00	29.64
10.00	29.71
11.00	29.80
13.00	29.59
15.00	28.57
17.00	29.31
19.00	29.64
21.00	29.49
25.00	30.41
29.00	31.00
33.00	31.88
37.00	32.70
51.00	34.70

$$F = \frac{M_t}{M_\infty}$$

where F is equal to the fraction of mass released, and M_t and M_∞ are the cumulative mass of drug released in time t and at infinite time (can be assumed to be final time point for the study), respectively.

a. Using Peppas equation, comment on the drug release mechanism from the hydrogel-based systems. Assume initial drug at the start of loading to be 6 mg.

b. Calculate the encapsulation efficiency of the system if 3.3 mg of hydrogels were used for the loading study.

c. What are the advantages of using such hybrid NPs over conventional hydrogels?

d. This drug release study has been performed as a function of temperature. Are there any guidelines from the system to choose this temperature in order to have maximum release?

10.10 The image below shows magnetic resonance imaging (MRI) being used for investigating the degradation of manganese oxide NPs (MONPs) inside the pig's eye. The tissue can be characterized by different relaxation times, as shown in the figure below.

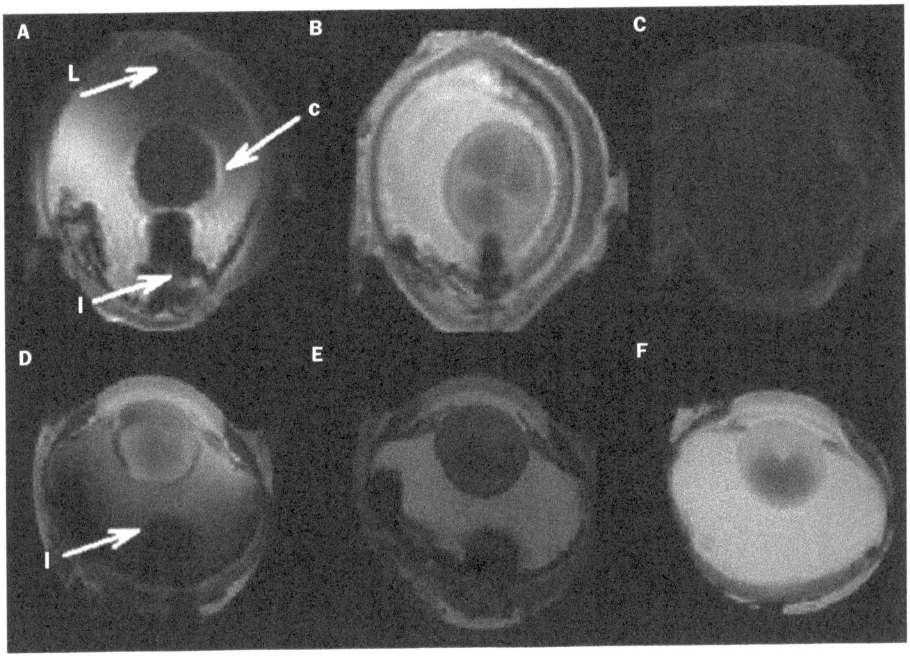

Figures A and B represent T_1w images of the pig's eye injected with MONPs acquired immediately and after 34 hours, respectively. Figure C is the T_1w image of the control eye taken immediately after injection. Similarly, figures D and E are the T_2^* w images immediately and after 34 hours of injection, respectively. Figure F is the T_2^* w image of the control eye taken after 34 hours of injection of MONPs.

a. Describe the changes in contrast that occur among the T_1 and T_2^* w images.

b. Explain the contrast change for T_1w images from figures A and B to figure C with respect to contrast agents.

c. Similarly, explain the contrast change for T_2^* w images from figures D and E to figure F with respect to contrast agents.

10.5 References

1. Mukherjee, B., Nanosize drug delivery system. *Current Pharmaceutical Biotechnology.* **2013,** *14*(15), 1221.
2. Torchilin, V. P., Passive and active drug targeting: drug delivery to tumors as an example. In: *Drug Delivery, Handbook of Experimental Pharmacology,* Schäfer-Korting, M. ed. Berlin, Heidelberg: Springer, 2010, Vol. 197, pp. 3–53.
3. Maeda, H.; Wu, J.; Sawa, T.; Matsumura, Y.; Hori, K., Tumor vascular permeability and the EPR effect in macromolecular therapeutics: a review. *Journal of Control Release.* **2000,** *65*(1–2), 271–284.
4. Brede, C.; Labhasetwar, V., Applications of nanoparticles in the detection and treatment of kidney diseases. *Advances in Chronic Kidney Disease.* **2013,** *20*(6), 454–465.
5. Liechty, W. B.; Peppas, N. A., Expert opinion: responsive polymer nanoparticles in cancer therapy. *European Journal of Pharmaceutics and Biopharmaceutics.* **2012,** *80*(2), 241–246.
6. Nagayasu, A.; Uchiyama, K.; Kiwada, H., The size of liposomes: a factor which affects their targeting efficiency to tumors and therapeutic activity of liposomal antitumor drugs. *Advanced Drug Delivery Reviews.* **1999,** *40*(1–2), 75–87.
7. Nagayasu, A.; Shimooka, T.; Kinouchi, Y.; Uchiyama, K.; Takeichi, Y.; Kiwada, H., Effects of fluidity and vesicle size on antitumor activity and myelosuppressive activity of liposomes loaded with daunorubicin. *Biological and Pharmaceutical Bulletin.* **1994,** *17*(7), 935–939.
8. Truong, N. P.; Whittaker, M. R.; Mak, C. W.; Davis, T. P., The importance of nanoparticle shape in cancer drug delivery. *Expert Opinion on Drug Delivery.* **2015,** *12*(1), 129–142.
9. Abraham, S. A.; McKenzie, C.; Masin, D.; Ng, R.; Harasym, T. O.; Mayer, L. D.; Bally, M. B., In vitro and in vivo characterization of doxorubicin and vincristine coencapsulated within liposomes through use of transition metal ion complexation and pH gradient loading. *Clinical Cancer Research.* **2004,** *10*(2), 728–738.
10. Alexis, F.; Pridgen, E.; Molnar, L. K.; Farokhzad, O. C., Factors affecting the clearance and biodistribution of polymeric nanoparticles. *Molecular Pharmaceutics.* **2008,** *5*(4), 505–515.
11. Choi, S. H.; Yoon, J. J.; Park, T. G., Galactosylated poly(*N*-isopropylacrylamide) hydrogel submicrometer particles for specific cellular uptake within hepatocytes. *Journal of Colloid and Interface Science.* **2002,** *251*(1), 57–63.
12. Byrne, J. D.; Betancourt, T.; Brannon-Peppas, L., Active targeting schemes for nanoparticle systems in cancer therapeutics. *Advanced Drug Delivery Reviews.* **2008,** *60*(15), 1615–1626.
13. Yang, H. W.; Hua, M. Y.; Liu, H. L.; Huang, C. Y.; Wei, K. C., Potential of magnetic nanoparticles for targeted drug delivery. *Nanotechnology, Science and Applications.* **2012,** *5*, 73–86.
14. Colletier, J. P.; Chaize, B.; Winterhalter, M.; Fournier, D., Protein encapsulation in liposomes: efficiency depends on interactions between protein and phospholipid bilayer. *BMC Biotechnology.* **2002,** *2*, 9.
15. Ellens, H.; Bentz, J.; Szoka, F. C., pH-induced destabilization of phosphatidylethanolamine-containing liposomes: role of bilayer contact. *Biochemistry.* **1984,** *23*(7), 1532–1538.
16. Guo, X.; Szoka, F. C., Jr., Steric stabilization of fusogenic liposomes by a low-pH sensitive PEG–diortho ester–lipid conjugate. *Bioconjugate Chemistry.* **2001,** *12*(2), 291–300.

17. Zhang, J. X.; Zalipsky, S.; Mullah, N.; Pechar, M.; Allen, T. M., Pharmaco attributes of dioleoylphosphatidylethanolamine/cholesterylhemisuccinate liposomes containing different types of cleavable lipopolymers. *Pharmacological Research.* **2004,** *49*(2), 185–198.
18. Needham, D.; Dewhirst, M. W., The development and testing of a new temperature-sensitive drug delivery system for the treatment of solid tumors. *Advanced Drug Delivery Review.* **2001,** *53*(3), 285–305.
19. Chang, H. I.; Yeh, M. K., Clinical development of liposome-based drugs: formulation, characterization, and therapeutic efficacy. *International Journal of Nanomedicine.* **2012,** *7*, 49–60.
20. Gaucher, G.; Dufresne, M. H.; Sant, V. P.; Kang, N.; Maysinger, D.; Leroux, J. C., Block copolymer micelles: preparation, characterization and application in drug delivery. *Journal of Controlled Release.* **2005,** *109*(1–3), 169–188.
21. Park, E. K.; Kim, S. Y.; Lee, S. B.; Lee, Y. M., Folate-conjugated methoxy poly(ethylene glycol)/poly(epsilon-caprolactone) amphiphilic block copolymeric micelles for tumor-targeted drug delivery. *Journal of Controlled Release,* **2005,** *109*(1–3), 158–168.
22. Kawano, K.; Watanabe, M.; Yamamoto, T.; Yokoyama, M.; Opanasopit, P.; Okano, T.; Maitani, Y., Enhanced antitumor effect of camptothecin loaded in long-circulating polymeric micelles. *Journal of Controlled Release.* **2006,** *112*(3), 329–332.
23. Adams, M. L.; Kwon, G. S., Relative aggregation state and hemolytic activity of amphotericin B encapsulated by poly(ethylene oxide)-block-poly(*N*-hexyl-L-aspartamide)-acyl conjugate micelles: effects of acyl chain length. *Journal of Controlled Release.* **2003,** *87*(1–3), 23–32.
24. Forrest, M. L.; Zhao, A.; Won, C. Y.; Malick, A. W.; Kwon, G. S., Lipophilic prodrugs of Hsp90 inhibitor geldanamycin for nanoencapsulation in poly(ethylene glycol)-b-poly(epsilon-caprolactone) micelles. *Journal of Controlled Release.* **2006,** *116*(2), 139–149.
25. Forrest, M. L.; Won, C. Y.; Malick, A. W.; Kwon, G. S., In vitro release of the mTOR inhibitor rapamycin from poly(ethylene glycol)-b-poly(epsilon-caprolactone) micelles. *Journal of Controlled Release.* **2006,** *110*(2), 370–377.
26. Huh, K. M.; Lee, S. C.; Cho, Y. W.; Lee, J.; Jeong, J. H.; Park, K., Hydrotropic polymer micelle system for delivery of paclitaxel. *Journal of Controlled Release.* **2005,** *101*(1–3), 59–68.
27. Kwon, G. S., Polymeric micelles for delivery of poorly water-soluble compounds. *Critcal Reviews in Therapeutic Drug Carrier Systems.* **2003,** *20*(5), 357–403.
28. Fu, Y.; Kao, W. J., Drug release kinetics and transport mechanisms of non-degradable and degradable polymeric delivery systems. *Expert Opinion on Drug Delivery.* **2010,** *7*(4), 429–444.
29. Kamaly, N.; Yameen, B.; Wu, J.; Farokhzad, O. C., Degradable controlled-release polymers and polymeric nanoparticles: mechanisms of controlling drug release. *Chemical Reviews.* **2016,** *116*(4), 2602–2663.
30. Narasimhan, B., Mathematical models describing polymer dissolution: consequences for drug delivery. *Advanced Drug Delivery Reviews.* **2001,** *48*(2–3), 195–210.
31. Siepmann, J.; Siepmann, F., Mathematical modeling of drug delivery. *International Journal of Pharmaceutics.* **2008,** *364*(2), 328–343.
32. Siepmann, J.; Peppas, N. A., Higuchi equation: derivation, applications, use and misuse. *International Journal of Pharmaceutics.* **2011,** *418*(1), 6–12.

33. Narasimhan, B., Mathematical models describing polymer dissolution: consequences for drug delivery. *Advanced Drug Delivery Reviews.* **2001,** *48*(2–3), 195–210.
34. Peppas, N. A.; Narasimhan, B., Mathematical models in drug delivery: How modeling has shaped the way we design new drug delivery systems. *Journal of Controlled Release.* **2014,** *190*, 75–81.
35. Narasimhan, B.; Peppas, N. A., Molecular Analysis of Drug Delivery Systems Controlled by Dissolution of the Polymer Carrier. *Journal of Pharmaceutical Sciences* **1997,** *86* (3), 297–304.
36. Hopfenberg, H. B., Controlled release from erodible slabs, cylinders, and spheres. In: *Controlled Release Polymeric Formulations.* United States of America, Editor: Robert F. Gould. American Chemical Society: 1976; Vol. 33, pp. 26–32.
37. Hopfenberg, H. B., Controlled release from erodible slabs, cylinders, and spheres. *ACS Symposium Series.* **1976,** *33*, 26–32.
38. Weissleder, R.; Pittet, M. J., Imaging in the era of molecular oncology. *Nature.* **2008,** *452*(7187), 580–589.
39. Huang, Y.; He, S.; Cao, W.; Cai, K.; Liang, X. J., Biomedical nanomaterials for imaging-guided cancer therapy. *Nanoscale.* **2012,** *4*(20), 6135–6149.
40. Krahne, R.; Morello, G.; Figuerola, A.; George, C.; Deka, C.; Manna, I., Physical properties of elongated inorganic nanoparticles. *Physics Reports—Review Section of Physics Letters.* **2011,** *501*(3–5), 75–221.
41. Sun, C.; Lee, J. S.; Zhang, M., Magnetic nanoparticles in MR imaging and drug delivery. *Advanced Drug Delivery Reviews.* **2008,** *60*(11), 1252–1265.
42. Mao, X.; Xu, J.; Cui, H., Functional nanoparticles for magnetic resonance imaging. *Wiley Interdisciplinary Reviews: Nanomedicine and Nanobiotechnology.* **2016,** *8*(6), 814–841.
43. Khawaja, A. Z.; Cassidy, D. B.; Al Shakarchi, J.; McGrogan, D. G.; Inston, N. G.; Jones, R. G., Revisiting the risks of MRI with Gadolinium-based contrast agents-review of literature and guidelines. *Insights Imaging.* **2015,** *6*(5), 553–558.
44. Wei, H.; Bruns, O. T.; Kaul, M. G.; Hansen, E. C.; Barch, M.; Wisniowska, A.; Chen, O.; . . . Bawendi, M. G., Exceedingly small iron oxide nanoparticles as positive MRI contrast agents. *Proceedings of the National Academy of Science of the United States of America.* **2017,** *114*(9), 2325–2330.
45. Jokerst, J. V.; Gambhir, S. S., Molecular imaging with theranostic nanoparticles. *Accounts of Chemical Research.* **2011,** *44*(10), 1050–1060.
46. Tang, H.; Wu, E. X.; Ma, Q. Y.; Gallagher, D.; Perera, G. M.; Zhuang, T., MRI brain image segmentation by multi-resolution edge detection and region selection. *Computerized Medical Imaging and Graphics.* **2000,** *24*(6), 349–357.
47. Sipkins, D. A.; Cheresh, D. A.; Kazemi, M. R.; Nevin, L. M.; Bednarski, M. D.; Li, K. C., Detection of tumor angiogenesis in vivo by alphaVbeta3-targeted magnetic resonance imaging. *Nature Medicine.* **1998,** *4*(5), 623–626.
48. Blanco-Andujar, C.; Walter, A.; Cotin, G.; Bordeianu, C.; Mertz, D.; Felder-Flesch, D.; Begin-Colin, S., Design of iron oxide-based nanoparticles for MRI and magnetic hyperthermia. *Nanomedicine (London).* **2016,** *11*(14), 1889–1910.
49. McDonagh, B. H.; Singh, G.; Hak, S.; Bandyopadhyay, S.; Augestad, I. L.; Peddis, D.; . . . Glomm, W. R., L-DOPA-coated manganese oxide nanoparticles as dual MRI contrast agents and drug-delivery vehicles. *Small.* **2016,** *12*(3), 301–306.

50. Miller, M. A.; Gadde, S.; Pfirschke, C.; Engblom, C.; Sprachman, M. M.; Kohler, R. H.; . . . Weissleder, R., Predicting therapeutic nanomedicine efficacy using a companion MR imaging nanoparticle. *Science Translational Medicine.* **2015,** *7*(314), 314ra183.
51. Ramanathan, R. K.; Korn, R. L.; Raghunand, N.; Sachdev, J. C.; Newbold, R. G.; Jameson, G.; . . . Fitzgerald, J. B., Correlation between ferumoxytol uptake in tumor lesions by MRI and response to nanoliposomal irinotecan in patients with advanced solid tumors: a pilot study. *Clinical Cancer Research.* **2017,** *23*(14), 3638.
52. Qiu, J.-D.; Huang, H.; Liang, R.-P., Biocompatible and label-free amperometric immunosensor for hepatitis B surface antigen using a sensing film composed of poly (allylamine)-branched ferrocene and gold nanoparticles. *Microchimica Acta.* **2011,** *174*(1–2), 97.
53. Tassa, C.; Shaw, S. Y.; Weissleder, R., Dextran-coated iron oxide nanoparticles: a versatile platform for targeted molecular imaging, molecular diagnostics, and therapy. *Accounts of Chemical Research.* **2011,** *44*(10), 842–852.
54. Biran, I.; Yu, X.; Walt, D. R., Optrode-based fiber optic biosensors (bio-optrode). In: *Optical Biosensors,* 2nd ed., Ligler, F.; Taitt, C. (eds.), , Amsterdam: Elsevier, 2008, pp. 3–82.
55. Tan, W.; Shi, Z. Y.; Kopelman, R., Development of submicron chemical fiber optic sensors. *Analytical Chemistry* **1992,** *64*(23), 2985–2990.
56. Tan, W.; Shi, Z. Y.; Smith, S.; Birnbaum, D.; Kopelman, R., Submicrometer intracellular chemical optical fiber sensors. *Science.* **1992,** *258*(5083), 778.

CHAPTER 11

Applications in Water Management

After the discussions on biomedical applications in the previous chapter, we now turn to environmental applications of nanomaterials. This chapter covers two important applications—wastewater treatment and hydrological tracers in water management.

11.1 Wastewater Treatment

Urbanization in the present world has escalated pollution of water, both at surface and ground level, and in parallel worldwide demand of quality water to meet human needs is increasing. As a result, the supply of clean water is becoming an intrinsic global issue. To overcome this scarcity of clean potable water, novel technological advancement for integral water management is necessary. So far different strategies have been adopted for the treatment of contaminated water.

Traditionally, wastewater treatment employs three steps, schematically shown in Fig. 11.1—primary, secondary, and tertiary treatment processes, to remove contaminants from the affected impure water.

In the first stage of water treatment known as *primary water treatment*, processes involve physical separation of solids, oil, and grease through screening out of insoluble solids as well as sedimentation and flocculation. Screened-out wastewater stream is kept for several hours in a primary sedimentation tank where heavy solids settle at the bottom due to gravity and light-weight impurities such as oil and grease, and tiny solid silt particles float to the top of the aqueous phase and are subsequently skimmed off. Flocculation forms clogs of suspended solid particles by using different chemicals known as *flocculating agents*. After removing almost 60–65 percent of solids and 30–35 percent of biochemical oxygen demand (BOD) of the impurities through primary treatment processes, the wastewater stream is subjected to secondary treatment procedures.

Secondary treatment processes involve the use of mainly biodegradation technologies to remove 80–85 percent of BOD and solid contaminants through activated sludge process and biofiltration, and using stabilization tank. In activated sludge process, microorganisms are mixed with wastewater sludge stream and oxygen, whereby the microorganisms consume biodegradable soluble waste organic matter as their food and break them down into flocs. Biofilters could be sand filters, trickling filters, or contact filters. These filters ensure the removal of the remaining sediment containing variable heavy organic load that has not been removed through primary treatment. The effluent from the filter is then passed through a sedimentation tank, which is also known as

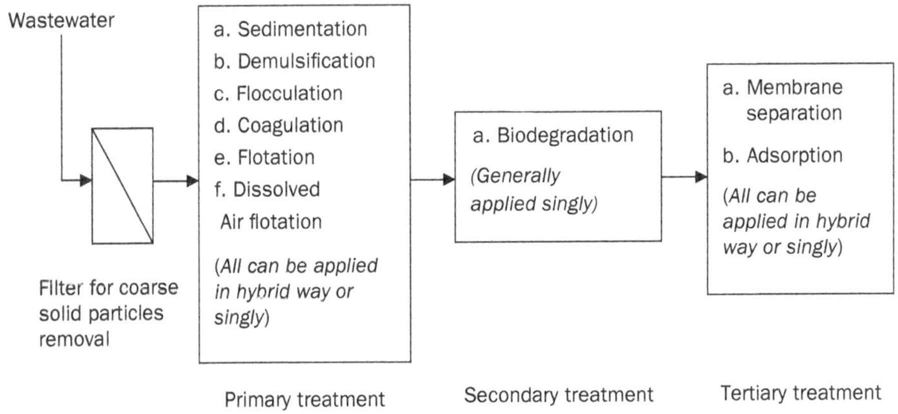

FIGURE 11.1 Diagrammatic presentation of different stages of water treatment.

secondary clarifier or *secondary settling tank*. In the secondary clarifier, solids settle down at the bottom. Part of the settled solid is subjected to solid management process and part is recycled back to the aeration tank as the carbon nutrition source for the population of microorganisms in the activated sludge process.

In the final stage, i.e., tertiary treatment, the wastewater is allowed for further purification to meet the final standards of the treated water stream chemically, physically, or biologically. Tertiary treatments, also called *advanced treatments*, are sometimes coupled with primary or secondary treatment processes or used instead of secondary processes to meet up the quality standard of the purified stream. Tertiary treatment processes can be microfiltration (MF), ultrafiltration (UF), reverse osmosis (RO), adsorption, ion-exchange, disinfection, etc. Progressive filtration technologies such as UF, MF, and RO remove residual components from secondary treated effluents. Disinfection with the application of UV, chlorine, or ozone can produce potable quality water. In some cases, the polluted water streams are subjected to a preliminary treatment process before the primary treatment procedure. In the preliminary treatment step, the raw waste water is subjected to a filtration to screen out coarse solid particles and large materials to ensure maintenance and efficiency of subsequent operations.

Nanotechnology could be a progressive contemporary solution for wastewater treatment, reproducing purified water from wastewater stream with improved treatment efficacy. The incredibly small sizes of the nanomaterials give an unparalleled edge over other conventional materials used for water purification with a larger surface area.

There are different conventional technologies adopted by several water treatment fraternities attributing to the purification of contaminated water. Some of these schemes are screening, gravity separation, sedimentation, oxidation, solvent extraction, distillation, dissolved air flotation, coagulation and flocculation screening, filtration, reverse osmosis, evaporation, distillation, crystallization, electrolysis, ion exchange, electrodialysis, adsorption, centrifugal and membrane separation, fluidization, neutralization and bioremediation techniques, and so on. All of these conventional methods based on physical and chemical principles have their own limitations in practical applications and do not ensure effective separation leading to an improved effluent quality. These conventional processes somehow get restricted due to toxic sludge generation, pH sensitivity, corrosion issues, and also economic feasibility. For instance, excess use of chemicals in the coagulation process restricts its application as it generates toxic sludge. Membrane processes have a disadvantage of economic viability, whereas biological processes have their own limitations with long time of operation. In this scenario, nanotechnology-based wastewater treatment processes are attracting researchers as an alternate tertiary treatment technology, with their effective, economic, and multifunctional operational procedures. Durability and high surface-to-volume ratio to interact with pollutants provide the nanotechnology an edge over conventional techniques. This forms one of the bases of applying nanotechnology to water treatment processes. We will discuss in this chapter about how nanotechnology can benefit water treatment through removal of contaminants by using adsorption, nanofiltration, and other relevant techniques.

Undesired materials of the wastewater stream can be adsorbed on the highly active surfaces of tiny NPs, leading to waste removal. Alternatively, they can be used as nanocatalysts as well as nanosieves for filtration processes. Nanofiltration could be an alternative to RO water purifier to produce drinking water ensuring lesser water wastage. It also softens the hard water and removes undesired toxic heavy metals and organic matters from the polluted water stream. Let us now take an overview of adsorption processes in water treatment.

11.1.1 Adsorption

Adsorption is commonly employed as a secondary or tertiary route to remove organic and inorganic impurities in water management. Adsorption is a surface phenomenon, where adsorbate (atoms, ions, or molecules of gas, liquid, or dissolved solids) adheres to the surface of the adsorbent. You are already familiar with the process of adsorption from discussions in Chaps. 5 and 9.

During adsorption, the solution containing adsorbable solute comes in contact with a highly porous solid adsorbent surface. Due to the intermolecular forces of attraction between the adsorbate and the surface, there starts a concentration buildup or deposition of adsorbate molecules over the adsorbent surface. In the adsorption process, equilibrium is reached when the concentrations of adsorbed adsorbate and adsorbate in solution become constant. Adsorption isotherm depicts the relationship between amounts of adsorbate adsorbed and remaining adsorbate in solution at equilibrium. Following are different adsorption models used widely to study adsorption kinetics: Langmuir, Freundlich, Halsey, Henderson, Smith, Elovich liquid film diffusion, intraparticle diffusion, and Lagergren. You have already studied the Langmuir adsorption model in Chap. 5—one of the most commonly used adsorption models for adsorption on flat surfaces. Similar concepts are used in this chapter to follow the succeeding discussion.

A high surface area of adsorbent is preferred to facilitate a high adsorption capacity and surface reactivity. Inadequate selectivity, lower surface area, or lesser active site reduces the efficiency of conventional adsorbents. Nanoadsorbents provide enhanced effectiveness along with higher specific surface area and associated sorption sites, surface chemistry, and adjustable pore size. The adsorption technology is quite effective for removal of inorganic pollutants such as arsenic, lead, copper, cobalt, cadmium, chromium, and selenium from the polluted water.

The activity of nanoadsorbents in an adsorption process is dependent on their inherent properties and external functionalization. Based on the activity, these can be classified into different groups, as shown in Fig. 11.2. Physicochemical properties as

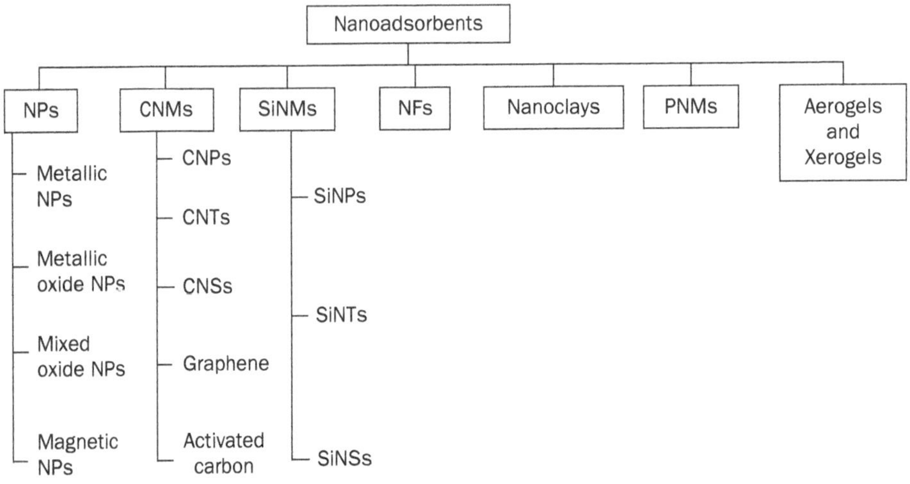

FIGURE 11.2 Classification of nanoadsorbents.

well as material characteristics of nanoadsorbents are also associated with their size, composition, and surface structure. Next, some of the nanoadsorbents are discussed in greater detail.

Nanoparticles (NPs)

NPs whose size range is 1–100 nm hold high efficacy to purify contaminated water. Tiny size, higher surface area, high reactivity, catalytic potential, and easy disengagement, including higher number of active sites to approach several contaminants, make NPs an ultimate choice as an adsorbent. These characteristics are the reasons behind the high adsorption capacities of NPs.

Though naturally occurring metals present in the environment are not harmful, metals exposed to environment from anthropogenic activities lead to environmental pollution affecting the sources of water. According to the World Health Organization, exposure of Ni, Zn, and other uncontrolled metals in water from anthropogenic activities will become a major concern in the near future due to their adverse health effects. These metal ions are removed from water through adsorption and during this adsorption process, the concentration of metal ions adsorbed over the surface of the adsorbents rises higher than the concentration of those ions in water. Sometimes adsorbents cannot bear this huge load of contaminants and expose them back into the water. This phenomenon is named *exhaustion*. For being a better alternative, adsorbents have to bear the load of the adsorbed contaminants and restrict this exhaustion.

Though at one point all adsorbents get exhausted, some studies have shown that exhaustion limits of NPs are very high.[1] Zinc oxide (ZnO), titanium oxide (TiO_2), alumina (Al_2O_3), manganese oxide (MnO), and magnesium oxide (MgO) are very effective metallic oxide NPs for heavy metal removal in the wastewater purification procedure.

ZnO is an extensively used catalyst in the photodegradation of several organic contaminants as well as in removal of dangerous toxic gases. Disposal of spent-out ZnO is considered as nonhazardous solid waste disposal—this widens application of ZnO in industrial wastewater treatment. Anatase form of TiO_2 NPs shows very high photocatalytic activity, due to its higher potential energy of photogenerated electrons. Most widely used semiconductor photocatalyst TiO_2 is known for its nontoxicity, longer life, higher surface area, along with photochemical stability. Additionally, its low cost makes it the most frequently used photo-catalyst in the photo-oxidation process for treatment of wastewater streams. It produces highly reactive oxidant OH^- for disinfection of microorganisms, bacteria, fungi, algae, viruses, and so on. During long simulated solar exposure, TiO_2 can diminish several waterborne pathogens such as protozoa, fungi, *Escherichia coli,* and *Pseudomonas aeruginosa.*[2] TiO_2 NPs are also used for decomposition of organic contaminants in wastewater, heavy metal adsorption such as Pb, Cd, and Zn, and dye removal from wastewater streams. The limitation of TiO_2 is that it only activates under UV light. One way to overcome this is by expanding its optical absorbance toward visible-light region with the use of transition metals[3] and anionic non-metals such as nitrogen, carbon, sulphur, or fluorine doping.[4] pH-resistant highly porous adsorbent aluminum oxide (Al_2O_3) NPs are very efficient in removing Cd, Zn, and other heavy metal ions from wastewater stream. MgO and Mg NPs are efficient biocides for gram-positive and gram-negative bacteria and bacterial spores.[5]

Metallic NPs of Ag have strong antimicrobial activity against fungi, bacteria, and viruses. Ag does not have toxic effects on humans, which makes its use a potent choice for water treatment. The Environmental Protection Agency has registered nano-Ag as swimming pool algaecides since 1954 and drinking water filters since 1970.[6]

Different mixed oxide NPs are used for the betterment of water purification. Fe/Mn oxide, e.g., is used for Cr removal in water treatment.

Iron (Fe) and iron oxides such as maghemite (Fe_2O_3) and magnetite (Fe_3O_4) are examples of magnetic NPs used in water treatment especially because their magnetic properties make separation easier. Both Fe_3O_4 and Fe_2O_3 NPs are known to possess superparamagnetic properties. Such materials are characterized by large magnetic moments, with faster response to the external magnetic field and negligible remanence and coercivity.[7] Hence, iron oxides are the most popular and one of the cheaper alternatives for use as nanoadsorbents. These are capable of removing several heavy metals from impure water including As, Pb, Cu, Cr, Ni, and Co in comparatively short contact times than conventional adsorbents.[8] Fe_2O_3 is an eco-friendly material that can be used directly in a contaminated environment with minor chance of secondary contamination.[9, 10] These magnetic NPs are very useful in groundwater purification, where they can be directly injected to an affected stream and after the adsorption process rather easily recovered using magnetic field. In drinking water purification, they are very effective in recognizing and targeting bacteria. Although metallic and metal oxide NPs can function as effective nanoadsorbents, this field has always seen huge potential in new materials development with even advanced properties.

Carbonaceous Nanomaterials (CNMs)

CNMs are nowadays very popular for removal of organic solutes, heavy metals, and dyes from impure water streams because of their easy preparation process, high sorption capacity, higher stability, nontoxicity, and abundant availability. Activated carbons are easily available from carbonaceous sources, such as coal, wood, coconut shells, etc. and are efficient in removal of inorganic and organic contaminants including As, Cr, Hg, Cu, and Pb.[11] Carbon nanotubes (CNTs) are a class of extensively discussed nanomaterials having potential to remove heavy metals and various organic contaminants from wastewater via adsorption.[12] Though poor dispersion and difficult separation restricted their applications in the primary days of water treatment, researchers have introduced modified multi-wall carbon nanotubes (MWCNTs) to overcome previous drawbacks.[13] While several studies have reported efficient Pb, Cu, Mn, and dye removal using MWCNTs, single-wall carbon nanotubes can also remove aromatic hydrocarbons such as benzene and toluene effectively. CNTs produce oxidative stress, which creates disturbances to cell membranes and diminishes a wide range of microorganism population, including viruses and bacteria such as *E. coli* and *Salmonella*[14] from wastewater streams. Graphenes comprise single- or multi-layered carbon atoms along with a 2D structure and are known to possess remarkable electrical, mechanical, and thermal properties.[15] Magnetic graphene composite adsorbents are very good As removers from water streams, while graphene oxides show their efficacy in Co, Cd, Pb, and Cr removal.[16] Generally graphene oxides have higher sorption capabilities compared to activated carbons and CNTs.

Silicon Nanomaterials

Nano silica (SiO_2) is used in photochemical degradation of organic contaminants. It is known to diminish biological toxins and pathogenic microorganisms and help in physical separation of traces of many metal ions and other contaminants from effluent wastewater streams. Grafted materials are the most used form of nanomaterials, prepared by simple attachment of NPs to an organic polymer matrix or an oxide matrix via covalent bonds. SiO_2 is the widely used oxide matrix due to its low cost, easy accessibility, wide-ranging uses, and various usable grafts. As silicon is more polarized than carbon, SiNTs offer stronger van der Waals interactions to adsorbed molecules, facilitating higher adsorption.

Nanofibers (NFs)

Polymeric electrospun NFs are very efficient heavy metal ion, dye, and organic pollutant removers in wastewater treatment. High selectivity and porosity, small pore size, long length, interconnectivity, and interstitial and huge specific surface area make electrospun NFs an efficient candidate for membrane filtration as well. Positively charged NFs are effective in trapping negatively charged viruses.

Nanoclays

Nanoclays are effective candidates in heavy metal ion, dye, oil, and grease removal in wastewater treatment. Montmorillonite is an example of a natural nanoclay efficient in heavy metal trace removal, whereas polymeric synthetic nanoclays are efficient in oil and grease removal.

Polymer-Based Nanomaterials (PNMs)

Organic polymers have excellent support matrix for NPs with highly porous structure, favorable mechanical behavior, and flexible functional groups. All of these along with innate feature of polymeric matrix and NPs make PNMs competent for effective wastewater treatment technology offering high adsorption capacity, chemical stability, expanded pH range, and high chemical stability. Chitin and chitosan are mostly used nontoxic, biodegradable, high adsorptive sustainable polymers for PNM fabrication. Though chitin is effective in adsorbing heavy metals, dyes, and antibiotics from wastewater, chitosan shows higher efficiency in adsorption owing to the presence of free amino groups. Amino and hydroxyl groups present in chitosan serve as active sites for water pollutants.[17] Generally, cross-linked chitosan is insoluble at low pH and it can be used over a wide pH range.[18] PNMs are recently becoming popular for their uses in chromatography, electrode production, and hydrogen storage. Polymer-based nano beads are highly effective in boron removal from extremely saline wastewater.[19]

Aerogels and Xerogels

Aerogels and xerogels are apt for heavy metal removal, oil removal, and organic contaminant removal. They have properties similar to that of an ideal adsorbent—large surface area, high porosity, and internal pore volume. Amorphous silica and γ-alumina xerogels are the most used porous materials in the sol-gel process. γ-Alumina granules with ideal mechanical properties and good pore structure are very popular in preparation of novel catalysts and adsorbents. These adsorbents also show high thermal and chemical stability, high reactivity for metal ions, and organic pollutant gases. Aerogels have higher surface area, porosity, and pore volume compared to xerogels. With great textural properties, adjustable surface chemistry and hydrophobicity, and much higher surface area per unit volume, aerogels are competent adsorbents in wastewater treatment. Silica aerogels can adsorb a wide range of miscible and immiscible organics, including ethanol, toluene, chlorobenzene, and trichloroethylene, two times more by weight, than activated carbon.[20] Hydrophobic silica aerogels can adsorb about 3.5 times crude oil of their own weight[21] and can be fully recovered after regeneration. Biodegradable cellulose aerogels are also good in oil adsorption. Carbon aerogel is reported as good sorbent in removal of U, Cd, Pb, Hg, Cu, Ni, Mn, and Zn.[22, 23]

Another field of applications where nanomaterials are becoming increasingly important is in membrane separations, and especially in the field of water treatment using membranes.

11.1.2 Membrane Separation

Membrane separation can be visualized as a pressure-driven filtration technique where the membrane acts as a selective barrier to restrict the passage of pollutants. The membrane itself is a semi-permeable, thin-layered sheet that works as a sieve during the filtration process, holding back the retentate—the contaminant—while allowing the permeate, the less contaminated water to pass through. Membrane separation ensures high-quality treatment of wastewater within a smaller footprint area.

The major limitation of any filtration process is the concentration polarization phenomenon. During filtration, as the process proceeds, the concentration of the retentate macromolecules increases around the vicinity of the membrane. Due to this concentration buildup of macromolecules over the membrane surface, the pores of the membrane get blocked. Further, the macromolecules get adsorbed over the membrane surface, reducing the permeate flux. These finally lead to permanent damage of the membrane and shorten the life of the expensive membranes. Nanotechnology gives the opportunity of inclusion of functional nanomaterials in membrane production, with manipulation of membrane structural properties, selectivity, porosity and roughness, and hydrophilicity while minimizing the chances for fouling. Thermally and mechanically stable modified membranes are also efficient in balancing permeability and selectivity.

Nanomembranes

Nanomembranes are presently a lucrative option for membrane filtration processes in water treatment. Researchers are currently using hydrophilic polymeric layer of polyvinyl alcohol (PVA) or chitosan to modify the membrane surface to ensure longer membrane life. Nanosheets made of carbon-based nanomaterials such as graphene, graphene oxide, and reduced graphene oxide are being widely used for highly hydrophilic membrane preparation. This allows only water to go through the membrane, holding back other liquids and gases. Both zeolite-based and graphene-based RO membranes have been known to perform with high permeate flux while maintaining good ion-rejection rate. Incorporation of nanocatalyst TiO_2 in the membranes increases hydrophilicity of the membranes and ensures efficient control of microorganisms and organic pollutants in pharmaceutical and agricultural wastewater.

Many natural and synthetic polymers have been electrospun into nanofibers to produce electrospun nanofiber membranes (ENMs) as such a process is less expensive, easy to control, and requires a lesser footprint than the conventional membrane preparation processes. ENMs effectively remove heavy metals and microorganisms from wastewater streams ensuring desalination as well. Electrospun polysulfone membranes have reported removal of ammonia and suspended solids and reduced COD from biologically treated wastewater. Electrospun polyvinylidene fluoride has shown up to 90 percent of microparticles rejection. The electrospun membranes are also efficient in removal of toxic heavy metals such as Ni, Cd, Cu, and Cr.[24] Amine cellulose acetate/silica nanofiber membranes have reported Cr removal from wastewater, where PAN/$FeCl_3$ composite membranes have been shown to convert Cr(IV) to Cr(III) along with removal of Cr from wastewater.[25] Chitosan ENMs are used to remove Pb and Cu.[26] Nanofibrous membranes can achieve almost 100 percent desalination, at lower operational pressure, high temperature, and with much lower energy consumption. The electrospun nanofibrous membrane also poses antimicrobial activity for both bacteria and viruses.

Polymerizable liquid-crystalline (LC) molecules and block copolymers have uniformly sized nanopores and sub-nanopores. Self-assembly is another promising process of producing stable functional materials and thin films from these materials.[27] Self-assembled membranes are also making their way into water treatment in order to remove ions and organic molecules. LC materials can produce 1D, 2D, and 3D nanochannels, and 1D, 2D, and 3D nanopores can be generated from block copolymers. Architectures and sizes of the self-assembled nano-channels and nanopores depend on the molecular structures and intermolecular interactions. The pore diameter of porous films of block copolymers ranges from 5 to 50 nm,[28] which is in the range required for UF and MF membranes for separating larger solutes. High-density arrays of well-defined nanopores of block copolymers offer high hydraulic permeability and separation selectivity while being incorporated into a membrane. The pore diameter of nanochannels generated from LC materials is of 0.5–2 nm.[29] These nanochannels function as nanofiltration (NF) and reverse osmosis (RO) systems to separate small molecules and ions. Subnanopores of thermotropic ionic LC materials are extremely potent as membranes for selective rejection of salts and hazardous components from seawater or brackish water.[30]

Aquaporin-based membranes are biologically motivated nanomembranes. Aquaporins are pore-forming protein channels and are omnipresent in living cells. Under definite conditions, they form selective water channels to reject most ionic molecules. The characteristics of high water permeability and selective rejection make them an apt material to build novel high flux biomimetic membranes for water treatment. Highly unstabilized aquaporins are incorporated in vesicles which are embedded in polymer matrix or onto polymeric substrates (such as nanofiber membranes) for stabilization and mechanical strength. Aquaporin Inside™ (Aquaporin A/S, Copenhagen, Denmark) is the first commercially available aquaporin-embedded biomimetic membrane that can withstand pressures up to 10 bar, and allows water flux rate >100 L/(h-m^2). Though recent studies on membrane nanotechnologies have indicated that aquaporin-based bioinspired membranes offer the best for revolutionary performance, to be commercialized in water treatment processes, aquaporin-based membranes have to be competitive with conventional membranes in terms of stability and long life.[31]

Nanofiltration

Membrane separation offers a wide variety of processes depending on the desired level of treated effluent quality. For example, microfiltration (MF) can remove particulate and colloidal forms of contaminants, whereas it can also be integrated as part of a membrane bioreactor (MBR) in a biological treatment system. Membrane technologies can be applied in combination with several physical, chemical, and biological processes to produce better quality of effluent. Coupling of MBR technology with activated sludge processes or an anaerobic biodegradation process with MF or ultrafiltration (UF) is one of the examples of that. Hence, a wide range of membrane-assisted technologies is available for wastewater treatment, and these come with their own advantages and disadvantages depending on effluent quality, energy consumption, and system complexity.

Nanofiltration (NF) is a recently developed concept of membrane technology for removal of finer components than MF and UF and closely resembles RO membrane separation. Having pore sizes of 1–5 nm, NF membranes are generally used for water softening and desalination while also holding back ions and organic macromolecules. NF membranes are widely used for treatment of surface and groundwater. Removal of larger particulate matter through prior treatment of the water stream subjected to NF membrane is crucial to maintain a good life of the membrane, as its pore sizes are very tiny. Numerous advantages of NF such as low operating pressure, high permeate flux, high retention of multivalent anion salts and organic matter, and relatively low costs sometimes make NF a reasonable, alternative, and cost-intensive RO technique.

So far, we have discussed different nanomaterial-based alternatives that may be applied for wastewater treatment. Now, we will discuss applications of nanomaterials in yet another exciting field—hydrology.

11.2 Hydrological Tracers in Water Management

Hydrology is the science dealing with the distribution, circulation, and quality of water of the earth and typically focuses on water and water quality on or below the land surface, and in rivers and lakes. In terms of the hydrologic cycle, hydrology can be described as the portion of the cycle from precipitation to stream discharge or re-evaporation, and in terms of water quality, hydrology studies the relation of water with the natural environment and human interactions. To be able to follow and quantify water flow paths is key for the development of hydrological understanding and concurrently for water resources management as water is one of the most important resources for society.

Tracers are the most important tool hydrologists have for the development of hydrological understanding and water resources management. Tracers are substances that can be detected in the water at very low concentrations and allow following, or tracing, the flow of water.[32] Tracers can be of two types: environmental (tracers that are part of the hydrological system, also called *passive tracers*) and artificial (brought into the hydrological system intentionally for tracer study, also called *active tracers*).

Here, we focus on the artificial tracers only. Artificial tracers are, for example, dissolved salts (sodium, potassium, lithium, chloride, and bromide), organic solutes and fluorescent dyes (e.g., uranine, rhodamines), stable isotopes (e.g., deuterium), or particulates (e.g., bacteria such as *E. coli*, viruses, phages, synthetic microparticles, and DNA).[32-34]

All water substances interact with the environment during their transport albeit with very different rates. The substances that have no or very limited interaction with the environment are called *conservative substances.* Both conservative and non-conservative or even reactive tracers find widespread applicability in hydrological research. Evidently, it is more difficult to interpret non-conservative or reactive tracers; for example, if the tracer substance in the water is lost (depleted) or enriched due to dissolution during its transport, to interpret the tracer experiment all the reaction processes have to be known and quantified.

Water travels in the environment, for example, from precipitation in a catchment to the stream, via infinite flow paths and travel times, whereas the toolkit available to hydrologists to study these dynamic processes consists of a few tracers only giving limited insight into the hydrological dynamics. Furthermore, if one wants to apply a tracer on a larger scale it would be diluted vastly and so the detection limit becomes an important constraint. For better and accurate results in hydrological studies, a new generation of tracers are used, which are detectable at very low concentrations, have (near) identical behavior in the environment, and are available in vast amounts. Theoretically, DNA fulfills most of these criteria.

When it comes to work on DNA tracers, a rather complete overview, including the many advantages of using DNA, is given by Liao et al.[35] DNA in tracer tests for hydrology or water management purposes has been applied in the form of single strands[36-41] or double strands,[42] attached to larger particles,[43] covered by a protective cage,[44] or covered by a protective shell.[45-47] In this section, we focus on tracers that include DNA, which is protected by a cover or cage ("covered" DNA tracers), and are composed of either a DNA core or of a core of engineered nanomaterial. The objective of this section is not only to demonstrate their potential, but also to investigate their inherent limited tracer-like behavior, whereby we focus on their hypothetical transport mechanisms in the environment, and, more specifically, on their potential interactions with surfaces and other particles; because this type of tracer is rather new and difficult to prepare for hydrologists, data to substantiate claims related to their transport behavior is largely lacking. However, based on the advancements made in colloid transport theory in the last decades,[48-54] we present a realistic particle or colloid transport framework for DNA tracers, including the most important processes, interactions with solids, and transport models.

To our knowledge, there are four types of covered DNA tracers:

- The particle produced by Garnett et al.[44] has an approximate size of 35 nm and is composed of DNA that is circumferentially wrapped in a condensed form and stabilized by a combination of polyamidoamine (PAA), a cationic homopolymer, and a PEG-PAA-PEG [PEG stands for poly(ethylene)glycol] copolymer as a self-assembling system.
- The particle produced by Sharma et al.,[47] and subsequently reproduced by Dahlke et al.,[55] has an approximate size of 480 nm, is composed of a core of iron oxide nanoparticles plus DNA, and encapsulated by polylactic acid (PLA).
- The particle used by Grass et al.[45] was reproduced from Paunescu et al.[56] and has a silica core, surrounded by a layer of DNA, and protected by an outer cover of—again—silica.

- The particles used by Kianfar et al.[46] were both reproduced based on the Paunescu protocol and based on a protocol developed by Puddu et al.[57] The latter has a similar silica cover layer as the Paunescu particle, but the core is composed of nanoparticles of superparamagnetic magnetite with DNA wrapped around them. The particles of both Paunescu and Puddu have an approximate size of 180 nm.

We will first discuss the potential of covered DNA tracers via a brief overview of a—still modest—number of cases. Then, we will discuss the energies and forces governing colloidal particle–surface and particle–particle contact, including example energy profiles. Thereafter, you will learn about colloid filtration and the mechanistic prediction of colloid retention. After discussing continuum scale transport models, which are applicable to groundwater environments, you will briefly learn about transport models for surface water environments. At the end of our discussion, you will encounter challenges in this field and will learn how nanotechnology can solve some of these.

11.2.1 Potential of Covered DNA Tracers: Case Studies

Surface Water Injection

With the Garnett particle, an instantaneous tracer injection test over a 6.1 km stretch of the River Fruin roughly from Inverlauren to Dumfin, west of Loch Lomond in Scotland was carried out. Discharge of the river at the time of the test was unknown. We estimated it to be in the range of m^3/s. A few more details about the test are given in Garnett et al.[44]: only 2.5 mg of DNA (60–80 mer) suspended in 1 L was injected. Samples were taken at 10–20 minute intervals and 20 μL of sample was directly used in a qPCR (quantitative polymerase chain reaction) well without further treatment. Maximum DNA copy number per qPCR sample was therefore in the range of 10,000. Comparison with the dye showed a similar time to peak for appearance of the DNA tracer at the sampling site at 610 minutes (Fig. 11.3). No data on DNA tracer mass losses compared to rhodamine mass losses were reported.

With the Sharma particle, a number of different DNA tracers were injected in a glacier in Sweden to map complex flow paths.[55] Particles were collected 1–2 km downstream of

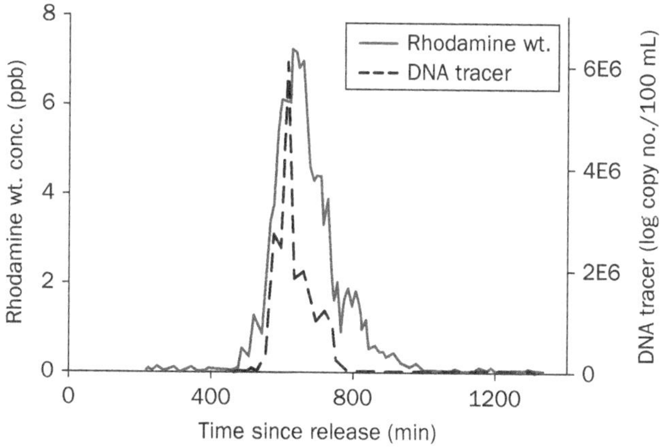

FIGURE 11.3 Instantaneous DNA tracer injection test in the River Fruin (Scotland).

the various injection points. Recovered DNA tracer mass ranged from 1 percent to 66 percent, while recovered fluorescent dye mass was 99 percent. Due to the cold environment, Dahlke et al.[55] rule out mass loss associated with microbial activity or photochemical degradation. They believe sorption of DNA tracers onto suspended sediments and loss of microtracer particles to sediment storage are the most probable sinks of DNA tracer mass.

Borehole to Borehole Tracer Test

In the United Kingdom, in a gravel aquifer the transport of the Garnett particle was tested in a forced gradient test. Thereto, groundwater was pumped from BHC (Fig. 11.4). DNA tracer and rhodamine were added to injection well BHA and monitored at BHB, some 3.5 m from BHA (Fig. 11.4). Samples were taken frequently and directly used in qPCR wells. Maximum DNA copy number concentrations in

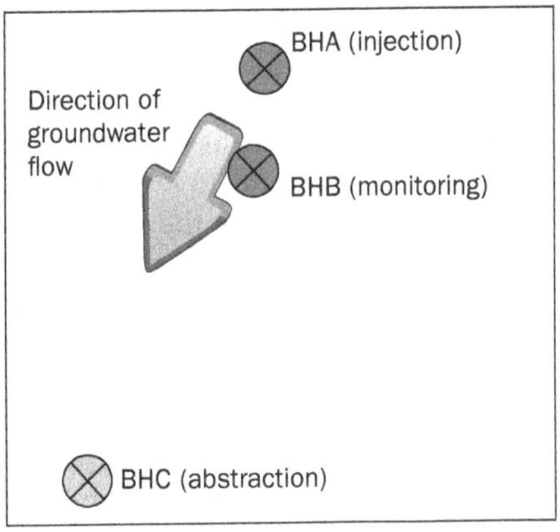

FIGURE 11.4 Layout of the test experiment.

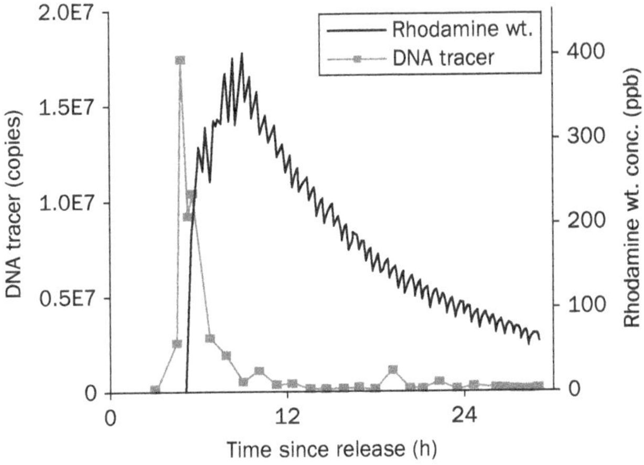

FIGURE 11.5 Borehole to borehole tracer trial.

samples were in the range of 18,000,000 (Fig. 11.5). Arrival of the DNA tracer was about 6 hours after injection and about 3 hours before rhodamine tracer arrival. In this case, no information exists about the tracer mass recovery and mass recovery compared to rhodamine.

Column Experiments

With the Sharma particle[47] and the Paunescu particle,[46,56] a number of column experiments were carried out with saturated sand and column heights in the order of 5–15 cm. Results indicated relative breakthrough concentrations in the order of 0.1–1.0, depending on grain size of the sand, flow velocity, and chemistry of the water in which the particles were suspended.

11.2.2 Colloidal Forces Governing Particle–Surface and Particle–Particle Contact

Particle–surface or particle–particle interactions that control nano- and microparticle attachment or deposition have traditionally been described by the Derjaguin–Landau–Verwey–Overbeek (DLVO) theory of colloidal stability,[49] which considers the sum of van der Waals (VDW) and electrical double-layer (EDL) interactions.

The VDW attraction force is the force between atoms due to the spontaneous appearance of temporary dipoles resulting from an always moving electron distribution within an atom. Just by chance, such electron distribution can become unevenly distributed, causing all other electron distributions in atoms in a lattice to become unevenly distributed. The attractive force between these temporary dipoles is the van der Waals force. The body-to-body interaction of the VDW force is determined by the Hamaker constant, which is different for each compound or mineral and for each fluid in which the compounds or minerals make contact. If two interacting dipoles are a distant apart, the time required for the electromagnetic signal of one unit to reach the neighbor, polarize the local electron cloud, and get the return signal is long compared to the lifetime of the original dipole. In those retarded cases, a reduction factor is used of the form $(1 + bh / \lambda)^{-1}$ where λ is a characteristic wavelength and b is a constant.[58] The characteristic wavelength for water is 10^{-5} cm and is a correction for the reduction due to retardation, because of transmission through a medium of dielectric constant not unity.[59]

The EDL interaction results from the presence of electric charge at the surface of a particle or colloid. The first layer is the positive or negative charge at the surface of a grain or particle due to lattice imperfections, ions adsorbed onto the surface, and/or hydrochemical or geochemical interactions (e.g., weathering). The second layer is composed of ions attracted to the surface charge. It is loosely associated with the first layer and made up of ions that move freely in the fluid under the influence of electric attraction and thermal motion. An important parameter in determining EDL energy is the Debye length, which is a measure of the net electrostatic effect in solution of a charged particle and how far its electrostatic effect persists. Besides a number of constants, the Debye length mainly depends on the ionic strength of a solution.

Additionally, non-DLVO forces such as steric, hydrophobic, and hydration forces can also play an important role in colloid attachment. Especially steric forces resulting from adsorbed layers of polymers or surfactants that might lead to steric repulsion can play a role in aquatic environments as most particles adsorb natural organic matter.[49] Such steric repulsion will become more important at higher ionic strengths due to the reduction of the EDL energy (reduced Debye length). The profile of total interaction

Parameter	Value		
Dielectric constant of water	80		
Permittivity of free space [C/Vm]	8.85E-12		
Radius [nm]	Garnett 24	Sharma 240	Paunescu 90
Surface potential [mV]	Garnett −24	Sharma −37	Paunescu −24
Surface potential of grain [mV]	−15		
Counterion valence	2		
Ionic strength [mol/m³ = mmol/L]	20		
Avogadro number	6E23		
Elementary charge [C]	1.06E-19		
Boltzmann constant [J/K]	1.38E-23		
Temperature [K]	278		
Debye length [nm]	3.2		
Hamaker Particle–Water–Silica [×1E-20 J]	Garnett 1	Sharma 1	Paunescu 0.7
Characteristic wavelength [cm]	1E-5		

Table 11.1 Parameter Values Used to Prepare the Total Energy Profiles of Fig. 11.6

energy as a function of particle–particle or particle–surface distance is influenced by properties of the colloids and the surrounding medium (i.e., particle size, zeta potential, Hamaker constant, and electrolyte composition). Aggregation or deposition occurs if the particles can overcome the energy barrier and reach the deep primary energy minimum at short separation distances. If the conditions of the system are repulsive, e.g., due to a high zeta potential, the particle requires sufficient kinetic energy to overcome the energy barrier to reach the primary energy minimum. An overview of key equations to evaluate particle–surface and particle–particle interactions is given by Petosa et al.[49] (Table 11.1). We implemented the equations [their Eqs. (3)–(8)] in a spreadsheet and presented the total energy interaction profiles for the Garnett, Sharma, and Paunescu particles in case of interaction with a large silica surface (Fig. 11.6).

Figure 11.6 shows a hypothetical situation whereby the DNA tracers encounter a silica surface in fresh water with an ionic strength of 20 mmol/L. We imagine water with divalent ions, such as Ca^{2+} and Mg^{2+} (counterion valence = 2; Table 11.1). Particle radii were taken from respective articles, and we used the zeta-potential instead of the surface potential of the particles and the silica surface. Zeta-potentials were also taken from respective articles. Hamaker constants were estimated based on the tables of Petosa et al.[49] and Hunter (2001).[60]

From Fig. 11.6, it is clear that all particles have to overcome an energy barrier in order to reach the primary minimum at very short separation distances of 0.1–0.2 nm. Also, the well-known secondary minimum at some distance from the collector surface due to the exponential reduction of the EDL and the reciprocal increase of the VDW with distance

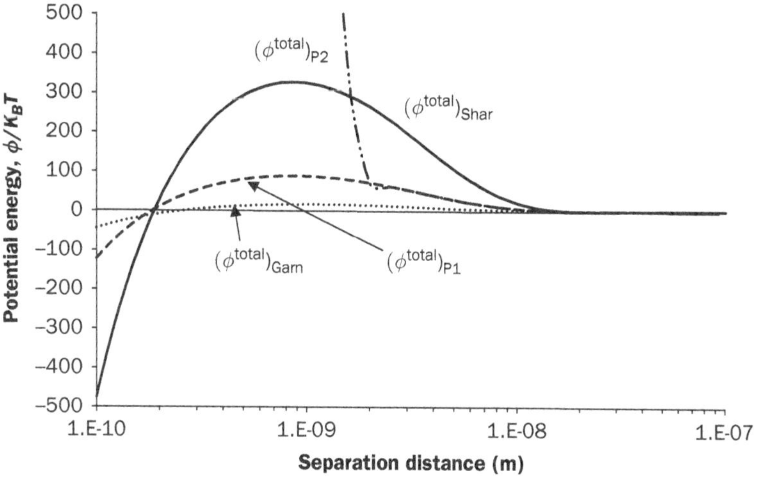

FIGURE 11.6 Total energy profiles of the Garnett ("Garn"), Sharma ("Shar"), and Paunescu ("P1" or "P2") particles in case of interaction with a (large) silica collector grain. Other conditions and parameter values are listed in Table 11.1 and discussed in the text.

appears to be absent for these conditions. A secondary minimum would, for instance, appear if the ionic strength would increase and the surface potentials of the particle and the silica surface would be identical. To show the importance of steric repulsion, we add a hypothetical case ("P2" in Fig. 11.6), whereby we assume that only the collector surface is covered with a layer of polymers equally spaced at 0.5 nm from each other and with a film thickness of 2 nm. As a result, the primary minimum disappears and repulsion remains. In case of particle–particle interactions, the energy equations change, and so do the total energy profiles. However, total energy patterns (primary energy minimum—barrier, no secondary minimum) are similar to the one discussed here.

11.2.3 Colloid Filtration and Mechanistic Prediction of Retention

Differences between molecular (e.g., solute) and colloidal transport processes in porous media drive differing modeling approaches to describe their transport.[48] While solutes have a rather random motion and no deterministic trajectory, colloids exhibit relatively limited diffusive (=Brownian) motion and possess mainly deterministic trajectories.[48]

There are more than 10 different colloid filtration theories (CFTs) that simulate these deterministic trajectories and the likelihood of colloids reaching collector surfaces.[48] Most CFTs have two major components: (1) a model that describes colloid trajectory and attachment and (2) correlation equation(s) that usually includes many variables, which then approximate(s) the results of the mechanistic model. In this way, it is possible to predict transport.

The most important parameter of such models is the "collector contact efficiency," η, defined as the fraction of colloids able to contact the collector in a given system.[48] We are not going into much detail regarding the various models that describe colloid trajectory and attachment (see Molnar et al.[48] for an overview); in most CFTs, transport of colloids from the pore fluid to the vicinity of a filter grain (collector) is typically described by three mechanisms: interception, gravitational sedimentation, and Brownian diffusion (see study by Tufenkji and Elimelech; Fig. 11.7).[61]

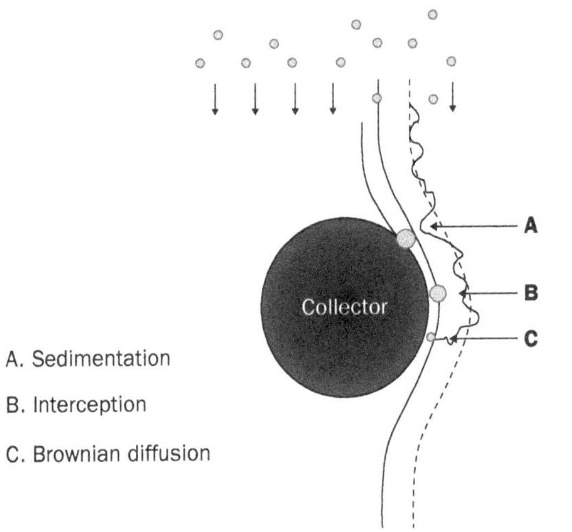

FIGURE 11.7 Typical deposition mechanisms in colloid filtration theory.

Transport of particles by interception occurs when a particle moving along a streamline comes into contact with the collector due to its finite size. Gravitational sedimentation refers to the settling of particles with densities greater than that of the fluid onto the collector surface. Smaller particles undergo Brownian motion (diffusion), which can result in contact with the collector grains.[61]

Yao et al.[62] presented the first CFT model and suggested that the three transport mechanisms are additive. However, this classical CFT has serious limitations, since it does not consider the influence of hydrodynamic (viscous) interactions and the universal van der Waals attractive forces.[61] Rajagopalan and Tien[63] were the first to solve numerically the trajectory equation for colloids in the non-Brownian particle size range. This resulted in the so-called RT correlation equation, which has been used extensively to predict deposition (filtration) of colloidal particles in natural and technological systems.[61] However, also the RT equation has a serious limitation as it neglects the influence of hydrodynamic and van der Waals interactions on the deposition of particles that are dominated by Brownian diffusion. The predictive correlation equation, developed by Tufenkji and Elimelech after numerically solving the convective-diffusion equation with hydrodynamic interactions and universal van der Waals forces fully incorporated, overcomes these shortcomings:[61]

$$\eta_0 = 2.4A_S^{1/3}N_R^{-0.081}N_{Pe}^{-0.715}N_{vdW}^{0.052} + 0.55A_SN_R^{1.675}N_A^{0.125} + 0.22N_R^{-0.24}N_G^{1.11}N_{vdW}^{0.053} \quad (11.1)$$

where η_0 is the single collector contact efficiency, N_R is the aspect ratio, N_{Pe} is Peclet number characterizing the ratio of convective over diffusive transport, N_{vdW} characterizes the ratio of van der Waals interaction energy to the particle's thermal energy, N_{gr} is a measure for the gravitational potential of the particle, N_A represents the combined influence of van der Waals attraction forces and fluid velocity on the particle deposition rate due to interception, and N_G is the ratio of Stokes particle settling velocity to approach velocity of the fluid. For more details, the reader is referred to Tufenkji and

Elimelech.[61] Also, a spreadsheet can be downloaded from the website of Tufenkji or Elimelech to calculate the single collector contact efficiency.

This TE correlation equation actually makes clear that colloid transport can be predicted. The equation is based on the same information as presented in Table 11.1. In addition to that, only the constants—pore water flow velocity, bulk diffusion coefficient, particle density, fluid density, and fluid viscosity—have to be known. However, the TE correlation equation does not include ionic strength, nor does it include characteristic features of the DLVO theory (energy minima and barrier). Research shows that under favorable conditions, that is when the charge of colloid and collector surface of colloid are opposing and there is no energy barrier to reach the primary minimum as predicted by the DLVO theory, transport can indeed be predicted using such correlation equation.[48] In case of unfavorable attachment, that is when the charge of colloid and collector surface of colloid have a similar sign, and the colloid is able to overcome the energy barrier and reaches the primary minimum, the correlation equation has little predictive value. Since most surfaces in the natural environment are negatively charged including the DNA tracer surfaces, it is because of these unfavorable conditions that it is difficult to predict DNA tracer transport. Conversely, it is difficult, if not impossible, to independently determine mass balance components of DNA tracer injection tests. This is a serious limitation of the use of DNA tracers in the environment.

11.2.4 Continuum-Scale Transport Models

Continuum-scale models of the transport of species in groundwater solve mass balance equations at the macroscopic scale. Microscopic properties are averaged over a representative elementary volume (REV) and the averaged property (e.g., porosity, fluid pressure, and colloid concentration) is treated as a value that represents the property throughout the REV.[48, 64] Continuum models necessarily blur microscopic details and employ simplifying assumptions in the averaging process. The result is a set of partial differential equations that describe spatial and temporal rates of change of key macroscopic properties (e.g., fluid velocity and colloid concentration) at a practical scale. These equations are typically discretized and solved on a numerical grid that correlates REVs to nodes. The advection-dispersion equation (ADE), which describes the transport of a nonreactive solute tracer in groundwater, is a typical example and frequently used in colloid transport.[48, 53]

$$\theta\frac{\partial c}{\partial t} + \frac{\rho_b}{\theta}\frac{\partial S}{\partial t} = -\nabla \cdot (\mathbf{u}c - \mathbf{D}\nabla c) + q_c \tag{11.2}$$

where θ is the porosity, C is the concentration, ρ_b is the solid bulk density, S is the concentration at the solid surface, $\mathbf{u}$ is the Darcy velocity, $\mathbf{D}$ is the dispersion tensor, and q_c is a source or sink. Deposition on the solid surface can then be described by the first-order rate kinetics:

$$\frac{\rho_b}{\theta}\frac{\partial S}{\partial t} = k_a C \tag{11.3}$$

where k_a is the colloid deposition rate coefficient. This is related to the single-collector removal efficiency such as that defined in Eq. (11.4) by Tufenkji and Elimelech:[61]

$$k = \frac{3(1-\theta)v}{2d_c}\eta_0\alpha \tag{11.4}$$

where d_c is the diameter of the collector grains and α is the sticking efficiency, which is defined as the fraction of colloids reaching the surface that also stick to that surface. Note that there are different relationships between k and η depending on the CFT used.[48]

In case of favorable conditions, $\alpha \approx 1$, the attachment rate coefficient can be calculated. In case of unfavorable conditions, $\alpha < 1$, and as was already described above, so far there has not been a theoretical or approximation method to quantify the sticking efficiency. This parameter has to be independently determined with tests such as column experiments. In literature, more particle transport models have been proposed: dual deposition mode, single-site deposition remobilization, dual-site deposition remobilization, single-site deposition remobilization and blocking, etc.[52] Also, many different types of particle transport models have been incorporated in HYDRUS-1D, a Windows-based modeling environment for analysis of water flow and solute transport in variably saturated porous media, which was developed by the U.S. Soil Salinity Laboratory. HYDRUS-1D can be downloaded free of charge from the Internet.

11.2.5 Transport of DNA Tracers in Surface Water

The instantaneous injection experiment of the Garnett particle in the Fruin River (Figs. 11.3 to 11.5) was successful in the sense that a DNA signal was recorded at the same time as the rhodamine tracer. However, no attempt was made to make a DNA tracer mass balance of the experiment. The tracer mass balance enhances the application of DNA tracers in surface waters and is required for a variety of hydrological applications. Currently, we do not know the fate and behavior of DNA tracers in aquatic systems besides advection and dispersion, as it is clear from Fig. 11.3 that dispersion is the dilution of DNA tracer mass without loss. Other likely processes influencing the fate and behavior of these covered DNA tracers in aquatic systems include agglomeration or aggregation. This is a mass-conserving process whereby particles of similar properties interact to form larger clusters held together by attractive van der Waals forces.[65,66] In natural waters, homo-aggregation is unlikely to control particle–particle collisions as natural colloids are typically present in higher concentrations, perhaps even orders of magnitudes higher than DNA tracers. Within complex media such as rivers and lakes, hetero-aggregation of DNA tracers with a variety of bio- and geocolloids of varying concentrations is more likely.[66] Hetero-aggregation could substantially modify DNA tracer surface chemistry and detection in downstream qPCR analyses. Collision frequencies between particles leading to aggregate formation are influenced by Brownian motion, fluid motion, and differential settling.[66] Successful collisions result in the formation of aggregates with increasing size and mass,[49] as described by the DLVO theory (see Sec. 11.2.2). Aggregation rates are influenced by particle collision frequency. A high attachment efficiency to suspended solids is likely to result in reduced transport.[66] Sedimentation could be an important removal mechanism for DNA tracers. Increasing particle size and density caused by colloidal interactions result in the formation of aggregates which are then subjected to settlement due to gravity.[65, 66]

Although we do not expect chemical and biological processes to influence the fate and behavior of DNA tracers, at present we cannot rule out that redox reactions, photochemical reactions, dissolution, precipitation, speciation, or biotransformation can have an influence on DNA tracer concentrations.

There are not many models that take these processes into consideration.[67] Also, there are not many studies known that provide the various rate parameter values. One such model is NanoDUFLOW,[68] which is part of the DUFLOW water quality modeling suite. Another model is INCA.[69] If only travel times have to be determined, an analytical solution of the well-known advection-dispersion differential equation, similar to Eq. (11.2), but then in one dimension and with different source and sink terms, fitted to the DNA tracer breakthrough curve seems most appropriate.

11.2.6 Challenges

As pointed out by Liao et al.,[35] the DNA tracer technique has a number of advantages: easily distinguishable from environmental background, theoretically unlimited number of tracers, simple to prepare, cost-effective, and environmentally friendly. However, the entire DNA multi-tracing technique has not been applied frequently, and it is still in its infancy. There are a number of important challenges to overcome.

In spite of the scale-up of material manufacturing, further reduction of cost, and improvement with regard to protection,[35] there are certain challenges as follows:

- *Mass balance problems:* What are the most important sinks in the environment and can the effect of these sinks be predicted or included in the DNA tracer transport models? Another perhaps even more fundamental question is: what is the dominant retention mechanism of DNA tracers in the environment? Is it indeed the first-order kinetic attachment without much detachment? Or, is another interaction model more appropriate? In addition to that, although it is clear from literature that equilibrium sorption is a thermodynamic impossibility in case of colloid transport,[50] there is the possibility of rapid attachment and detachment, giving rise to a kind of pseudo-equilibrium sorption mechanism that causes retardation of the DNA tracer breakthrough curve in a similar way as retardation of solutes (e.g., see Fig. 3 in Molnar et al.[48]). Such DNA tracer behavior might lead to erroneous travel time estimations when unnoticed.
- *Sample up-concentration problems*: To detect DNA tracer particles in dilute samples, a reduction in the sample volume has to take place. The easiest way to detect the DNA tracer particles in large samples (e.g., 5–100 L) seems to be by magnetically harvesting the particles in a continuous sample water stream without harvesting "collateral" particles, like those in environmental samples, omnipresent bio- and geocolloids. So far, in our labs we have not been able to up-concentrate (i.e., to increase concentration) samples of more than 100 mL in a convenient, reproducible, and quantitative way. Especially in river water samples, the concomitant harvesting of biocolloids and cleansing with a positive silica filter lead to an unacceptable high number of DNA tracer samples with an inhibited qPCR signal.

Finally, determining DNA concentrations on a qPCR apparatus is an essentially enzymatic method with a lot of manual pipetting actions, especially in typical small research laboratories. As already pointed out by Foppen et al.[37], minor temperature differences in the qPCR apparatus give rise to minor differences in the enzymatically determined DNA amplification fluorescence values, and, in addition to that, carrying our qPCR and determining relative concentrations require at most some 36 pipetting actions including pipetting 36 errors. For quantitative hydrology purposes this remains a concern. Recently, we acquired a PCR robot for our labs, which reduced the manual pipetting errors and, as a result, substantially improved the reproducibility of qPCR cut-off cycles in consecutive samples or series of samples. This in turn opened up a host of possibilities for focusing on DNA-based hydrological tracers.

This chapter has provided an overall perspective of how nanomaterials can make their way in water management—whether in wastewater treatment or in hydrological applications as tracers. The chapter recapitulated the fundamental principles of synthesis and functionalization of nanomaterials, which forms the basis of their end uses.

11.3 Study Questions

11.3.1 Concept-Based Questions

11.1 Describe the advantages of NF process over RO process.

11.2 Define exhaustion of an adsorbent. How do nanoparticle adsorbents rate against conventional adsorbents?

11.3 Explain the mechanism of the TiO_2-photocatalyzed oxidative degradation.

11.4 Discuss different water chemistry conditions that can affect the antimicrobial activities of Ag NPs.

11.5 Describe the characteristics of iron and iron-oxide nanoparticles that make them an apt adsorbent.

11.6 Discuss the advantages and disadvantages of using carbon nanotubes (CNTs) in wastewater treatment.

11.7 Explain concentration polarization phenomenon. Suggest possible methods to minimize the same.

11.8 What are hydrological tracers? Explain with examples the different types of tracers.

11.9 Compare and contrast conservative and non-conservative tracers.

11.10 What are the four types of covered DNA tracers discussed in this chapter?

11.11 List the advantages of using DNA tracers for hydrological studies.

11.12 Explain the typical deposition mechanisms in colloid filtration theory.

11.3.2 Research-Based Questions

11.13 Residual COD and conductivity limit the reuse of effluent of wastewater treatment plant of textile industries. In 2003 Bes-Piá et al.[70] studied activated sludge process in combination with NF and in combination with chemical oxidation using ozone and UV irradiation to reuse water in a printing, dyeing, and finishing textile plant. Wastewater coming from activated sludge process was prefiltered to remove suspended solids followed by the analysis of COD, conductivity, and pH of the prefiltered samples. NF experiments were performed in a laboratory plant with a membrane system of four plane membranes. The permeate flux variation, COD, and conductivity removal were studied. The ozonation experiments were also carried out in a laboratory plant consisting of three ozone generators and the redox potential was measured to control the oxidation reactions. Final effluent coming out of the NF process shows a COD reading of <50 ppm and conductivity of 0.39–0.51 mS/cm. Final effluent stream from ozonation and UV irradiation holds also <50 ppm COD but with a conductivity reading of 3.04 mS/cm.

Among the above-mentioned treated final effluent streams, which one will you prefer for reuse and why?

11.14 In 2014 Akhbarizadeh et al.[71] synthesized maghemite NPs to determine the capability of the maghemite nanoparticles to remove the most common contaminants of concern: Cu, Ni, Mn, Cd(II), and Cr(VI) in the acid mine drainage (AMD). In parallel, these 14-nm sized NPs were charged to simulated wastewater as well. Simulated wastewater of initial pH 6.14 was prepared by dissolving $CuSO_4{\cdot}5H_2O$, $NiSO_4{\cdot}6H_2O$, $MnSO_4{\cdot}H_2O$, $Cd(NO_3)_2{\cdot}4H_2O$, and KCr_2O_4 in distilled water. The concentrations of Cu, Ni, Mn, Cd(II), and Cr(VI) in the AMD sample were 92.58, 7.50, 33.89, 0.31, and 1.15 mg/L, respectively, with an initial pH 3.43 and 10°C temperature. The AMD sample might comprise some carbonates, sulfates, and other metal ions (such as As).

To compare the ability of maghemite NPs to remove metals, the process was tested at optimal conditions for the best adsorption, with simulated wastewater solution and AMD. 5 mg of maghemite was added to 100 mL of 50-ppm simulated wastewater and AMD and kept for 24 hours at room temperature. The tests were performed at three pH values—2.6, initial pH (6.14 for simulated solution and 3.43 for AMD), and 9.5. A solution of 0.01 M HCl and 0.01 M NaOH was used for pH adjustment. After the experiments, the following removal efficiencies were reported.

	Removal Efficiency (%)					
	@pH 2.5		@pH 3.43	@pH 6.14	@pH 9.5	
Metal Ions	**AMD**	**Simulated Wastewater**	**AMD**	**Simulated Wastewater**	**AMD**	**Simulated Wastewater**
Cd	26.35	18.15	17.66	26.64	59.04	98.85
Cr	65.79	86.56	84.11	81.73	4.52	9.40
Cu	13.60	3.13	4.23	90.22	96.03	99.77
Mn	12.05	22.92	2.36	80.21	96.21	99.71
Ni	8.74	11.74	10.11	77.41	88.71	99.51

Explain why there are differences in removal efficiencies of metals in between simulated wastewater and AMD.

11.4 References

1. Engates, K. E.; Shipley, H. J., Adsorption of Pb, Cd, Cu, Zn, and Ni to titanium dioxide nanoparticles: effect of particle size, solid concentration, and exhaustion. *Environmental Science and Pollution Research.* **2011,** *18*(3), 386–395.
2. Lonnen, J.; Kilvington, S.; Kehoe, S. C.; Al-Touati, F.; McGuigan, K. G., Solar and photocatalytic disinfection of protozoan, fungal and bacterial microbes in drinking water. *Water Research.* **2005,** *39*(5), 877–883.
3. Subramanian, V.; Wolf, E.; Kamat, P. V., Semiconductor–metal composite nanostructures. To what extent do metal nanoparticles improve the photocatalytic activity of TiO_2 films? *Journal of Physical Chemistry B.* **2001,** *105*(46), 11439–11446.

4. Amin, M. T.; Alazba, A. A.; Manzoor, U., A Review of removal of pollutants from water/wastewater using different types of nanomaterials. *Advances in Materials Science and Engineering.* **2014,** *2014*, 24.
5. Stoimenov, P. K.; Klinger, R. L.; Marchin, G. L.; Klabunde, K. J., Metal oxide nanoparticles as bactericidal agents. *Langmuir.* **2002,** *18*(17), 6679–6686.
6. Gehrke, I.; Geiser, A.; Somborn-Schulz, A., Innovations in nanotechnology for water treatment. *Nanotechnology, Science and Applications.* **2015,** *8*, 1–17.
7. Sonti, S. V.; Bose, A., Cell separation using protein-A-coated magnetic nanoclusters. *Journal of Colloid and Interface Science.* **1995,** *170*(2), 575–585.
8. Gupta, V. K.; Tyagi, I.; Sadegh, H.; Shahryari-Ghoshekandi, R.; Makhlouf, A. S. H.; Maazinejad, B.; , Nanoparticles as adsorbent: a positive approach for removal of noxious metal ions: a review. *Science, Technology and Development.* **2015,** *34*(3), 195–214.
9. Li, Y.-H.; Ding, J.; Luan, Z.; Di, Z.; Zhu, Y.; Xu, C.; . . . Wei, B., Competitive adsorption of Pb^{2+}, Cu^{2+} and Cd^{2+} ions from aqueous solutions by multiwalled carbon nanotubes. *Carbon.* **2003,** *41*(14), 2787–2792.
10. Li, Y.-H.; Wang, S.; Luan, Z.; Ding, J.; Xu, C.; Wu, D., Adsorption of cadmium(II) from aqueous solution by surface oxidized carbon nanotubes. *Carbon.* **2003,** *41*(5), 1057–1062.
11. Zare, K.; Gupta, V. K.; Moradi, O.; Makhlouf, A. S. H.; Sillanpää, M.; Nadagouda, M. N.; . . . Kazemi, M., A comparative study on the basis of adsorption capacity between CNTs and activated carbon as adsorbents for removal of noxious synthetic dyes: a review. *Journal of Nanostructure in Chemistry.* **2015,** *5*(2), 227–236.
12. Ren, X.; Chen, C.; Nagatsu, M.; Wang, X., Carbon nanotubes as adsorbents in environmental pollution management: a review. *Chemical Engineering Journal.* **2011,** *170*(2), 395–410.
13. Daneshvar Tarigh, G.; Shemirani, F., Magnetic multi-wall carbon nanotube nanocomposite as an adsorbent for preconcentration and determination of lead (II) and manganese (II) in various matrices. *Talanta.* **2013,** *115*, 744–750.
14. Savage, N.; Diallo, M. S., Nanomaterials and water purification: opportunities and challenges. *Journal of Nanoparticle Research.* **2005,** *7*(4), 331–342.
15. Moradi, O.; Gupta, V. K.; Agarwal, S.; Tyagi, I.; Asif, M.; Makhlouf, A. S. H.; . . . Shahryari-ghoshekandi, R., Characteristics and electrical conductivity of graphene and graphene oxide for adsorption of cationic dyes from liquids: kinetic and thermodynamic study. *Journal of Industrial and Engineering Chemistry.* **2015,** *28*, 294–301.
16. Sadegh, H.; Ali, G. A. M.; Gupta, V. K.; Makhlouf, A. S. H.; Shahryari-ghoshekandi, R.; Nadagouda, M. N.; . . . Megiel, E., The role of nanomaterials as effective adsorbents and their applications in wastewater treatment. *Journal of Nanostructure in Chemistry.* **2017,** *7*(1), 1–14.
17. Vunain, E.; Mishra, A. K.; Mamba, B. B., Dendrimers, mesoporous silicas and chitosan-based nanosorbents for the removal of heavy-metal ions: a review. *International Journal of Biological Macromolecules.* **2016,** *86*, 570–586.
18. Bhatnagar, A.; Sillanpää, M., Applications of chitin- and chitosan-derivatives for the detoxification of water and wastewater—a short review. *Advances in Colloid and Interface Science.* **2009,** *152*(1), 26–38.
19. Singh, P. P.; Ambika, Environmental remediation by nanoadsorbents-based polymer nanocomposite (Chapter 10). In: *New Polymer Nanocomposites for Environmental Remediation*, Hussain, C. M.; Mishra, A. K., eds. Elsevier, 2018, pp. 223–241.

20. Maleki, H.; Hüsing, N., Aerogels as promising materials for environmental remediation—a broad insight into the environmental pollutants removal through adsorption and (photo)catalytic processes (Chapter 16). In: *New Polymer Nanocomposites for Environmental Remediation*, Hussain, C. M.; Mishra, A. K., eds. Elsevier, 2018, pp. 389–436.
21. Adebajo, M. O.; Frost, R. L.; Kloprogge, J. T.; Carmody, O.; Kokot, S., Porous materials for oil spill cleanup: a review of synthesis and absorbing properties. *Journal of Porous Materials.* **2003,** *10*(3), 159–170.
22. Coleman, S. J.; Coronado, P. R.; Maxwell, R. S.; Reynolds, J. G., Granulated activated carbon modified with hydrophobic silica aerogel-potential composite materials for the removal of uranium from aqueous solutions. *Environmental Science & Technology.* **2003,** *37*(10), 2286–2290.
23. Gabelich, C. J.; Tran, T. D.; Suffet, I. H., Electrosorption of inorganic salts from aqueous solution using carbon aerogels. *Environmental Science & Technology.* **2002,** *36*(13), 3010–3019.
24. Nasreen, S. A.; Sundarrajan, S.; Nizar, S. A.; Balamurugan, R.; Ramakrishna, S., Advancement in electrospun nanofibrous membranes modification and their application in water treatment. *Membranes.* **2013,** *3*(4), 266–284.
25. Taha, A. A.; Wu, Y.-n.; Wang, H.; Li, F., Preparation and application of functionalized cellulose acetate/silica composite nanofibrous membrane via electrospinning for Cr(VI) ion removal from aqueous solution. *Journal of Environmental Management.* **2012,** *112*, 10–16.
26. Teng, M.; Wang, H.; Li, F.; Zhang, B., Thioether-functionalized mesoporous fiber membranes: sol–gel combined electrospun fabrication and their applications for Hg^{2+} removal. *Journal of Colloid and Interface Science.* **2011,** *355*(1), 23–28.
27. Werber, J. R.; Osuji, C. O.; Elimelech, M., Materials for next-generation desalination and water purification membranes. *Nature Reviews Materials.* **2016,** *1*, 16018.
28. Warkiani, M. E.; Bhagat, A. A. S.; Khoo, B. L.; Han, J.; Lim, C. T.; Gong, H. Q.; Fane, A. G., Isoporous micro/nanoengineered membranes. *ACS Nano.* **2013,** *7*(3), 1882–1904.
29. Zhou, M.; Kidd, T. J.; Noble, R. D.; Gin, D. L., Supported lyotropic liquid-crystal polymer membranes: promising materials for molecular-size-selective aqueous nanofiltration. *Advanced Materials.* **2005,** *17*(15), 1850–1853.
30. Henmi, M.; Nakatsuji, K.; Ichikawa, T.; Tomioka, H.; Sakamoto, T.; Yoshio, M.; Kato, T., Self-organized liquid-crystalline nanostructured membranes for water treatment: selective permeation of ions. *Advanced Materials.* **2012,** *24*(17), 2238–2241.
31. Kunduru, K. R.; Nazarkovsky, M.; Farah, S.; Pawar, R. P.; Basu, A.; Domb, A. J., Nanotechnology for water purification: applications of nanotechnology methods in wastewater treatment (Chapter 2). In: *Water Purification*, Grumezescu, A. M., ed. Academic Press, 2017, pp. 33–74.
32. Leibundgut, C.; Seibert, J., Tracer hydrology. In: *Treatise on Water Science*, Wilderer, P., ed. Elsevier, 2011, Vol. 2, pp. 215–236.
33. Leibundgut, C.; Maloszewski, P., *Tracers in Hydrology*. John Wiley & Sons, 2011.
34. Abbott, B. W.; Baranov, V.; Mendoza-Lera, C.; Nikolakopoulou, M.; Harjung, A.; Kolbe, T.; . . . Campeau, A., Using multi-tracer inference to move beyond single-catchment ecohydrology. *Earth-Science Reviews.* **2016,** *160*, 19–42.
35. Liao, R.; Yang, P.; Wu, W.; Luo, D.; Yang, D., A DNA tracer system for hydrological environment investigations. *Environmental Science & Technology.* **2018,** *52*(4), 1695–1703.

36. Foppen, J. W.; Orup, C.; Adell, R.; Poulalion, V.; Uhlenbrook, S., Using multiple artificial DNA tracers in hydrology. *Hydrological Processes.* **2011,** *25*(19), 3101–3106.
37. Foppen, J. W.; Seopa, J.; Bakobie, N.; Bogaard, T., Development of a methodology for the application of synthetic DNA in stream tracer injection experiments. *Water Resources Research.* **2013,** *49*(9), 5369–5380.
38. Aquilanti, L.; Clementi, F.; Landolfo, S.; Nanni, T.; Palpacelli, S.; Tazioli, A., A DNA tracer used in column tests for hydrogeology applications. *Environmental Earth Sciences.* **2013,** *70*(7), 3143–3154.
39. Aquilanti, L.; Clementi, F.; Nanni, T.; Palpacelli, S.; Tazioli, A.; Vivalda, P. M., DNA and fluorescein tracer tests to study the recharge, groundwater flowpath and hydraulic contact of aquifers in the Umbria-Marche limestone ridge (central Apennines, Italy). *Environmental Earth Sciences.* **2016,** *75*(7), 626.
40. Sabir, I. H.; Torgersen, J.; Haldorsen, S.; Aleström, P., DNA tracers with information capacity and high detection sensitivity tested in groundwater studies. *Hydrogeology Journal.* **1999,** *7*(3), 264–272.
41. Sabir, I. H.; Haldorsen, S.; Torgersen, J.; Alestrom, P.; Gaut, S.; Colleuille, H.; . . . Alestrom, P., Synthetic DNA tracers: examples of their application in water related studies. *IAHS Publication(International Association of Hydrological Sciences).* **2000,** 262, 159–165.
42. Pang, L.; Robson, B.; Farkas, K.; McGill, E.; Varsani, A.; Gillot, L.; . . . Abraham, P., Tracking effluent discharges in undisturbed stony soil and alluvial gravel aquifer using synthetic DNA tracers. *Science of the Total Environment.* **2017,** *592*, 144–152.
43. Pang, L.; Farkas, K.; Bennett, G.; Varsani, A.; Easingwood, R.; Tilley, R.; . . . Lin, S., Mimicking filtration and transport of rotavirus and adenovirus in sand media using DNA-labeled, protein-coated silica nanoparticles. *Water Research.* **2014,** *62*, 167–179.
44. Garnett, M. C.; Ferruti, P.; Ranucci, E.; Suardi, M. A.; Heyde, M.; Sleat, R., Sterically stabilized self-assembling reversibly cross-linked polyelectrolyte complexes with nucleic acids for environmental and medical applications. *Biochemical Society Transactions.* **2009,** *37*(4), 713–716.
45. Grass, R. N.; Schälchli, J.; Paunescu, D.; Soellner, J. O.; Kaegi, R.; Stark, W. J., Tracking trace amounts of submicrometer silica particles in wastewaters and activated sludge using silica-encapsulated DNA barcodes. *Environmental Science & Technology Letters* **2014,** *1*(12), 484–489.
46. Kianfar, B.; Bogaard, T.; Foppen, J. W., Environmental factors affecting the transport of DNA-tagged particle tracers in saturated porous media. *Geophysical Research Abstracts.* **2018,** *20*, EGU2018-16963-3.
47. Sharma, A. N.; Luo, D.; Walter, M. T., Hydrological tracers using nanobiotechnology: proof of concept. *Environmental Science & Technology.* **2012,** *46*(16), 8928–8936.
48. Molnar, I. L.; Johnson, W. P.; Gerhard, J. I.; Willson, C. S.; O'Carroll, D. M., Predicting colloid transport through saturated porous media: a critical review. *Water Resources Research.* **2015,** *51*(9), 6804–6845.
49. Petosa, A. R.; Jaisi, D. P.; Quevedo, I. R.; Elimelech, M.; Tufenkji, N., Aggregation and deposition of engineered nanomaterials in aquatic environments: role of physicochemical interactions. *Environmental Acience & Technology.* **2010,** *44*(17), 6532–6549.
50. Praetorius, A.; Tufenkji, N.; Goss, K.-U.; Scheringer, M.; von der Kammer, F.; Elimelech, M., The road to nowhere: equilibrium partition coefficients for nanoparticles. *Environmental Science: Nano.* **2014,** *1*(4), 317–323.

51. Bradford, S. A.; Torkzaban, S.; Leij, F.; Simunek, J., Equilibrium and kinetic models for colloid release under transient solution chemistry conditions. *Journal of Contaminant Hydrology.* **2015,** *181*, 141–152.
52. Goldberg, E.; Scheringer, M.; Bucheli, T. D.; Hungerbühler, K., Critical assessment of models for transport of engineered nanoparticles in saturated porous media. *Environmental Science & Technology.* **2014,** *48*(21), 12732–12741.
53. Salama, A.; Negara, A.; El Amin, M.; Sun, S., Numerical investigation of nanoparticles transport in anisotropic porous media. *Journal of Contaminant Hydrology.* **2015,** *181*, 114–130.
54. Torkzaban, S.; Bradford, S. A.; Vanderzalm, J. L.; Patterson, B. M.; Harris, B.; Prommer, H., Colloid release and clogging in porous media: effects of solution ionic strength and flow velocity. *Journal of Contaminant Hydrology.* **2015,** *181*, 161–171.
55. Dahlke, H. E.; Williamson, A. G.; Georgakakos, C.; Leung, S.; Sharma, A. N.; Lyon, S. W.; Walter, M. T., Using concurrent DNA tracer injections to infer glacial flow pathways. *Hydrological Processes.* **2015,** *29*(25), 5257–5274.
56. Paunescu, D.; Puddu, M.; Soellner, J. O.; Stoessel, P. R.; Grass, R. N., Reversible DNA encapsulation in silica to produce ROS-resistant and heat-resistant synthetic DNA "fossils". *Nature Protocols.* **2013,** *8*(12), 2440.
57. Puddu, M.; Paunescu, D.; Stark, W. J.; Grass, R. N., Magnetically recoverable, thermostable, hydrophobic DNA/silica encapsulates and their application as invisible oil tags. *ACS Nano.* **2014,** *8*(3), 2677–2685.
58. Lotito, V.; Zambelli, T., Approaches to self-assembly of colloidal monolayers: a guide for nanotechnologists. *Advances in Colloid and Interface Science.* **2017,** *246*, 217–274.
59. Schenkel, J.; Kitchener, J., A test of the Derjaguin–Verwey–Overbeek theory with a colloidal suspension. *Transactions of the Faraday Society.* **1960,** *56*, 161–173.
60. Hunter, R. J., *Foundations of Colloid Science.* Oxford, UK: Oxford University Press, 2001.
61. Tufenkji, N.; Elimelech, M., Deviation from the classical colloid filtration theory in the presence of repulsive DLVO interactions. *Langmuir.* **2004,** *20*(25), 10818–10828.
62. Yao, K.-M.; Habibian, M. T.; O'Melia, C. R., Water and waste water filtration. Concepts and applications. *Environmental Science & Technology* **1971,** *5*(11), 1105–1112.
63. Rajagopalan, R.; Tien, C., Trajectory analysis of deep-bed filtration with the sphere-in-cell porous media model. *AIChE Journal.* **1976,** *22*(3), 523–533.
64. Bear, J.; Cheng, A. H. D.; Batu, V.; Chen, C.; Kuang, X.; Jiao, J. J., Methods to derive the differential equation of the free surface boundary. *Groundwater.* **2010,** *48*(4), 486–493.
65. Hartmann, N. I. B.; Skjolding, L. M.; Hansen, S. F.; Baun, A.; Kjølholt, J.; Gottschalk, F., *Environmental Fate and Behaviour of Nanomaterials: New Knowledge on Important Transfomation Processes.* Copenhagen K: Danish Environmental Protection Agency, 2014. Environmental Project, No. 1594.
66. Shevlin, D.; O'Brien, N.; Cummins, E., Silver engineered nanoparticles in freshwater systems: likely fate and behaviour through natural attenuation processes. *Science of the Total Environment.* **2017**. doi: 10.1016/j.scitotenv.2017.10.123.
67. Markus, A. A.; Parsons, J. R.; Roex, E. W.; de Voogt, P.; Laane, R. W., Modelling the release, transport and fate of engineered nanoparticles in the aquatic environment—a review. *Review of Environment Contamination and Technology.* **2017,** *243*, 53–87.
68. Kooi, M.; Besseling, E.; Kroeze, C.; van Wenzel, A. P.; Koelmans, A. A., Modeling the fate and transport of plastic debris in freshwaters: review and guidance. In: *Freshwater Microplastics.* Wagner M, Lambert S. Cham, Switzerland: Springer, 2018, pp. 125–152.

69. Nizzetto, L.; Bussi, G.; Futter, M. N.; Butterfield, D.; Whitehead, P. G., A theoretical assessment of microplastic transport in river catchments and their retention by soils and river sediments. *Environmental Science: Processes & Impacts.* **2016,** *18*(8), 1050–1059.
70. Bes-Piá, A.; Mendoza-Roca, J. A.; Roig-Alcover, L.; Iborra-Clar, A.; Iborra-Clar, M. I.; Alcaina-Miranda, M. I., Comparison between nanofiltration and ozonation of biologically treated textile wastewater for its reuse in the industry. *Desalination.* **2003,** *157*(1), 81–86.
71. Akhbarizadeh, R.; Shayestefar, M. R.; Darezereshki, E. Competitive removal of metals from wastewater by maghemite nanoparticles: a comparison between simulated wastewater and AMD. *Minewater and the Environment.* **2013**. 1–7.

CHAPTER 12

Applications in Energy Harvesting

In the 21st century, technological advancements have made significant impacts on our daily lives—for instance, the increasing use of wireless mobile devices, and the use of artificial intelligence in building automated systems and transportation. Energy spectrum shown in Fig. 12.1 depicts the energy required to operate nano/micro-electronics to macro-electrical systems.[1] These advancements have improved our quality of life, but have escalated the need for energy. Growing world population and rapid economic expansion are also resulting in increase in consumption of energy and hence the increase in energy demand. It is known that natural resources (oil, natural gas, and coal) are not endless and the increasing demand for energy cannot be fulfilled by these resources. Furthermore, the energy generation from fossil fuels is not environment friendly because of greenhouse gas emissions. It has been estimated that to meet the energy demand, the energy supply should be raised to more than double by 2050. Therefore, it is critical that researchers from across disciplines develop alternative but sustainable and environmentally benign energy technologies.

Energy can be produced from renewable resources such as sunlight, wind, waves, and geothermal heat and used in different sectors including transportation, electricity generation, and heating water. However, when the energy produced from renewable resources is transformed to another form or used for practical applications, much of the energy is lost. In fact, the amount of energy lost is much higher during the transmission and distribution stages than the amount actually used. This unusable form of energy is released into the environment in the form of heat or light. In order to develop sustainable and environment-friendly energy technologies, it was necessary to discover novel techniques to capture energy that is unusable for any practical purpose. This necessity led to the discovery of a technique called *energy harvesting*.

Energy harvesting can be defined as a process involving electronically trapping and accumulating the energy that is considered to be unused or waste for practical purposes from a variety of sources. Sometimes the process of capturing residual energy generated as a byproduct of industrial processes or natural sources can also be referred to as energy harvesting.

Examples of common energy harvesting sources include mechanical energy resulting from vibration, stress, and strain; thermal energy from furnaces, combustion engines, and other heating sources; even biological, solar energy from all forms of light sources, ranging from lighting, light emissions, and the sun; electromagnetic energy that is captured via inductors, coils, and transformers; wind and fluid energy resulting from air and liquid flow; chemical energy from naturally recurring or biological

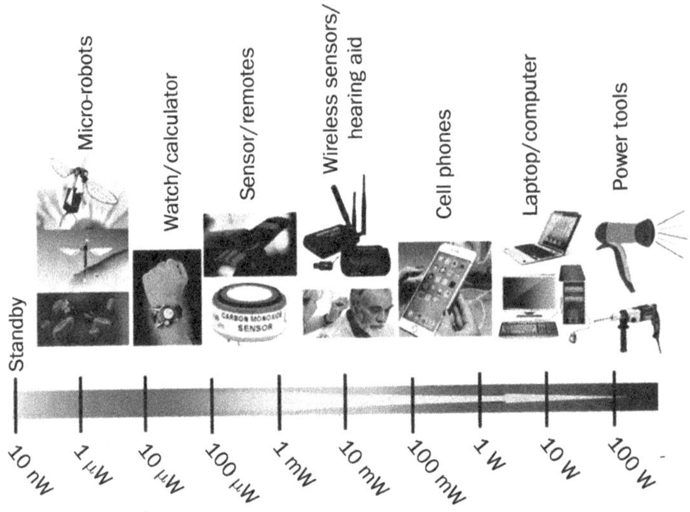

FIGURE 12.1 Range of power consumptions for a wide spectrum of electrical devices and their manifestations.[1]

processes; and huge amounts of radio frequency energy in the environment because of ubiquitous radio transmitters and television broadcasting.

Energy harvesting is a cross-disciplinary topic. New design principles for high-performance energy harvesting materials can be learned from nanotechnological fabrication approaches such as top-down and bottom-up.[2-5] These approaches enable the generation of size-, shape-, and composition-controlled nanomaterials showing a great promise in the field of energy harvesting technology. Nanomaterials exhibit unusual physical properties because of dramatic increase in the surface-area-to-volume ratio compared to bulk materials. Progress has also been made in designing next-generation functional artificial materials by the self-assembly of nanoscale building blocks.[6] Such artificial self-assembled materials exhibit enhanced collective physical properties because of the synergy between their individual counterparts. The properties of self-assembled materials can be tailored by tuning the size and shape of nanoscale building blocks and their organization within the self-assembled materials.

The aim of this chapter is to provide an overview of recent development in the materials design for rapidly expanding energy harvesting field. The chapter is divided into sections on heat harvesting, photovoltaic and solar energy, magnetic harvesting, and mechanical harvesting. Each topic starts with an introduction followed by materials design principles and the role of nanomaterials in the rapidly evolving field of energy harvesting. Therefore, this chapter will provide a thorough understanding of different types of nanomaterials that may be used for energy harvesting applications.

12.1 Harvesting Energy from Heat

12.1.1 Overview

It is well known that out of the primary energy that we consume, only one-third is effectively used, and the majority of energy is lost as waste heat (Fig. 12.1).[7] In the 20th century, humankind learned to achieve unprecedented control over charge carriers

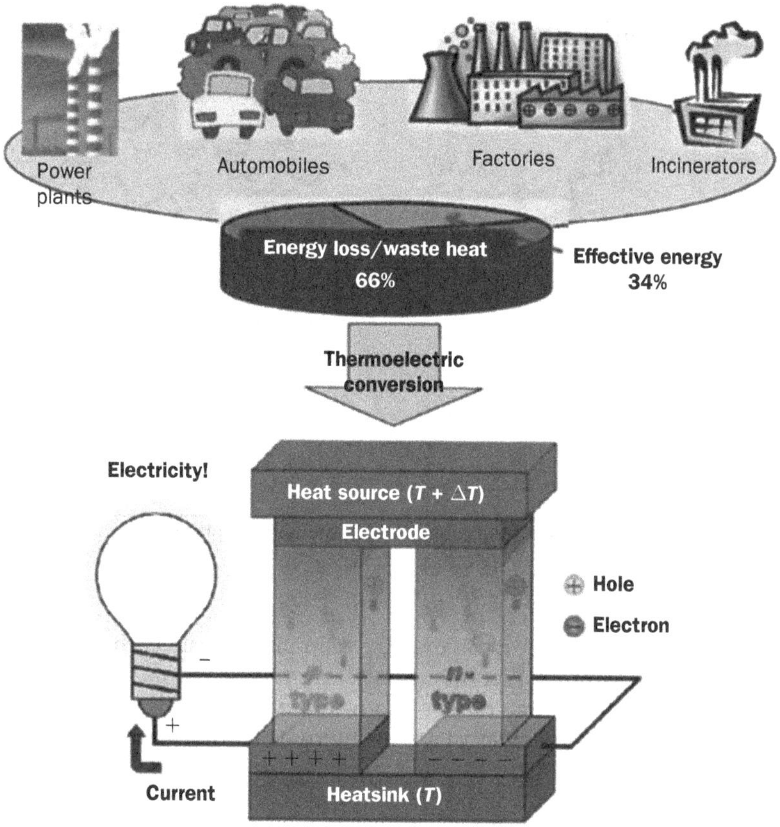

FIGURE 12.2 Schematic showing waste heat and thermoelectric power generation. Carrier diffusion to the lower temperature side generates electricity.[12]

(electrons), spins, and photons; however, advanced control over phonons and thermal energy is still not satisfactory and remains one of the important scientific challenges for the 21st century.

The majority of waste heat is estimated to have temperatures below 150°C, and thermoelectrics, which can compactly convert heat to electricity through solid-state devices using the Seebeck effect without scaling, are quite promising. The Seebeck effect is the phenomenon where electrical voltage is generated by the diffusion of charge carriers when a temperature difference is applied to a material (Fig. 12.2). In a thermoelectric device, two materials, one n and one p type, are connected to the heat source and the heat sink. Electrons are transferred due to different temperatures, resulting in an electrical potential. Conversely, a temperature difference can be induced when a potential is applied. One difficulty in achieving high conversion efficiency in thermoelectric materials is that the figure of merit (ZT), defined by the following equation:

$$ZT = (S^2\sigma T)/\kappa \tag{12.1}$$

is comprised of competing requirements in physical properties, namely, optimally designing a material with a large Seebeck coefficient (S), as an insulator, but with high

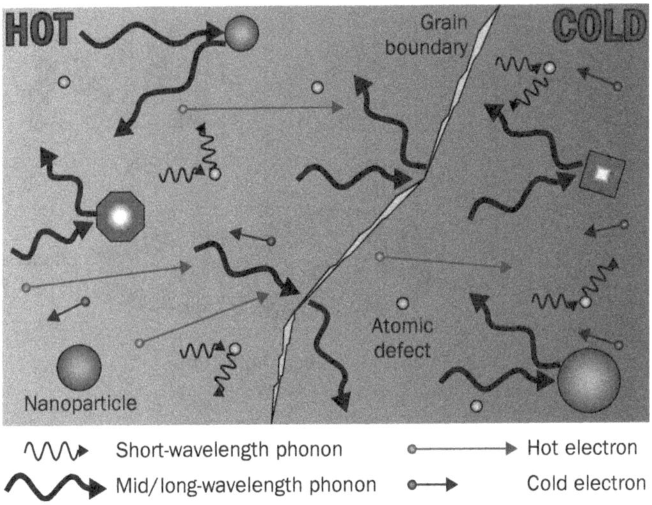

FIGURE 12.3 Sketch showing the probable mechanisms of phonon scattering, including electron transport of hot and cold electrons.[13]

electrical conductivity (σ), like a metal, yet simultaneously with low thermal conductivity (κ); T is temperature.[8,9] A dimensionless number ZT is often used as a figure of merit for the thermoelectric material. For conventional materials $ZT < 1$ at $T = 300$ K. The higher the ZT, the more efficient are the thermoelectric-based devices.

To overcome the conducting electricity while blocking heat transport as much as possible, selective scattering of phonons has been shown to be an effective strategy to achieve high ZT.[10,11] The thermal conductivity is composed of heat transport by the electrical carriers, κ_{el}, and that carried by the lattice, κ_{lat}, where:

$$\kappa = \kappa_{el} + \kappa_{lat} \tag{12.2}$$

According to Wiedemann–Franz law, κ is proportional to the electrical conductivity: $\kappa_{el} \propto \sigma$. Since high σ is desirable, it is necessary to find ways to lower κ without overly harming the electrical conductivity. This can be strategically carried out by utilizing the fact that for the range of materials suitable for thermoelectrics, the phonon mean free path (MFP) is typically longer than the carrier MFP. If materials can be modified such that phonons are selectively scattered, this can enhance the ZT. Figure 12.3 shows the possible mechanism of phonon scattering.

12.1.2 Nanomaterials for Heat Energy Harvesting

Nanostructures are able to lower their thermal conductivity relative to the bulk. Such a reduction is expected to increase the thermoelectric figure of merit (ZT) in the superlattice systems. Superlattices in general are not good candidates for applications when considering a large scale. To overcome this, nanocomposites would be a better candidate for use in applications.

Designing multi-scale nano-microstructures, ranging from point defects to volume defects, which have different phonon scattering behaviors, can be effective at reducing κ_{lat}.[14] Point defects have a frequency (ω) dependence of the scattering rate (the inverse of the phonon relaxation time), of $\tau_{PD}^{-1} \propto \omega^4$, meaning they are particularly effective in scattering high-frequency phonons. Line defects have relaxation time dependencies of $\tau_{DS}^{-1} \propto \omega^1$ and $\tau_{DS}^{-1} \propto \omega^3$ for dislocation strain fields and dislocation cores, respectively, and therefore scatter mid-frequency phonons well. Two-dimensional defects, such as grain boundaries, are typically frequency independent with $\tau_{GB}^{-1} \propto \omega^0$ and scatter low-frequency phonons. Volume defects can have $\tau_{NP}^{-1} \sim \omega^0 + \omega^4$, and therefore can scatter phonons over a wide range. Strategic nano-microstructural control can thus lead to large enhancement of ZT.

While such nano-microstructuring with phonon selective scattering has led to significant enhancements in ZT for a wide range of materials, finding new broadly applicable principles to enhance the power factor, $S^2\alpha$, the denominator of ZT, has been more elusive. Various principles regarding utilizing resonance levels, band convergence, modulation doping, and magnetism have been proposed.[15-17] Quantum materials have the potential to overcome the traditional bounds of conventional materials.

Quantum-sized effects were first predicted by Hicks and Dresselhaus in semiconductor nanostructures, such as quantum-well structures, superlattices, nanowires, and quantum dots, as a potential way to significantly enhance the power factor through modification of the band structure.[18] As interesting quantum systems with particular surface states, topological insulators (TIs) have embodied many high-performance thermoelectric systems such as Bi_2Te_3, Bi_2Se_3, Sb_2Te_3, Sn-doped PbTe, and PbSe.[19] TIs possess a bulk bandgap and highly conductive gapless surface states that are protected by time-reversal symmetry and therefore do not backscatter, and they possess high-mobility charge carriers. Nanograin materials with different grain-size distributions were pointed out as a way to tune the contribution of the topological states and thereby tune the thermoelectric properties.

A huge challenge for thermoelectrics has been achieving widespread power-generation applications. The development of flexible materials such as organic and hybrid materials enables wearable thermoelectric applications.[20] While not as extensively investigated compared to their inorganic counterparts, organic thermoelectric materials were developed from a relatively early time. A large enhancement in performance can be achieved through carrier tuning of PEDOT:Tos (poly(3,4-ethylenedioxythiophene):tosylate anion), and this created intense interest in these materials.[21] Organic materials have an advantage over inorganic materials in the general low cost of materials and scalable processing, as well as their lightweight and flexible nature, which enable wearable applications. Their relatively low temperature stability is also not as critical an issue when dealing with wearable or moderate temperature applications below 150°C, where the majority of unutilized (waste) thermal energy exists. However, the general long-range structural disorder in organic materials largely hinders electrical conduction, typically outweighing the benefit gained from their low thermal conductivity. Therefore, it is necessary to realize a significant improvement in their thermoelectric figure of merit to make them suitable for applicable devices. This may also be a solution to the difficulty in finding n-type organic thermoelectric materials with good thermoelectric performance. The flexibility and low thermal conductivity of organic materials and the good electrical properties, that is, high power factor, of inorganic materials make inorganic–organic hybrid composites an obvious choice for such materials. Inorganic–organic hybrid composites have reportedly exhibited high power factor, such as about 900 $\mu W/mK^2$ for a TiS_2 or organic superlattice material. The overall design and advanced synthesis techniques of organic and hybrid materials are optimization strategies to achieve high ZT.

12.2 Photovoltaics

12.2.1 Overview

There are approximately 120,000 terawatts (TW) of solar power irradiating on Earth. The global consumption is merely 15 TW. Therefore, harnessing solar energy via photovoltaics (PV) technology can be used as source for supplying energy in sustainable and environment friendly fashion.

Photovoltaics involves converting solar radiation into electrical energy through semiconductor materials. Solar energy harvesting involves transferring light into electricity through certain types of chemical reactions. The energy production usually starts with the light being absorbed by assembled materials. Subsequently, electrons are excited and transferred to an acceptor. The most common materials used today are amorphous silicon, cadmium telluride, copper indium gallium selenide, copper indium gallium sulfide, and single-crystal and polycrystalline silicon.[22] There are a wide variety of nanomaterials used as solar materials for energy harvesting, including nanotubes,[23] quantum dots,[24] nanocrystals,[25] among others. The advancement of NP synthesis in well-defined geometrical shapes could result in innovative designs for solar energy conversion.

The power conversion efficiency, η (conversion of solar energy into electric power), is depicted by:

$$\eta = \frac{FFJ_{SC}V_{OC}}{P_{inc}} \tag{12.3}$$

where P_{inc} is the incident solar energy, V_{OC} is the open circuit voltage, J_{SC} is the short-circuit current density, and FF is the fill factor. The short-circuit current density J_{SC} is the current density derived at zero bias voltage, while the open circuit voltage V_{OC} is the maximum voltage that a solar cell can generate. The physical processes in a solar device encompass the following:

1. *Irradiation of sunlight:* A typical solar spectrum covers a wide range of wavelengths from ultraviolet (UV) to infrared (IR), resembling to the spectrum of black body radiation at the temperature of about 5700 K. The short-wave radiation, UV light is filtered out by ozone due to atmospheric absorption and very little reaches the surface. IR light is readily absorbed by water and carbon dioxide molecules and converted into heat energy. Therefore, IR radiation is responsible for warming of the Earth's surface.
2. *Light absorption:* The photoactive layer in the solar cell device absorbs sunlight and excites the electron from ground energy level to high energy states, generating hole-electron pair, called an *exciton*. Therefore, the materials should possess appropriate bandgap to prevent non-radiative relaxation process. Semiconductors are the most widely used materials because these materials possess moderate bandgap compared to metals (zero bandgap) and insulators (high bandgap) to enable the energy conversion.
3. *Hole-electron separation:* In solar cell devices, the charge carriers (holes and electrons pair) can be separated by a built-in internal field. As a result, the electron drifts to n-type layer and the hole into p-type layer. A proper alignment of electronic energy band structures or fermi level at the charge separation junction is very important in dissociating the hole-electron pair to lead them in

the opposite direction. In addition, an efficient charge separation also requires low electron-hole recombination rates.

4. *Transport of photo-generated charge carriers to electrodes:* Carrier transporting layers are required to achieve high mobility as well as long diffusion path lengths of electron and holes to ensure an efficient collection of charge carriers.
5. *Energy dissipation in the circuit:* The collected electrons dissipate their energy in the external circuit and return to the solar cell, and annihilate with holes.

Photovoltaics have significantly evolved through three generations:

1. *The first generation of crystalline p–n junction silicon solar cells:* Although relatively high in efficiency, these cells involve high manufacturing and installation costs.[26,27]
2. *The second generation of thin-film solar cells, e.g., amorphous silicon,[28] poly-crystalline silicon,[29] cadmium telluride,[30,31] and copper indium gallium selenide:[32]* These solar cells are significantly less expensive than single-crystalline PV cells due to reduced materials and processing costs, and increased manufacturing throughput. The efficiencies of these cells are lower than the first generation solar cell.
3. *The third generation of solution-processed solar technologies that aim to lower costs, while maintaining high solar energy conversion efficiencies:* These technologies provide the advantage that large area and flexible structures with roll-to-roll coating/printing can be accomplished. However, the technical challenge is to fabricate PV devices over the entire solar spectrum as well as to overcome the fundamental power-loss mechanism.

12.2.2 Nanomaterials for Photovoltaics

The broad solar spectrum spans from about 280 nm in the UV region to about 2500 nm in the mid-IR region. The efficiency of solar cell devices depends on the matching of their light absorption spectrum to the irradiated solar spectrum on the Earth's surface. Semiconducting materials hold great potential to be used as light-absorbing materials and harvest the whole solar spectrum. The absorption properties can be tuned by tailoring the bandgap of the semiconducting materials. Advances in nanotechnology have enabled the synthesis of semiconductor nanoparticles (NPs), e.g., quantum dots (QDs). The main advantage of using semiconductor nanomaterials based on QDs is the tunability in the bandgap of semiconductor QDs, which can be tailored by controlling the type, size, shape, and composition or alloying of the material. This determines the portion of solar spectrum the materials can absorb.

The wet chemical colloidal synthetic approach is widely used for the preparation of different types of semiconductor QDs. This approach employs three components: (i) a high boiling point solvent (benzyl ether, octadecene, and dichlorobenzene), (ii) capping ligands or surfactant or reducing agents [olyelamine, oleic acid, and trioctylphosphine oxide (TOPO)], and (iii) organometallic precursors. A high boiling solvent is needed to obtain the required temperature for the crystallization of inorganic NPs. The ligands or surfactants act as reducing agent controlling the reduction of precursor as well as capping ligands to control the growth of NPs. The organometallic precursor provides elements (Cd and Te for CdTe) to grow semiconductor QDs.

There are two ways for introducing organometallic precursor to the reaction flask: the *heat-up* method and the *hot-injection* method. The latter has already been introduced in Chap. 5, but will be revisited here for the purpose of comparison. The *heat-up* method involves the addition of precursor to flask containing the solvent and ligands at room temperature followed by a slow heating of the reaction solution to the boiling temperature of the solvent for preparing NPs. In the *hot-injection* method, the precursor is usually introduced to the hot boiling reaction mixture of solvent and ligands. Semiconductor QDs are generally prepared by *hot-injection* method. The rapid injection of precursor to boiling reaction mixture leads to the immediate burst of nucleation, followed by a growth without further nucleation. To produce monodisperse semiconductor QDs, it is important to separate the nucleation and growth processes. The size of QDs can be adjusted by changing the ratio of precursor to ligands, temperature, reaction time, and nature of the solvent and ligands. Anisotropic QDs can be prepared by introducing ligands of different chemistries into the reaction mixture. The ligands preferentially bind to the particular facets depending on the surface energy of growing crystals and inhibit the growth of absorbing facets. The *hot-injection* method has also been used to produce hybrid NP systems, for example, monodisperse plasmonic semiconductor NPs and lanthanide-doped up-conversion $NaYF_4$:YB^{3+}/ER^{3+} NPs. The *heat-up* method has widely been used for the synthesis of plasmonic, magnetic, transparent conducting NPs (ITO NPs), up-conversion/down-conversion NPs, and hybrid (plasmonic-magnetic) NPs. The advantage of using the *heat-up* method is that the approach allows the synthesis of NPs in large sizes because of Ostwald ripening. As you already know by now that Ostwald ripening involves the growth of large particles at the expense of energetically less stable smaller particles. The NPs synthesized by both the approaches are dispersible in organic solvents such as toluene, hexane, and chloroform due to the attachment of hydrophobic long-chain capping molecules to their surfaces. This solution-based processing of NPs provides an opportunity to fabricate NP-based solar devices. The disadvantage of using long-chain hydrophobic molecules is that these inhibit the charge transfer between the NPs due to their insulating nature. This influences the performance of NP-based solar cell devices. Therefore, further surface modification strategies are required to replace the long-chain molecules with short-chain molecules or conducting organic molecules, techniques that you have studied in Chap. 5.

Semiconductor QDs or NPs prepared by hot colloidal synthesis have been utilized to design next-generation multi-junction tandem or rainbow solar cells with impressive performance. For example, IV–VI semiconductor lead sulfide (PbS) QDs prepared in different sizes were used to the tune the absorption onset from 3000 to 600 nm. The multi-junction solar cell is prepared from a single material system; the top layer consists of 2.6-nm PbS QDs with absorption onset at 600 nm; the second layer is composed of 3.6-nm PbS QDs with absorption onset at 1070 nm, and the bottom layer is made from 7.2-nm PbS QDs with absorption onset at 1750 nm. The QDs in different sizes allow the engineering of bandgap to optimize the absorption at a particular wavelength of the solar spectrum.

Inorganic semiconductor or polymer composite provides another opportunity to build efficient hybrid solar cells. Such hybrid solar cell device is composed of semiconductor QDs (CdSe, PbS, and PbSe) in a hole-conducting conjugated polymer. These hybrid solar cells offer several advantages compared to previous generation solar cells, including (i) low fabrication cost due to simple solution processing, (ii) light weight and flexible substrate due to polymer matrix (suitable for roll-to-roll production), and (iii) high electron mobility, size-dependent optical bandgap, and carrier multiplication.

There are three approaches to fabricate semiconductor and polymeric hybrid materials. The first approach is based on the ligand exchange of insulating capping ligands with functionalized conducting polymers. In a second approach, conjugated polymers can be directly tethered to bifunctional ligands attached to semiconductor NPs. The bifunctional ligands act as stabilizing ligands as well as provide the surface chemistry to anchor other molecules. The third approach is based on the direct growth of semiconducting NPs within conjugated polymers, allowing the intimate contact between the NPs and polymer.

12.3 Mechanical Energy Harvesting

12.3.1 Overview

Mechanical energy is generated by the position and motion of an object. There are materials and devices that convert this mechanical energy into other types of energy, such as electrical, heat, magnetic, etc. For example, the amount of produced electrical energy is given by the following equation:

$$E_{elec} = \eta_{elec}\eta_{conv}\eta_{abs}E_{source} \tag{12.4}$$

where η_{abs} is the proportion of natural mechanical energy absorbed by the system, η_{conv} is the proportion of the absorbed energy converted into electrical energy, η_{elec} is the electrical efficiency, and E_{source} is the energy available from the environment.

Mechanical energy is harvested from different sources such as vibrations (engines and human body), shocks (balls) or strain (shoe soles), and worn or washed clothes. The most common type of mechanical energy being harvested is mechanical vibration using resonant mechanical structures. With piezoelectric materials, the mechanical energy harvested from vibrations can be converted into electrical energy.

Piezoelectric materials vary greatly in properties: lead zirconate titanate (PZT), polymers polyvinylidene fluoride (PVDF), and crystalline (quartz). The conversion efficiency is given by the coupling coefficient κ as defined in the following equation:

$$\kappa = \frac{\text{Electrical energy}}{\text{Mechanical energy}} \tag{12.5}$$

The larger the coupling coefficient, the larger is the conversion efficiency, i.e., conversion of mechanical energy into electrical energy.

A piezoelectric material cannot be chosen based on this coefficient; mechanical and dielectric losses should also be taken into account. Crystalline and ceramic materials possess strong coupling coefficient ($\kappa = 0.7$) and are thus more efficient for converting mechanical energy into electrical energy. However, these materials are more susceptible to ageing. Polymeric materials show high mechanical resistance but their coupling coefficient is low ($\kappa = 0.2$). Hybrid materials made from ceramic and polymers can be designed to achieve both the properties of individual materials, but the use of these materials is restricted because of the presence of lead in their composition and environmental standards. Furthermore, the Curie temperature (above which piezoelectric properties disappear) limits their use. The efficiency of mechanical energy harvesting also depends on the design of piezoelectric materials employed in the device fabrication. Therefore, different structures of materials have been investigated to achieve improved energy conversion efficiency. One example is the design of piezoelectric polymers into wave shapes. The buckled ribbons enhance the piezoelectric effect significantly.

12.3.2 Nanomaterials for Mechanical Energy Harvesting

The advances in the strategies for synthesizing nanomaterials have fostered the development of new materials to harvest mechanical energy, i.e., the nanogenerator. PZT nanofibers can be produced by electrospinning and their size can be varied by changing the concentration of polymer solution. A soft polymer (polydimethylsiloxane, PDMS) was applied on top of the PZT nanofibers. Nanofibers were polled by an externally applied electric field across the electrodes. When the applied pressure was transferred to the PZT nanofibers through the PDMS matrix, it resulted in charge generation due to the combined tensile and bending stresses in the PZT nanofibers. A voltage difference between the two adjacent electrodes was thereby induced due to this separation of charge. The interdigitated electrodes could enhance the power output of the nanogenerator. The piezoelectric nanofibers between each pair of adjacent electrodes served as unit cells, and each cell was connected in parallel.

Mechanical energy harvesting can be done through hybrid composite, lead-free materials in linear or nonlinear vibration, wind energy, and locomotion. Efforts have been made to model the efficiency of such devices. Thin-film materials are potentially very attractive for Pb-free piezoelectric harvesting.

12.4 Magnetic Energy Harvesting

12.4.1 Overview

Among various forms of renewable energy such as sunlight, heat, and mechanical vibration, magnetic fields are also considered as a potential resource because magnetic fields are produced as a result of current flow. Most of the locations these days possess stray magnetic fields arising from the electric power transmission infrastructure of buildings, overhead power lines, industrial machinery, and consumer electronic devices. Such fields are considered to be noisy or harmful to the human body. The electric transmission cable generates low magnetic field (<1 mT $= 10$ G at a distance of 10 mm of 50 A current-conducting cable) at a fixed frequency of 50–60 Hz. This magnetic field can be captured and harvested by the use of electromagnetic devices employing coils and magnetic cores and operating on the principle of Faraday's induction law as well as by the incorporation of composite materials-based magneto-mechano-electric generators (Fig. 12.4).

The most widely used device to harvest magnetic field, i.e., traditional transformer, operates on the principle of Faraday's induction. The energy conversion efficiency can be optimized by altering the shape of the coil geometry, core materials, and winding protocol. In the core materials, cast iron is used because it significantly reduces losses from eddy currents. However, the traditional electromagnetic transformers have limitations related to weight, size, power density, installation, and conversion efficiency. Furthermore, when the electromagnetic generators are installed in an electrical system, the electric current flowing in the system becomes larger than those in normal operation and influences the quality of the power transmission when the power generated is large. Therefore, this system can be assumed to be stealing electricity from system, not harvesting waste energy. The development of alternative devices based on the discovery of new materials to harvest clean energy from magnetic noise fields is required.

12.4.2 Magnetic Nanomaterials

Magnetic materials play an important role in the development of efficient magnetic energy harvesting devices. There is a need for establishing manufacturing strategies

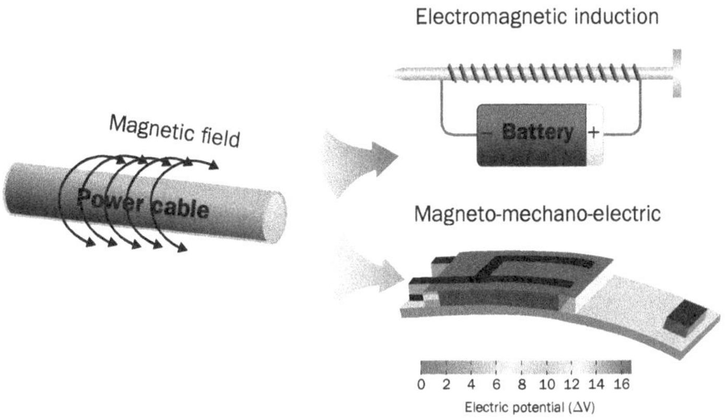

Figure 12.4 Magnetic energy harvesting from electromagnetic devices using Faraday's induction principle.[1]

for synthesizing a wide range of magnetic nanomaterials in different sizes, shapes, and composition. One particularly interesting route is the thermal decomposition of metallic precursor at high temperature in an inert atmosphere. This method relies on the heating up approach involving the slow heating of reaction mixture containing metal oleate or acetylacetonate, surfactant or ligand (oleic acid or oleylamine), and high boiling solvent (octadecene, diphenyl ether) to the reaction temperature (about 320°C). After refluxing the reaction at this temperature for 30–60 minutes, the metal oxide NPs can be synthesized. The size of magnetic NPs can be tailored by controlling the reaction temperature, solvent, and ratio of precursor to surfactant. The main advantage of this approach is that the method can be scaled up to produce high-quality magnetic nanomaterials with uniform size distribution. The shape or morphology of magnetic NPs can be varied by changing type of surfactant and its concentration in the reaction solution.

Another interesting route to synthesize magnetic nanomaterials is spark erosion. Spark erosion consists of generating spark discharges between the electrodes immersed in a dielectric fluid. After the generation and breakdown of complex plasma, the melted target materials are ejected from the electrodes and quenched in the dielectric medium. In dielectric liquid, ejected materials condense into particles. The spark erosion is a very versatile technique to synthesize NPs from a wide range of materials with particle sizes from micron to several nanometers. The same strategy can also be used to synthesize hollow spherical particles of metal (nickel, iron-cobalt) and the ferromagnetic shape memory alloy ($Fe_{83}Ga_{17}$) in different sizes from micrometers to nanometers when liquid nitrogen is used as the dielectric liquid (Fig. 12.5). Upon replacing the dielectric liquid nitrogen with argon, solid particles can be synthesized (Fig. 12.5).

These synthetic approaches also allow the synthesis of composite materials, i.e., the composite materials formed from the combination of magnetoelectric, ferromagnetic, and piezoelectric materials. These composite materials (such as CoFeB) will allow the energy transformation between magnetic and piezoelectric forms leading to higher conversion efficiency.

12.5 Summary

Advancement in the nanomaterials fabrication strategies has revolutionized the development of next-generation energy harvesting technologies. The progress can be seen in

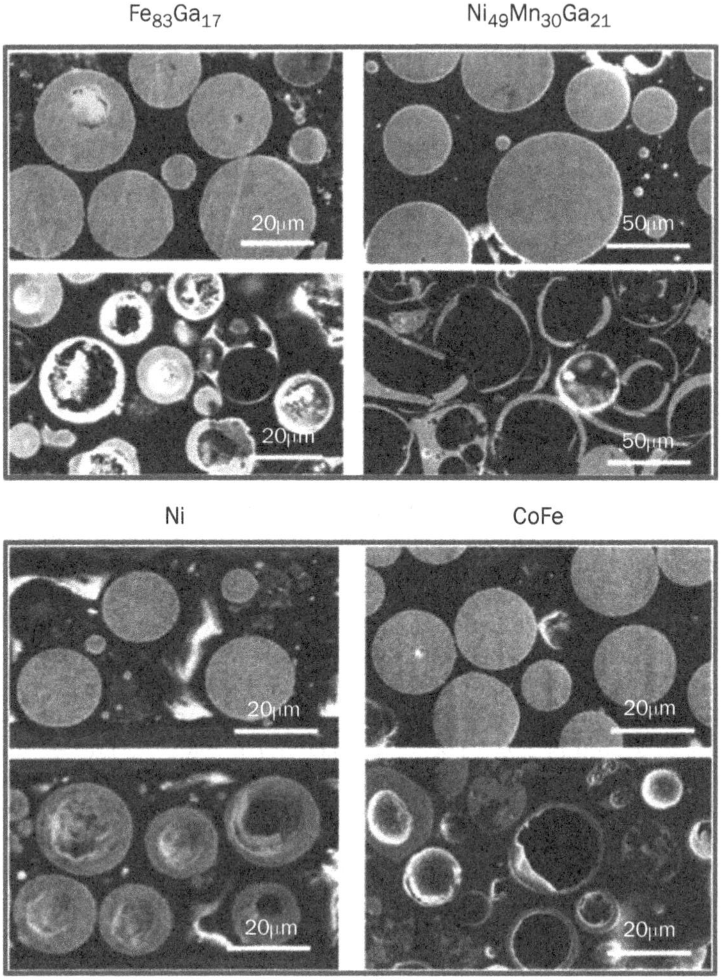

FIGURE 12.5 SEM images of solid particles prepared in liquid argon (upper panel) and hollow particles prepared in liquid nitrogen (lower panel).

wireless sensors, personal mobiles, Internet of things, and transport areas. An immensely important aspect of energy harvesting is its sustainability from a critical materials perspective, as research success transitions into widespread practical applications. To develop a high-performance thermoelectric materials–based device, the materials should incorporate the more abundant elements such as that in the first transition metal series in the periodic table. This is necessary to provide long-term adaptation of thermoelectric technology for practical purposes. Nanotechnology-based approaches have also provided numerous opportunities to develop third-generation solar cells based on semiconductor QDs with impressive energy conversion efficiencies at low fabrication costs. By controlling the size, shape, and composition of semiconductor QDs, light from broad solar spectrum can be harvested enabling development of tandem solar cell devices. Despite the significant progress, the world-record efficiencies of current semiconductor nanomaterials-based solar devices are still low (about 10 percent). Therefore, new advances in light harvesting using solar devices are required, such as development of new IR light absorber, *in situ* generation of NPs avoiding the use of insulating capped layer of ligands, the design of new conductive ligands, photon up-conversion and down-conversion, and printing and spray fabrication technologies. The magnetic energy technique is less developed compared to other energy harvesting technologies. Thus, the development of functional nanomaterials is required to achieve a drastic performance improvement in the energy conversion efficiency. The combination of two or more types of energy transfer is very important. This can be achieved by fabricating composite materials from the materials possessing magnetoelectric, piezoelectric, and electromagnetic properties or development of organic or inorganic hybrid energy platforms.

Finally, portable energy devices should be developed so that energy harvested under day light or in different geographic locations can be stored and transported to regions where energy is required. The new design criteria for synthesis of nanomaterials should be developed with dual functionality such as harvesting energy from different sources and their storage.

In an ideal situation, in a few years from now, a single nanodevice fabricated based on the knowledge gained through this textbook may be able to perform several of these functions (and maybe even more) simultaneously. Multifunctional NPs may pave the way for catering to a host of applications, all functional from a single platform! Imagine the DNA-based hydrological tracers, discussed in Chap. 11, capable of storing terabytes of data in less than a microliter solution of the NPs, thereby harvesting energy in a sustainable way and hence benefiting human society.

12.6 Study Questions

12.1 What is meant by energy harvesting? List some of the sources from where energy may be harvested.

12.2 Define figure of merit for thermoelectric materials.

12.3 Compare and contrast *hot-injection* and *heat-up* methods for synthesis of semiconductor quantum dots (QDs).

12.4 What is the fundamental principle of magnetic energy harvesting?

12.5 Illustrate the physical processes in a solar device.

12.6 Explain spark erosion and mention its advantages over other conventional methods.

12.7 What is meant by photovoltaics? Set up an equation showing the power conversion efficiency for photovoltaics.

12.7 References

1. Annapureddy, V.; Palneedi, H.; Hwang, G. T.; Peddigari, M.; Jeong, D. Y.; Yoon, W. H.; . . . Ryu, J., Magnetic energy harvesting with magnetoelectrics: an emerging technology for self-powered autonomous systems. *Sustainable Energy & Fuels.* **2017,** *1*(10), 2039–2052.

2. Gregorczyk, K.; Knez, M., Hybrid nanomaterials through molecular and atomic layer deposition: top down, bottom up, and in-between approaches to new materials. *Progress in Materials Science.* **2016,** *75,* 1–37.
3. Merkel, T. J.; Herlihy, K. P.; Nunes, J.; Orgel, R. M.; Rolland, J. P.; DeSimone, J. M., Scalable, shape-specific, top-down fabrication methods for the synthesis of engineered colloidal particles. *Langmuir.* **2010,** *26*(16), 13086–13096.
4. Vasquez, Y.; Henkes, A. E.; Bauer, J. C.; Schaak, R. E., Nanocrystal conversion chemistry: a unified and materials-general strategy for the template-based synthesis of nanocrystalline solids. *Journal of Solid State Chemistry.* **2008,** *181*(7), 1509–1523.
5. Wang, X.; Zhuang, J.; Peng, Q.; Li, Y. D., A general strategy for nanocrystal synthesis. *Nature.* **2005,** *437*(7055), 121–124.
6. Boles, M. A.; Engel, M.; Talapin, D. V., Self-assembly of colloidal nanocrystals: from intricate structures to functional materials. *Chemical Reviews.* **2016,** *116*(18), 11220–11289.
7. Akhbarizadeh, R.; Shayestefar, M. R.; Darezereshki, E. J. Competitive removal of metals from wastewater by maghemite nanoparticles: a comparison between simulated wastewater and AMD. *Minewater and the Environment.* **2013,** 1–7.
8. Bes-Piá, A.; Mendoza-Roca, J. A.; Roig-Alcover, L.; Iborra-Clar, A.; Iborra-Clar, M. I.; Alcaina-Miranda, M. I., Comparison between nanofiltration and ozonation of biologically treated textile wastewater for its reuse in the industry. *Desalination.* **2003,** *157*(1), 81–86.
9. Beekman, M.; Morelli, D. T.; Nolas, G. S., Better thermoelectrics through glass-like crystals. *Nature Materials.* **2015,** *14*(12), 1182–1185.
10. Khan, A. U.; Kobayashi, K.; Tang, D. M.; Yamauchi, Y.; Hasegawa, K.; Mitome, M.; Xue, Y. M.; Jiang, B. Z.; . . . Mori, T., Nano-micro-porous skutterudites with 100% enhancement in ZT for high performance thermoelectricity. *Nano Energy.* **2017,** *31,* 152–159.
11. Biswas, K.; He, J. Q.; Blum, I. D.; Wu, C. I.; Hogan, T. P.; Seidman, D. N.; Dravid, V. P.; Kanatzidis, M. G., High-performance bulk thermoelectrics with all-scale hierarchical architectures. *Nature* **2012,** *489* (7416), 414–418.
12. Mori, T.; Priya, S., Materials for energy harvesting: at the forefront of a new wave. *MRS Bulletin.* **2018,** *43*(3), 176–180.
13. Vineis, C. J.; Shakouri, A.; Majumdar, A.; Kanatzidis, M. G., Nanostructured thermoelectrics: big efficiency gains from small features. *Advanced Materials.* **2010,** *22*(36), 3970–3980.
14. Liu, Z. H.; Mao, J.; Liu, T. H.; Chen, G.; Ren, Z. F., Nano-microstructural control of phonon engineering for thermoelectric energy harvesting. *MRS Bulletin.* **2018,** *43*(3), 181–186.
15. Heremans, J. P.; Jovovic, V.; Toberer, E. S.; Saramat, A.; Kurosaki, K.; Charoenphakdee, A. . . Snyder, G. J., Enhancement of thermoelectric efficiency in PbTe by distortion of the electronic density of states. *Science.* **2008,** *321*(5888), 554–557.
16. Pei, Y. Z.; Shi, X. Y.; LaLonde, A.; Wang, H.; Chen, L. D.; Snyder, G. J., Convergence of electronic bands for high performance bulk thermoelectrics. *Nature.* **2011,** *473*(7345), 66–69.
17. Zebarjadi, M.; Joshi, G.; Zhu, G. H.; Yu, B.; Minnich, A.; Lan, Y. C.; . . . Chen, G., Power factor enhancement by modulation doping in bulk nanocomposites. *Nano Letters.* **2011,** *11*(6), 2225–2230.

18. Hicks, L. D.; Dresselhaus, M. S., Effect of quantum-well structures on the thermoelectric figure of merit. *Physical Review B.* **1993,** *47*(19), 12727–12731.
19. Moore, J. E.; Balents, L., Topological invariants of time-reversal-invariant band structures. *Physical Review B.* **2007,** *75*(12), doi: 10.1103/PhysRevB.75.121306.
20. Tian, R. M.; Wan, C. L.; Hayashi, N.; Aoai, T.; Koumoto, K., Wearable and flexible thermoelectrics for energy harvesting. *MRS Bulletin.* **2018,** *43*(3), 193–198.
21. Bubnova, O.; Khan, Z. U.; Malti, A.; Braun, S.; Fahlman, M.; Berggren, M.; Crispin, X., Optimization of the thermoelectric figure of merit in the conducting polymer poly(3,4-ethylenedioxythiophene). *Nature Materials.* **2011,** *10*(6), 429–433.
22. Jacobson, M. Z., Review of solutions to global warming, air pollution, and energy security. *Energy and Environmental Science.* **2009,** *2*(2), 148–173.
23. Joshi, P.; Xie, Y.; Mwaura, J.; Ropp, M.; Galipeau, D.; Qiao, Q., Carbon NPs for counter electrode catalyst in dye-sensitized solar cells. *MRS Proceedings.* **2011,** *1102,* 1102-LL01-05.
24. Dissanayake, N.; Hatton, R. A.; Giusca, C.; Lutz, T.; Curry, R.; Silva, R., Organic:PbS-nanocrystal:fullerene hybrid photovoltaics. *MRS Proceedings.* **2011,** *1102,* 1102-LL07-06.
25. Wang, C.-T.; Yeh, C.-H., Quantum dot solar cells based on CdSe-assembled titania semiconductors. *MRS Proceedings.* **2011,** *1102,* 1102-LL04-05.
26. Powell, D. M.; Winkler, M. T.; Choi, H. J.; Simmons, C. B.; Needleman, D. B.; Buonassisi, T., Crystalline silicon photovoltaics: a cost analysis framework for determining technology pathways to reach baseload electricity costs. *Energy & Environmental Science.* **2012,** *5*(3), 5874–5883.
27. Green, M. A., The path to 25% silicon solar cell efficiency: history of silicon cell evolution. *Progress in Photovoltaics.* **2009,** *17*(3), 183–189.
28. Carlson, D. E.; Wronski, C. R., Amorphous silicon solar cell. *Applied Physics Letters.* **1976,** *28*(11), 671–673.
29. Slunjski, R.; Capan, I.; Pivac, B.; Le Donne, A.; Binetti, S., Effects of low-temperature annealing on polycrystalline silicon for solar cells. *Solar Energy Materials & Solar Cells.* **2011,** *95*(2), 559–563.
30. Wu, X. Z., High-efficiency polycrystalline CdTe thin-film solar cells. *Solar Energy.* **2004,** *77*(6), 803–814.
31. Shen, K.; Bai, Z. Z.; Deng, Y.; Yang, R. L.; Wang, D. Z.; Li, Q.; Wang, D. L., High efficiency CdTe solar cells with a through-thickness polycrystalline CdTe thin film. *RSC Advances.* **2016,** *6*(57), 52326–52333.
32. Panthani, M. G.; Akhavan, V.; Goodfellow, B.; Schmidtke, J. P.; Dunn, L.; Dodabalapur, A.; . . . Korgel, B. A., Synthesis of CuInS2, CuInSe2, and Cu(InxGa1-x)Se2 (CIGS) nanocrystal "Inks" for printable photovoltaics. *Journal of the American Chemical Society.* **2008,** *130*(49), 16770–16777.

Index

Note: Page numbers followed by *f* denote figures; by *t*, tables.

D

E

G

H

M

O

P

Q

R

S

T

U